# Global Congress on Process Safety 2014



**Topical Conference at the 2014 AIChE Spring Meeting and 10th Global Congress on Process Safety**

## New Orleans, Louisiana, USA
## 30 March - 3 April 2014

**ISBN: 978-1-63439-058-3**

**Printed from e-media with permission by:**

Curran Associates, Inc.
57 Morehouse Lane
Red Hook, NY 12571

CURRAN ASSOCIATES INC.
proceedings.com

**Some format issues inherent in the e-media version may also appear in this print version.**



Printed by Curran Associates, Inc. (2014)



AIChE
3 Park Avenue
New York, NY 10016-5991

Phone: (203) 702-7660
Fax: (203) 775-5177

www.aiche.org

**Additional copies of this publication are available from:**

Curran Associates, Inc.
57 Morehouse Lane
Red Hook, NY 12571 USA
Phone: 845-758-0400
Fax: 845-758-2634
Email: curran@proceedings.com
Web: www.proceedings.com

# TABLE OF CONTENTS

# 354012 Possible vs. Practical: Engineers Must Lead the Development of Practical Technologies

**Abstract:**

Chemical Engineering is need today more than ever. There is a strong societal desire for sources of energy and feedstocks that can sustainably support a high standard of living for all. Thermodynamic, material and economic challenges stand in the way. Most of the general public does not have the basic understanding of the energy flows, chemical transformations, and the scale of the systems that support our current way of life. These concepts are fundamental to chemical engineering. Chemical engineers should be on the vanguard of separating hype from truly practical options yet occationally we let society down. This talk will present several examples where the fundamentals of chemical engineering can be used to access the practicality of proposed solutions to energy and water issues, as well as explain why the venture capital model struggles in the materials market.

# 349805 How to Better Manage the Combined Effects of Barrier (IPL) Impairment and Ongoing Work Hazard Risks


**Abstract:**

In recent years there has been a growing recognition that the way we identify and manage cumulative risk in our offshore installations needs to be modified to address the frequency of change of the sources of risk. As one of those main sources of risk, monitoring and maintaining the condition of our independent layers of protection (ILP's) has become a major challenge given the number of protection and mitigation systems on a typical offshore installation. This is especially true for aging assets where maintaining integrity of these systems has become a never ending battle in more mature basin areas. Adding to the complexity of this challenge is the necessity to account not only for the condition of these systems but also for the ongoing day to day work on the plant. We need to understand how this impacts on the effectiveness of protection layers, the new hazards which it introduces and be in a position to judge whether the resultant cumulative risk posed by a work plan is acceptable or not. This paper will explain this challenge in detail and talk about ways in which technology can help manage such risks. It will discuss the operations reality of day to day work on a plant, the not so uncommon conditions of protection layers and the necessity to defeat or remove many of these systems in order to carry out maintenance and repair work.

Rigorous quantitative risk assessments techniques do not lend themselves to the dynamically changing conditions of day to day operations. The assessment of these risks and the judgement of what activities are safe to execute therefore relies on the experience and competence of the people at the front line. There are now emerging, new technology platforms where much improved tools can be used to help support these "in field" decisions. The paper will explore how these tools can be used on a day to day basis and how as an industry we can start to use more meaningful real time data to provide a better picture of the sources and magnitude of risk. The status of protection layers can be shown graphically alongside ongoing or planned work to show their potential impact on each other. As a result, decisions on what is safe to execute today or to plan for tomorrow can be made with much better insight on risk

For nearly 12 years our software systems have been used for managing the risks associated with work offshore (in the UKCS North Sea our systems are used on 90% of platforms). They have been used to help operations people manage task hazard risk as well as to account for risks associated with non performing equipment and impaired independent protection layers.

# 349102 Comparative Consequence Analysis Between LNG Import and Export Terminals through the Use of PHAST and GIS

**Abstract:**

According to the US Energy Information Administration (EIA), the natural gas consumption in the US was 1,726 billion cubic feet in January 2013, and the demand has been increasing steadily around the world as natural gas becomes the fuel of choice for electric power providers. In order to meet this demand, the US economic interest was focused on the Liquefied Natural Gas (LNG) import terminals installation. However, since 2009, changes in domestic shale production and other commercial changes, the focus in the US has shifted from the construction and operation of LNG import terminals to the creation of new export terminals in pursuit of exporting LNG. Based on this shifting focus, risk assessments for LNG terminals have also shifted towards potential hazards caused by flammable and cryogenic characteristics of LNG such as vapor cloud, flash fires, pool fires and Boiling Liquid Expanding Vapor Explosion. Potential hazards associated with LNG export terminals during liquefaction and storing processes require more in-depth study.

The implementation of Computational Fluid Dynamics (CFD) modeling plays an important role in identifying the contrasting consequences of incidents associated with potential scenarios at LNG export terminals. It is intended to determine the exclusion zones for these scenarios through the utilization of CFD in order to estimate the areas where people, property, or the environment would be more severely affected. Additionally, this project seeks to generate high accuracy consequence predictions by including the implementation of a Geographic Information System (GIS) which will allow the use of geo-spatial factors. This computational approach will help to develop a more robust analysis, which will predict the consequences of an incident. Not only will GIS facilitate the integration of topographical information into the model, but it will also provide a significant input for the emergency mitigation aspect. The location of emergency response personnel would be incorporated, and the time of response would be assessed. Consequently, a wider number of variables would facilitate the performance of the consequence analysis as realistically as achievable. The main goal of this project is the combination of these two computational tools (CFD and GIS) to fill a gap between the consequence estimation of LNG catastrophic events and the final representation of the results.

# 350020 Challenges of Explosion Risk Management in Artic Environments

**Abstract:**

Fully enclosed platforms utilized in artic climates can present challenges for managing explosion risk. Generally, modules on modern platforms are built with an "open" wall design. These open wall designs provide: (1) increased ventilation, which helps to prevent the accumulation of flammable gases within the module; (2) open pathways for flammable gases to dissipate; and (3) open pathways for the explosion to vent, thus reducing the overpressures. Due to the harsh conditions in artic environments, the outer walls of the platform are solid and enclosed, which prevents flammable gas from escaping during unintentional releases. Unlike open modules, nearly all mass flow leak rates (including the high frequency of occurrence, low mass flow rate leaks) can fill an entire process area with near stoichiometric gas concentrations at some time during the release. When ignited, these clouds can produce devastating explosion overpressures. Enclosed modules also increase the duration of time for which these clouds are present, which can increase the probability of ignition as mechanical ventilation is the only means of diluting and removing the gas vapor. This study will use FLACS CFD explosion and dispersion modeling to analyze the application of various mitigation techniques to help manage the explosion risk in these enclosed platforms. More specifically, the effectiveness of soft barriers/partitions (to reduce cloud sizes) as well as partially opening the outer enclosing walls (to reduce cloud sizes and the time that gas vapor may remain in the flammable range) will be evaluated. Exceedance curves will be produced for various geometry changes to analyze the $10^{-4}$ and $10^{-5}$ per year frequency of occurrence overpressures.

# Experimental study on the relationship between the charge amount of polypropylene granules and electrostatic discharges while silo loading

**Kwangseok Choi**
**National Institute of Occupation Safety and Health, Japan**
**1-4-6 Umezono, Kiyose, Tokyo, Japan**
**choiks@s.jniosh.go.jp**

**Tomofumi Mogami**
**Kasuga Denki INC**
**2-4 Shinkawasaki, Saiwai, Kawasaki, Kanagawa, Japan**

**Teruo Suzuki**
**Kasuga Denki INC**
**2-4 Shinkawasaki, Saiwai, Kawasaki, Kanagawa, Japan**

**Mizuki Yamaguma**
**National Institute of Occupation Safety and Health, Japan**
**1-4-6 Umezono, Kiyose, Tokyo, Japan**

Prepared for Presentation at
American Institute of Chemical Engineers
2014 Spring Meeting
10th Global Congress on Process Safety
New Orleans, LA
March 30 – April 2, 2014

UNPUBLISHED

# Experimental study on the relationship between the charge amount of polypropylene granules and electrostatic discharges while silo loading

**Kwangseok Choi**
**National Institute of Occupation Safety and Health, Japan**
**1-4-6 Umezono, Kiyose, Tokyo, Japan**
**choiks@s.jniosh.go.jp**

**Tomofumi Mogami**
**Kasuga Denki INC**

**Teruo Suzuki**
**Kasuga Denki INC**

**Mizuki Yamaguma**
**National Institute of Occupation Safety and Health, Japan**

**Keywords:** Electrostatic charges, Electrostatic discharges, Granules, Polypropylene, Silo

## Abstract

This paper is a report on the relationship between the charge amount (charge to mass ratio) of polypropylene (PP) granules and the electrostatic discharges that occur while loading a metal silo. The feedback control system was used in order to control the charge to mass ratio of PP granules. The electrostatic discharges inside the silo were also observed using a conventional image-intensifier system. The results obtained from the experiments show that (1) two kinds of electrostatic discharges were clearly observed inside a metal silo while loading PP granules, i.e., brush discharges and incendiary bulk surface discharges; (2) the number of brush discharges and incendiary bulk surface discharges increased with the increase in the charge to mass ratio of PP granules, almost reaching saturation; and (3) brush discharges and incendiary bulk surface discharges began to occur at the -1.16 μC/kg and -2.33 μC/kg points in the charge to mass ratio, respectively, in this study.

## 1. Introduction

A variety of combustible powders are currently being used in the food and pharmaceutical industries and in the materials used in electronic devices. Dust explosions and/or fires attributed to electrostatic discharges often occur in various processes of industries. There is no risk of ignition by electrostatic discharges with granules that are greater than 500 μm [1]. However, mixtures of granules and fine powders that are more concentrated than the low explosion limit are potentially hazardous in industrial situations. In fact, our previous studies observed visually the electrostatic discharges generated inside a metal silo while loading polypropylene granules. Some results obtained were noteworthy and, thus, published [2]. However, the question still remains as to what amount of charged polypropylene granules is needed to generate electrostatic discharges. Therefore, the main focus of this present paper is to investigate clearly the relationship between the charge amount of polypropylene granules and the electrostatic discharges that occur while loading a metal silo. A primary objective of this study is to communicate new information about safety management that will be important to individuals working in the field of accident prevention and mitigation.

## 2. Experimental

### *2.1 Pneumatic Powder Transport Facility and Sample Granules*

An actual-sized pneumatic powder transport facility was used in this study (Fig. 1). It consists of an earthed cylindrical silo (stainless steel; diameter, 1.5 m; body length, 3 m; capacity, 4.8 $m^3$), a pipeline (stainless steel; diameter, 0.1 m; total length, 23 m), an air blower (11 $m^3$/min) driven by an inverter motor, and an air conditioning unit controlling the temperature (30 ± 2 °C) and humidity (30 ± 2% RH) of the blowing air. The metal silo had a rotary valve (13 rpm) driven by an inverter motor to discharge sample granules from the bottom of the silo. The discharged sample granules were driven back into the silo from the top through a pipeline with a length of 20 m.

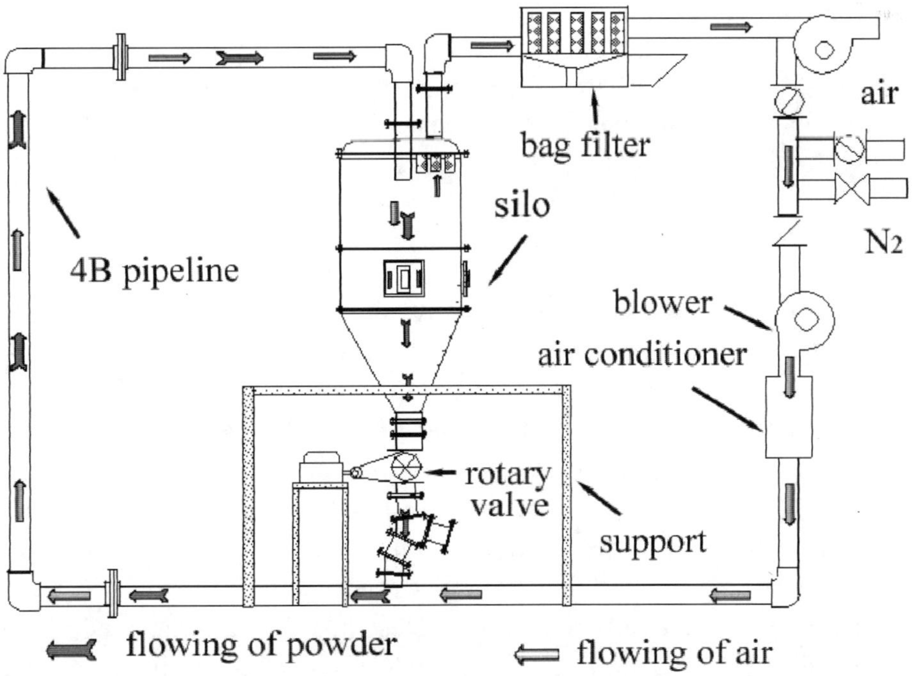


Fig. 1 Pneumatic powder transportation facility used in this experiment.

As a sample, 300 kg of polypropylene (PP, Fig. 2) granules was employed in this experiment, since PP has been widely used in chemical industries. The PP granules used in this study were the same used in our previous studies. The PP granules had a typical particle size of 2 to 3 mm. The apparent volume resistivity, R [Ω•m], of the PP granules was on the order of $10^{15}$ Ω•m.

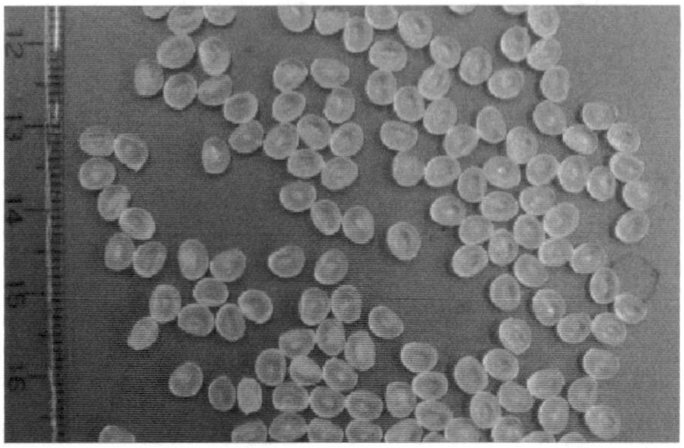

Fig. 2 Polypropylene (PP) granules used in the experiment.

### 2.2 *Observation of Electrostatic Discharges*

For observing electrostatic discharges inside the silo, an image-intensifier unit (Hamamatsu Photonics, Ltd.; C9016-02; max. gain, 1,000,000) was set on the windowpanes of the silo roof. The unit consists of an image-intensifier head, a remote controller, a CCD camera or a single-lens reflex camera, a wide-angle lens (18–35 mm), a power supply, an AC adapter, a PC, and a digital image recorder. The image intensifier was operated and controlled using the remote controller or a PC via the USB interface connector provided on the rear panel of the image-intensifier head. The image-intensifier unit was used with block-out curtains placed around the silo to prevent sunlight from entering.

### 2.3 *Controlling the charge amount of PP Granules*

The feedback control system was used to control the charge amount (charge-to-mass ratio, $q/m$) of PP granules. This feedback control system has been developed from our previous study and published [3]. The system is composed mainly of an ionizer (NKF10, Kasuga Denki Inc., flange-type, Fig. 3), an electrostatic air-blow field meter (KSF-0201, Kasuga Denki Inc., Fig. 4), a high-voltage power source, compressed air with zero relative humidity, and computer control equipment. The ionizer and the electrostatic air-blow field meter were attached to the end of the loading pipe inside the silo. The feedback control system imitates the process whereby the electrostatic air-blow field meter detects and monitors the electrostatic field $E_e$ [kV/m] that occurs as a result of the charged granule particles. In this way, the feedback of the monitored data to the ionizer causes the ionizer to generate ionized air until the charged powders have reached the desired value.

The relationship between the $E_e$ and the $q/m$ is shown in the following equation (1) [3]:

$$q/m = \frac{2\varepsilon_0 v \pi r E_e}{u}, \qquad \text{[Eq. 1]}$$

where $u$ [kg/s] is the flow rate of granules, $v$ [m/s] is the velocity of granules, $r$ [m] is the radius of the pipe for the electric field, and $\varepsilon_0$ [F/m] is the permittivity of free space ($8.854\times10^{-12}$ F/m).

It should be noted that $u$ is 0.42 kg/s, $v$ is 7 m/s, and $r$ is 0.05 m in this study.

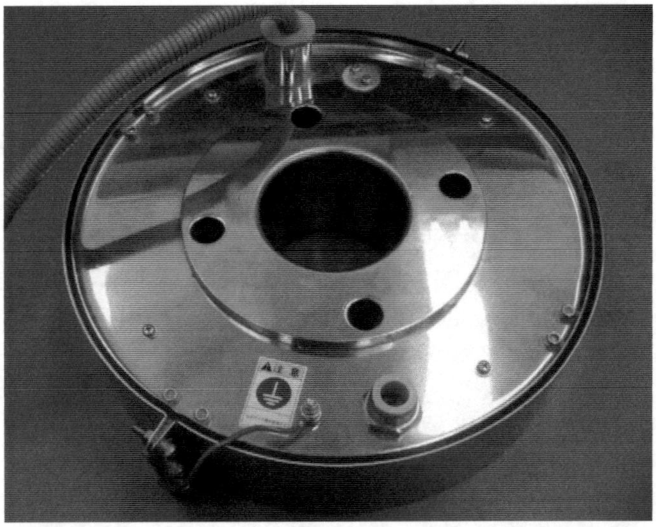

Fig. 3 Flange-type ionizer used in the experiment.

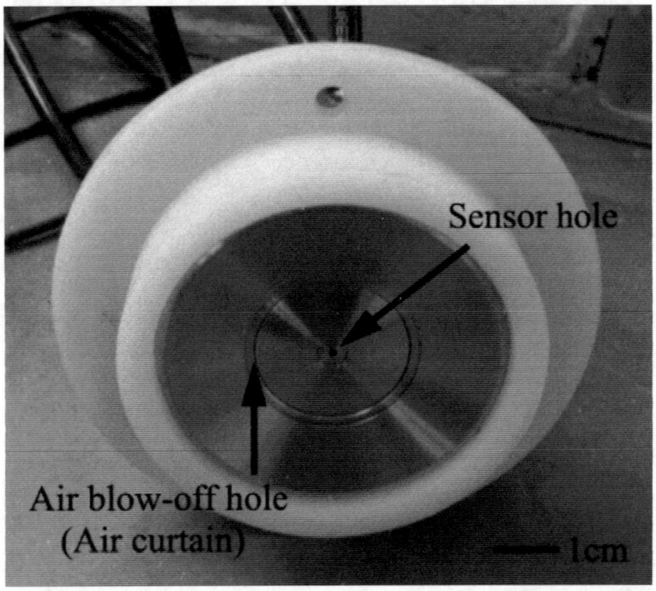


Fig. 4 Electrostatic air-blow field meter used in the experiment [4].

## 3. Results and Discussion

Two kinds of electrostatic discharges inside a metal silo while loading PP granules, i.e., brush discharges and incendiary bulk surface discharges known as cone discharges, were clearly observed as shown in Fig. 5. This fact was agreed upon in our previous studies, and their electrostatic hazards were discussed [2]. Specifically, it is well known that the energy released in incendiary bulk surface discharges depends on the silo diameter and the particle size of the products forming the powder heap. For silos with diameters in the range of 0.5 m to 3.0 m and powders with a median range of 0.1 mm to 3.0 mm, the energy released in bulk surface discharges can be estimated using the following equation (2) [1]:

$$W = 5.22\, D^{3.36}\, d^{1.46}, \qquad \text{[Eq. 2]}$$

where $W$ is the upper limit of the energy of the bulk surface discharge (mJ), $D$ is the diameter of the silo (m), and $d$ is the median of the particle size distribution (mm).

The $D$ of the silo and the $d$ of the PP were 1.5 m and 2 to 3 mm, respectively, in this study. Following from the above equations, $W$ was 56 mJ to 101 mJ. These energy values can easily ignite combustible powders, including fine particles.

The question still remains as to what amount of charged PP granules is needed to generate these electrostatic discharges. In this study, the PP granules were negatively charged using the feedback control system, since PP granules were negatively charged in the pneumatic powder transport facility. The charging control range for PP granules was from 0 to -12 μC/kg in the $q/m$.

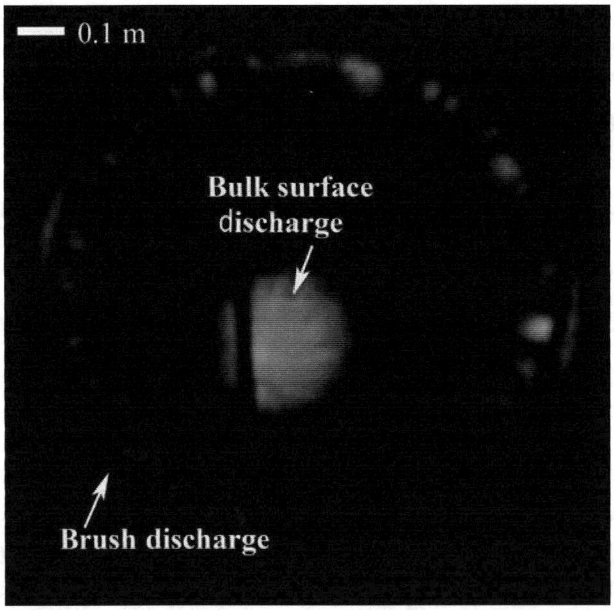


Fig. 5 Electrostatic discharges that occurred while loading PP granules.

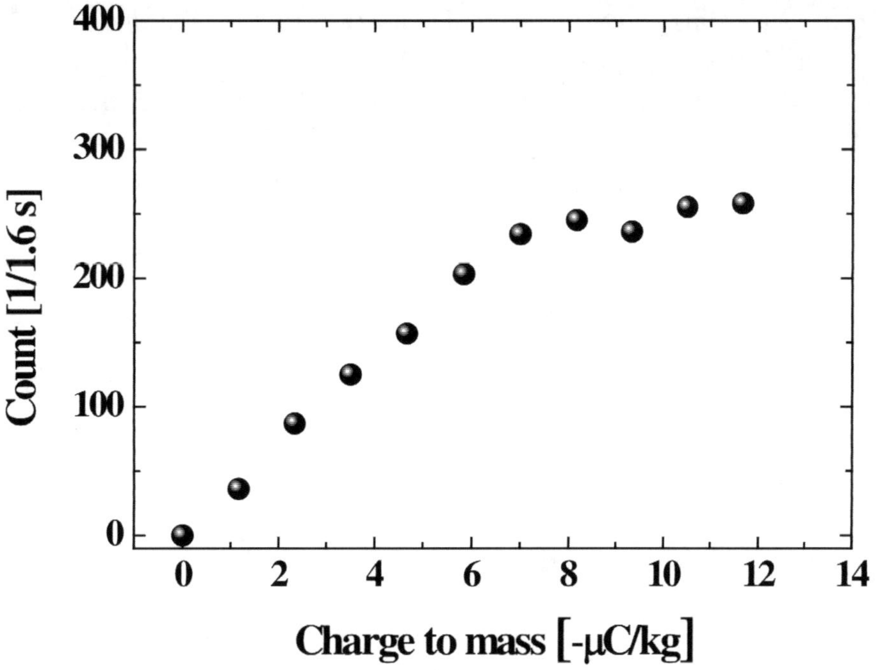


Fig. 6 Relationship between the amount of charged PP granules and the brush discharges that occurred while loading.

Figure 6 shows the relationship between the *q/m* of PP granules and the brush discharges that occurred while loading. Each dot represents the total number of brush discharges out of fifty random photographs. The exposure time was 2 s. Figure 7 shows the relationship between the *q/m* of PP granules and the incendiary bulk surface discharges that occurred while loading. Each dot represents the total number of bulk surface discharges that occurred during the five minutes. The results obtained from the experiments led to the following conclusions: (1) Neither brush discharges nor bulk surface discharges occurred when the *q/m* of PP granules was 0 μC/kg, as expected; (2) The generation of brush discharges was about ten times larger than that of bulk surface discharges; (3) The *q/m* significantly affected the generation of electrostatic discharges, with higher *q/m* resulting in the occurrence of more brush discharges and bulk surface discharges, almost reaching saturation; (4) Brush discharges and incendiary bulk surface discharges began to occur at the -1.16 μC/kg and -2.33 μC/kg points, respectively, in this study. This means that the threshold that must be exceeded to begin brush discharges is between greater than 0 and -1.16 μC/kg, and that of bulk surface discharges is between greater than -1.16 μC/kg and -2.33 μC/kg. It is well known that the threshold that can cause electrostatic charging problems with powders, especially electrostatic agglomeration and adhesion in process manufacturing, is 0.10 μC/kg (absolute value) [5]. However, the threshold related to electrostatic discharges has not been determined. Therefore, the findings mentioned above are noteworthy for communicating new information about electrostatic risk management that will be important for individuals working in the field of dust explosion (and/or fire) prevention and mitigation. Especially, it is noteworthy that bulk surface discharges, known as incendiary electrostatic discharges, did not occur when the *q/m* was -1.16 μC/kg in this study. If the real industry process is similar to the experimental condition in this paper, we suggest that the *q/m* of PP granules while loading must be kept below -1 μC/kg to prevent dust explosions and/or fires attributed to incendiary bulk surface discharge. Many different testing conditions, such as the kinds and sizes of granules and the size of the silo, may affect the electrostatic charging and discharging of granules; this needs to be examined in the future.

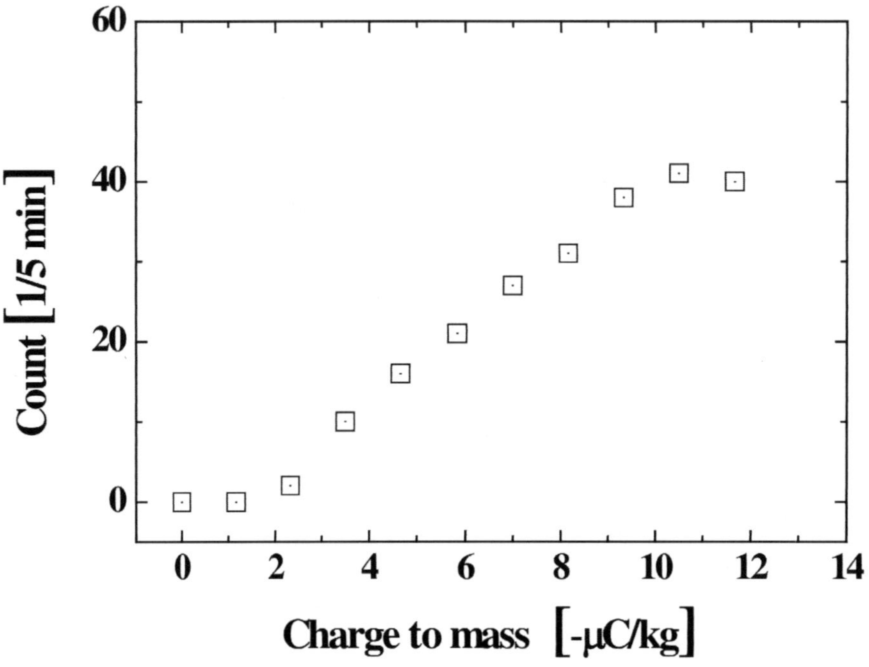


Fig. 7 Relationship between the amount of charged PP granules and the bulk surface discharges that occurred while loading.

## 4. Conclusion

To achieve a better understanding of electrostatic hazards, the relationship between the charge-to-mass ratio and electrostatic discharges generated inside a metal silo while loading polypropylene granules was investigated experimentally. The key results are summarized below:

(1) Two kinds of electrostatic discharges inside a metal silo while loading PP granules, i.e., brush discharges and incendiary bulk surface discharges, were clearly observed.
(2) Both brush discharges and incendiary bulk surface discharges increased with the increase in the charge-to-mass ratio of polypropylene granules, almost reaching saturation.

(3) Brush discharges and bulk surface discharges began to occur at the -1.16 μC/kg and -2.33 μC/kg points, respectively, in this study.

## 5. References


[1] IEC50404,"Electrostatic-Code of practice for the avoidance of hazards due to static electricity," 2003.

[2] K. Choi, T. Mogami, T. Suzuki, S. Kim, and M. Yamaguma,"Charge reduction on polypropylene granules and suppression of incendiary electrostatic discharges by using a novel AC electrostatic ionizer, " J. of Loss Prevention in the Process Industries, Vol. 26, NO.1, pp.255–260, 2013.

[3] T. Mogami, T. Suzuki, K. Choi, and T. Ikehata," Control of electrostatic charge for powder by using feedback control-type ionizer system, "J. of Loss Prevention in the Process Industries, Vol.23, NO.2, pp.237–241, 2010.

[4] K. Choi, S. Kim, J. Chung, "Experimental Study on Electrostatic Field of Polymer Powders in Freeboard Region of Fluidized Bed Reactor, " Advanced Science Letter, Vol. 19, pp.113-117, 2013.

[5] T.B. Jones and J.L. King: Powder Handling and Electrostatics, Lewis Publishers, p.34, 1986.

# 350342 Management of Change at Shell

**Abstract:**

Management of Change or lack thereof has been cited in many industrial accident investigation reports as a contributing factor to the cause of those accidents. Shell considers Management of Change to be a critical business control process and require that all businesses within Shell implement a Management of Change process. This paper will discuss how Shell Exploration and Production Company, SEPCo, implemented their Management of Change process and the steps associated with that process. This paper will also discuss what types of activities are covered by the Management of Change process. Lastly, this paper will look at the implementation challenges, lessons learned, key success factors, and Key Performance Indicators associated with Management of Change.

# Inherently Safer Design: Lessons Learned About the Principle of Simplification

**Andrew R. Carpenter**
**Russell A. Ogle**
**Brenton L. Cox**
**Sean J. Dee**
**Exponent**
**4580 Weaver Parkway**
**Warrenville, IL 60555**
**acarpenter@exponent.com, rogle@exponent.com, bcox@exponent.com, sdee@exponent.com**

Prepared for Presentation at
American Institute of Chemical Engineers
2014 Spring Meeting
10th Global Congress on Process Safety
New Orleans, LA
March 30 – April 2, 2014

UNPUBLISHED

# Inherently Safer Design: Lessons Learned About the Principle of Simplification

**Andrew R. Carpenter**
**Exponent**
**4580 Weaver Parkway**
**Warrenville, IL 60555**
**acarpenter@exponent.com**

**Russell A. Ogle**
**Brenton L. Cox**
**Sean J. Dee**
**Exponent**
**4580 Weaver Parkway**
**Warrenville, IL 60555**

**Keywords:** Inherently Safer Design; Simplification; hazardous release, fire, explosion

## Abstract

The American Institute of Chemical Engineers (AIChE), the Chemical Safety Board (CSB), and the Occupational Safety and Health Administration (OSHA) have all emphasized the importance of advancing inherently safer design concepts into chemical process plants. Incident investigations offer an important opportunity to identify, evaluate, and correct potential shortcomings in the design, construction, operation, and maintenance of a chemical process unit that has experienced a release.

This paper focuses on the inherently safer design principle of simplification. The case studies share the common theme of storing or handling liquids. Each case study illustrates how a design flaw led to unintended flow, reverse flow, or overfilling of a vessel which ultimately led to a fire, explosion, or hazardous release. Based on the incident investigation experiences of the authors, the paper illustrates how application of the simplification principle could have prevented these incidents.

## 1. Introduction

Process risk management has four generic strategies for reducing risk in chemical processes: Inherently Safer, Passive, Active and Procedural. Inherently safer design (ISD) eliminates potential process hazards through the use of safer materials or operating conditions. [1] ISD can be considered during initial design conceptualization and typically results in solutions with low operating costs, high reliability and low levels of complexity when compared with the three alternative risk reduction strategies. [2] While it is not always feasible to eliminate hazards, many professional organizations and regulatory groups have advocated that ISD should be considered before accepting process hazards and relying on safety systems for hazard management. [3-4] Therefore, ISD is commonly placed at the top of the risk reduction design solution hierarchy. [5-6]

This study focused on the examination of the principle of simplification within ISD through case studies. Simplification, along with Minimization, Substitution, and Moderation, are the four main approaches to designing inherently safer processes. In the context of ISD, simplification refers to the elimination of unnecessary complexity thereby reducing the probability of error or misinterpretation. Herein, four case studies are presented involving the storage and handling of liquids: Chlorine Release at a Commercial Water Park, Fire and Explosion at a Specialty Chemical Manufacturer, Chemical Release at an Ethanol Production Facility, and Mixing of Incompatible Wastes at a Waste Treatment Facility. All four cases highlight the role of the principle of simplification in creating inherently safer processes.

## 2. Methods of Risk Reduction and Design

The prevention of the progression of a hazardous event into undesired consequences can be controlled through various layers of protection. [7] These layers of protection can be grouped into three categories. At the outer edges of the layers of protection are response layers, which correspond to actions taken after a hazard event has taken place. Response layers intend to reduce the severity of the consequence of an event, but their effectiveness is often unknown. Underneath the response layers, are control layers which are intended to keep the hazards of a process within safe allowable operating limits. Control layers include basic controls, alarm systems, safety instrumented systems, automated shutdown systems, and physical protection. The impact of these systems can influence both likelihood and severity of an event several orders of magnitude. The design core lies at the core of the layers of protection. This design layer determines the inherent hazards within the process and is the most important layer because it has the capability to completely eliminate hazards through ISD. All the additional layers of protection around the design core served to mitigate risks and hazards that cannot be eliminated through ISD. Therefore, when performing a layer of protection analysis, inherently safety design is the preferred method for reducing risk when compared with adding independent protection layers.

## 3. Principles of Inherently Safer Design

Inherent safety is at the top level of the hierarchy of process risk management strategies, above passive, active and procedural approaches. ISD is the process of identifying and implementing solutions that result in the permanent and inseparable elimination of a process hazard. [8] Reduction of process hazards are also included in ISD, and are referred to as second order approaches to inherent safety.

ISD includes the following four design strategies:

- Minimization – The use of smaller quantities of hazardous materials.

- Substitution – The replacement of a hazardous material with a less hazardous material.
- Moderation – The concept of limiting the impact of an event by using less hazardous conditions, less hazardous forms of a material, or modifying the facility
- Simplification – The elimination of unnecessary complexity in the design of a facility, resulting in the reduction in the number of errors or an tolerance of process operation fluctuations.

While the concepts of minimization, substitution, and moderation appear simple to apply to a process, simplification can be much more intimidating. Simplification involves the deliberate questioning of the necessity of every aspect of a process and a broad knowledge of all potential alternatives. Simplification also requires individuals to discount the operational history or a process or unit and consider that safer approaches were either disregarded in the initial design, or not considered all together. The primary reason complexity exists in plant designs stems from the need to control hazards. [9] Unnecessary complexity can also result from the need to control hazards, the desire for technical elegance, failure to carry out a hazard analysis until too late in the design process, following standards or specification that are not applicable or necessary, or creating flexibility and redundancy. The strategy of simplification implies a reduction in the complexity of safeguards while simultaneously maintaining or increasing the level of risk reduction.

Understanding the appropriate definition of risk is paramount to the utilization of the concept of simplification in ISD. Risk is the product of an event likelihood and consequence severity. It is important to remember that process designs can be made inherently safer though strategies that affect both likelihood and consequence. Process equipment can often be designed to withstand overpressures and fires at much lower costs than complex safety instrumentation systems. This eliminates the likelihood of a loss of containment event. However, there are circumstances where a loss of containment event cannot be eliminated through process design. In those situations, hazard consequences can sometimes be mitigated by reducing potential consequences through siting. For example, if a vessel cannot be designed to withstand the overpressure generated from an explosion, consequences can be mitigated by ensuring the vessel is not located next to critical pieces of equipment, storage of additional hazardous materials, or occupied buildings. Thus, by being conscious of siting, the risk can be mitigated through reduction of an event consequence.

## 4. Case Studies: Storage and Handling of Liquids

The desire to reduce risk though complex control schemes can create new hazards that can be mitigated through inherently safer design. The handling and storage of hazardous chemicals at facilities is an example of a routine task that can become unnecessarily complex. These case studies illustrate three common accident scenarios that can arise: unintended transfers caused by siphoning, receiving vessel overflows, and unexpected accumulation in low points of the flow system. In each case, the principles of inherently safer design could have served to reduce the likelihood or severity of the associated incidents.

### *4.1 Chlorine Release at a Commercial Water Park*

This case study involved a chlorine gas release at a public swimming pool. Over one hundred patrons were exposed to the chlorine gas release. The chlorine gas was generated by the inadvertent mixing of sodium hypochlorite solution and muriatic acid (31% aqueous solution of hydrochloric acid). The swimming pool involved in this incident was a lazy river type pool that had a total volume of water greater than 1,000,000 liters. The water treatment system for the pool consisted of filtration, disinfection with bleach, and pH control with hydrochloric acid. The water treatment system operated continuously and had a design capacity of 2,400 liters per minute.

A maintenance contractor was performing some non-routine maintenance on the water treatment system. This activity required that he shut the system down. What the contractor did not know was that while the system was idle, both the concentrated bleach solution and the hydrochloric acid began to siphon from the feed tanks and drain into the mixing tee of the circulation pipe. The mixing tee was a short vertical pipe spool at the lowest point in the mechanical room. The orientation of the injection ports for the chemicals resulted in the lighter solution, muriatic acid, forming a stable layer above the denser solution of sodium hypochlorite. When the main circulation pump was restarted, the superchlorinated water was discharged into the pool area releasing chlorine gas.

Two design improvements would have resulted in an inherently safer design. The first improvement would be to lower the elevation of the chemical feed tanks to reduce the potential for siphoning. The second improvement would be to reverse the installation of the injection ports so that, in the event of siphoning, the heavier solution would enter the mixing tee above the less dense solution, thus encouraging the mixing of the two chemicals.

### *4.2 Fire and Explosion at a Specialty Chemical Manufacturer*

A specialty chemical manufacturer produced a very high purity chemical for the electronics and optics industries. The process involved several stages of batch distillation and batch vacuum distillation operating at near total reflux conditions to achieve the very high purity required. A typical batch distillation could run for several days producing only milliliters of high purity distillate product per hour. The distillate was water reactive, and when allowed to oxidize produced a highly unstable, shock sensitive solid.

A control system failure occurred in the evening while the facility was not staffed. The control system alerted the operators via email that the system had shut down, but the control system did not place the system into a safe state. The control system failure left the heating system in operation and opened the product withdrawal valve on the distillation system. The product withdrawal valve normally operates by opening and closing such that a very slow flow rate of product flows into a receiver. The distillation system typically processes a batch of 30 to 40 liters, while the receiving vessel was approximately 5 liters in volume. With the heating system on and the product withdrawal valve open the contents of the distillation system overflowed the receiving vessel. When the operators arrived at the site the following morning, they observed that the receiving vessel had overflowed and a significant volume of distillate had spilled onto the production floor. While attempting to clean the spill, the product ignited resulting in a fire that destroyed the entire facility and caused the rupture of several chlorine tanks resulting in an environmental release. Fortunately, the operators who were cleaning the spill had just exited the area prior to the fire and were not injured.

While the extent of damage precluded the ability to determine the specific nature of the control system failure, a simple design improvement would have resulted in an inherently safer design that would have prevented the release and the fire. The receiving vessel was significantly smaller in volume that the volume of material being processed. By increasing the size of the receiving vessel to contain the entire batch contents, the overflow condition resulting in the fire and chlorine release would not have occurred.

### *4.3 Chemical Release at an Ethanol Production Facility*

A corn based ethanol producer sold a variety of animal feed stocks from the residual products left over from the fermentation process. One of these products, wet-cake, is a combination of distillers dry grain (ddg) and corn syrup. The wet cake is produced by adding hot corn syrup at 180°F into the ddg as it exits a grain dryer using a configuration similar to that shown in Figure 1. Hot corn syrup was injected into the ddg conveyor system near the top of an inclined conveyor transition. This location allowed hot corn syrup to pool in the bottom of the conveyor. This build up would progress and eventually plug the vertical transition, preventing ddg from entering the inclined conveyor. To clear the blockage, the facility operator cut a porthole near the bottom of the inclined transition. When an operator attempted to clear the blockage by opening the porthole and rodding through the plug, a stream of hot corn syrup was released which resulted in significant burn injuries to the operator.

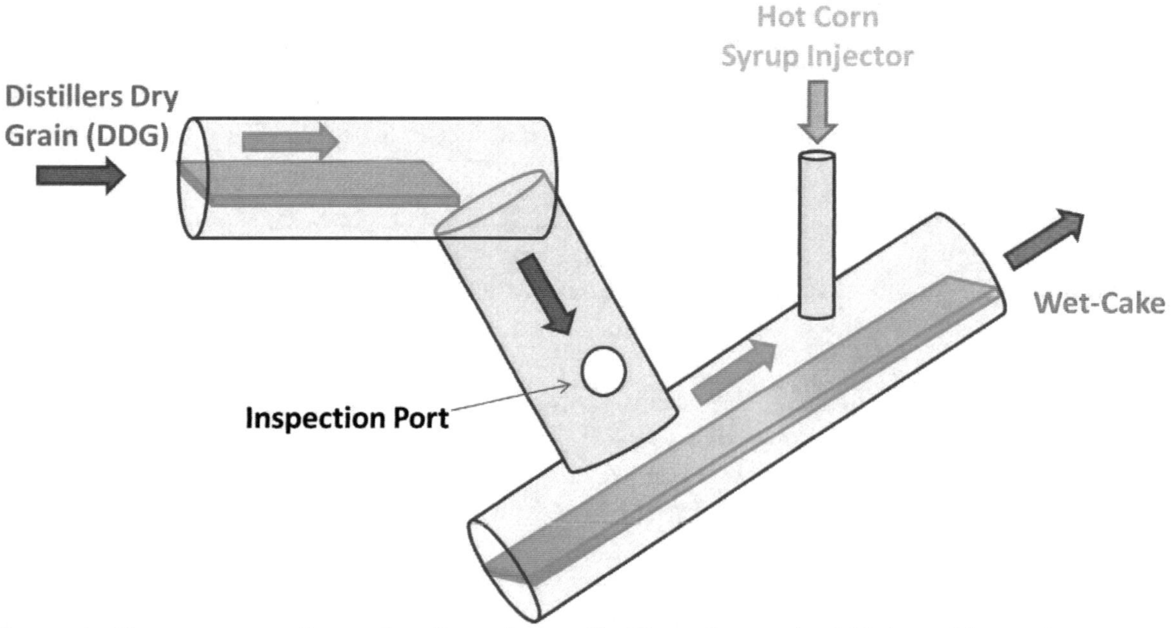


Figure 1: Conveyor configuration for mixing distillers dry grain (ddg) and heated corn syrup to produce wet-cake.

Two design improvements would have resulted in an inherently safer design. The first would be to place the conveyor inspection hatch at the top of the conveyor system such that an operator would not have to stand below the opening while attempting to clear a blockage. The second was to move the corn syrup injection point to the front of the horizontal conveyor that feeds into the inclined conveyor. This location would allow for the ddg and the corn syrup to mix and produce a wet cake and eliminate the pooling which was causing the blockage to occur.

## 5. Conclusion and Lessons Learned

The storage and handling of liquids requires the use of relatively simple technologies. But even simple technologies can experience failures. If the liquids are hazardous, the consequences of these potential failures must be controlled. These case studies illustrate three common accident scenarios that can arise: unintended transfers caused by siphoning, receiving vessel overflows, and unexpected accumulation in low points of the flow system.

In the case studies presented above, relatively minor changes could have mitigated, or potentially eliminated, the likelihood or severity of the hazards associated with each incident. In the first case study, both system layout and mixing point location alterations could have moderated or prevented the release of superchlorinated water. In the second case, a larger receiving vessel could have prevented the release of the hazardous chemical, and, in turn, the subsequent fire and explosion. In the third case, relocating the an inspection port would have kept the operator out of harm's way during a hazardous procedure, and relocating the injection point for the hot corn syrup may have eliminated the need for the procedure altogether.

Note that, while these alternatives may be obvious in hindsight, the alternatives, or even the hazards, may not have been apparent to the engineers at the time of the design. However, when designing or reviewing a system, let these incidents serve as a reminder to ask, "Are there modifications to the system that could reduce or eliminate any associated hazards?"

## 6. References


[1] Lees, F. P. *Loss Prevention in the Process Industries: Hazard Identification Assessment and Control.* 2nd Edition, Volume 1. 1996.

[2] CCPS. *Guidelines for Engineering Design for Process Safety.* Center for Chemical Process Safety. 2012.

[3] "CSB Releases New Safety Video on Inherently Safer Design and Technology: 'Inherently Safer: The Future of Risk Reducation' Examines how Industry Can Eliminate or Reduce Hazards," Jul 11, 2012, http://www.csb.gov/csb-releases-new-safety-video-on-inherently-safer-design-and-technology-inherently-safer-the-future-of-risk-reduction-examines-how-industry-can-eliminate-or-reduce-hazards/ (accessed Jan. 20, 2014).

[4] Hendershot, D. C. "Inherently Safer Design: The Fundamentals," CEP, January 2012.

[5] Crowl, D. A. *Understanding Explosions.* Center for Chemical Process Safety/AIChE. 2003.

[6] Johnson, R. W.; Rudy, Steven W.; Unwin, Stephen D. *Essential Practices for Managing Chemical Reactivity Hazards.* 2003.

[7] CCPS. *Guidelines for Safe and Reliable Instrumented Protective Systems.* Center for Chemical Process Safety. 2007.

[8] CCPS. *Inherently Safer Chemical Processes: A Life Cycle Approach.* Center for Chemical Process Safety. 2009.

[9] Kletz, T. A. *Process Plants: A Handbook for Inherently Safer Design.* 1998.

# On What Do You Base the Safety of Your Process?

**John C. Wincek**
**Croda, Inc.**
**Mill Hall, PA**
**john.wincek@croda.com**

Prepared for Presentation at
American Institute of Chemical Engineers
2014 Spring Meeting
10th Global Congress on Process Safety
New Orleans, LA
March 30 – April 2, 2014

UNPUBLISHED

# On What Do You Base the Safety of Your Process?

John C. Wincek
Croda, Inc.
Mill Hall, PA
john.wincek@croda.com

**Keywords:** Basis of Safety; Process Safety Information

## Abstract

Process Plants often require certain safeguards to satisfy management that the risk of operating the plant is sufficiently low. The required safeguards are based upon the credible consequences of a process incident, and the events that cause them. Safeguards can be preventative (preventing the incident from occurring), protective (e.g. safety relief valves) or mitigative (controlling and minimizing damage after the incident has occurred). In each category, the safeguards might be split into those that are "have to have" (i.e. the process will not be operated without this safeguard or a suitable replacement) and "nice to have" (i.e. those without which the process could run for defined periods of time). When the number of safeguards becomes large, it is difficult for the process operator to know them all, understand why they exist and what they do. It can be difficult to prioritize the safeguards when developing Asset Integrity plans. Management of Change (MOC) items can be difficult to evaluate without knowing if a "have to have" vs. a "nice to have" safeguard is impacted. This article will present a method of organizing and prioritizing safeguards. The key to this effort is determining which safeguards are required to meet a minimum level of safety. By identifying the subset of "have to have" safeguards, a concise Basis of Safety can be documented for the process. This document can then be used for operator training, maintenance and asset integrity prioritization, MOC review, and identification of safety-critical equipment such as sensors, regulators, and valves.

## 1. Introduction

Many companies subject their hazardous processes to one or more hazard studies before operation [1]. This might be a single Hazard and Operability (HazOp) study, another type of study, or several studies throughout design and construction. Existing facilities have often undergone one or more hazard studies as well. These studies are typically updated at some scheduled frequency to ensure any changes to the process risks have been reviewed and accounted for.

Hazard Study reports are typically large documents with summaries of findings, detailed worksheets and recommendations. There is also a significant amount of Process Safety Information (PSI) [2] for the process, including Piping and Instrumentation Drawings (P&IDs),

chemical safety data, equipment data and process technology data. Because there is so much information available it can be difficult for anyone, especially a front-line operator, to determine what information is most important for safely operating the process.

The Basis of Safety (BoS) document provides a summary of the most important information about the process. In a single, concise document, it contains a description of the process hazards, and the causes and consequences. This is followed by a list of the safety objectives that must be met to operate the process, and the minimum safeguards that must be in place to meet these objectives. Finally, it lists the Safe Operating Envelope for the process.

## 2. Process Knowledge

Anyone who operates, manages or interacts with a process should be familiar with the hazards and potential consequences it is capable of causing. This will help them understand the importance of operating the process safely. However to operate safely, one must know the potential causes of these incidents and the means to prevent them. The minimum safeguards that must be in place to meet the safety objectives must be known. For complex processes, this can be an overwhelming amount of information. When you add to this the safe operating limits of the process and other PSI such as chemical hazards, it easily can be too much information to expect employees to remember.

## 3. Equipment Safeguards

Safeguards are used to achieve the safety objectives for the process. For example, the safeguards used to meet the safety objective of "inerting with nitrogen" might include a vessel inerting procedure, a system for maintaining positive nitrogen pressure on the vessel, and nitrogen makeup when material is discharged from the tank. There might also be an oxygen meter in the vessel headspace, and a low pressure alarm on the tank. Safeguards can be classified into three categories: *Preventive*, *Protective* and *Mitigative* [3].

*Preventive* safeguards are those that preclude an incident from occurring through early detection and action. One example is a high level switch on a storage tank that prevents tank overflow by alerting an operator or automatically stopping the filling operation. Another example is a high temperature shutoff switch on a reactor, stopping reactant feeds to prevent a runaway reaction.

*Protective* safeguards usually prevent a serious incident, but often allow an incident with less severe consequences to occur. An example is a relief valve on a pressurized storage tank. In a fire impingement scenario, the relief valve likely prevents high pressure-induced tank failure by relieving excessive pressure. If the valve discharges to atmosphere, the valve opening allows a (hopefully less severe) release of material to the atmosphere at a planned, safer location.

*Mitigative* safeguards act to minimize the severity of an incident that has already occurred. Examples a tank dike, fire suppression systems, and water sprays on a chemical leak.

## 4. Minimum Required Safeguards

Some companies have established Quantitative Safety Risk Criteria as described by the Center for Chemical Process Safety [4]. These often specify a maximum allowable risk to employees in terms of fatal injuries. Other companies use varying methods of measuring risk. In many cases, however, companies do not simply stop when they have reached this level. Many additional safeguards are added based on industry practices, regulatory requirements and cost-effectiveness. While safeguards are important and must be maintained, one could separate them into "have-to-have" and "desirable-to-have" safeguards.

The "have-to-have" safeguards are those that are required to reach a minimum level of safety, such as those needed to meet the company's Quantitative Safety Risk Criteria. As will be shown later, this is typically a small subset of the existing safeguards that, without the benefit of the "desirable-to-have" safeguards, reduce the risk below whatever maximum allowable risk criteria are used.

One way to differentiate between "have-to-have" and "desirable-to-have" safeguards is through the use of Layer of Protection Analysis (LOPA). As LOPA practitioners know, not all safeguards can be counted as an Independent Protection Layer (IPL). The typical general requirements used to qualify IPLs are [5]:

1. It (solely) must be effective in preventing the consequences when it functions.
2. It must be independent of the initiating event and the components of any other IPL claimed for the scenario being evaluated.
3. It must be auditable (i.e. capable of being validated by some means)

When a scenario is evaluated using LOPA, many of the existing safeguards cannot be counted as contributing to the safety of the system because they do not meet the above rules. If LOPA is being used to demonstrate that the maximum risk criteria are met, and only safeguards that qualify as IPLs can be considered to reduce risk, then the safeguards claimed as IPLs must be sufficient to reduce the risk below the maximum risk criteria. These would be classified as the "have-to-have" safeguards. System safeguards not claimed as IPLs would be classified as "desirable-to-have."

As an example of this in use, consider the risk of overfilling the hydrocarbon storage tank in Figure 1. To prevent this from occurring, the following are in place:

1. An operator monitors the tank level using Level Transmitter A, and closes Valve B when the tank reaches 90% full.
2. The Basic Process Control System (BPCS) alarms at 92% full, as measured by Level Transmitter A.
3. The BPCS has an interlock that closes Valve C when Level Transmitter A reaches 93% full.
4. The BPCS closes Valve E if Level Switch D activates (at 94% full)
5. A Safety Instrumented System (SIS), independent of the BPCS, closes Valve F if Level Switch F is activated at the 95% full level.
6. The tank is contained in a diked area.

**Figure 1. Example Classification of Safeguards for a Hydrocarbon Storage Tank**

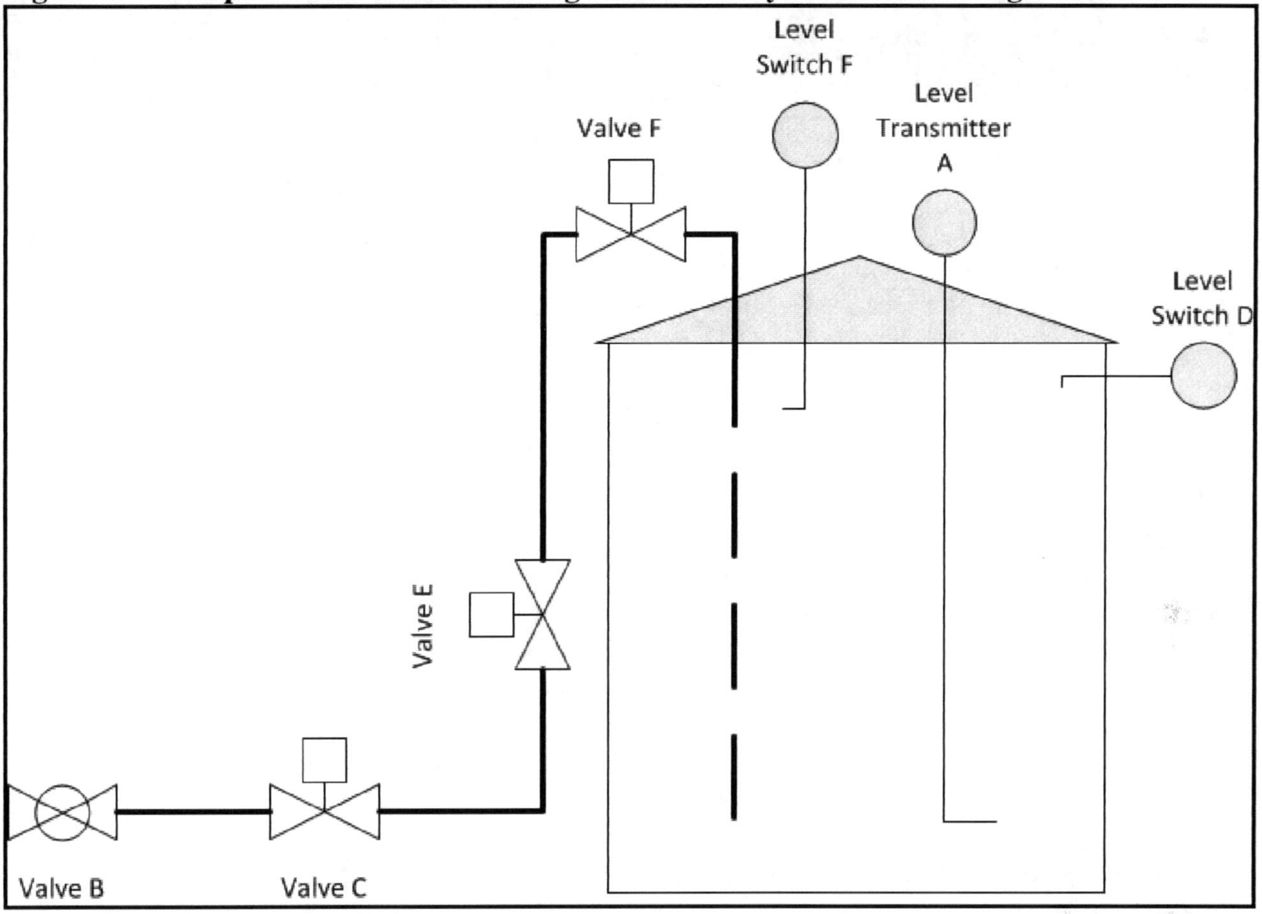


Table 1 shows which of the above could be considered IPLs in a scenario where the operator becomes distracted and does not close Valve B when required.

**Table 1. Qualification of Independent Protection Layers**

| Potential Safety Mechanism | IPL? |
|---|---|
| # 1 above could not be considered a safeguard, as it is not independent of the initiating event (i.e. it has already failed) | No |
| #2 relies on the operator (who initiated the event by failing to close Valve B) to take action. This is not independent of the initiating event; it could not be considered an IPL. | No |
| #3 and 4 rely on the BPCS to function; they are not independent of each other. Therefore, only one of them would count as an IPL. | Yes (but only one of them) |
| #5 - The SIS is independent of the BPCS, its sensors and valves – it would qualify as an IPL. | Yes |
| #6 - The dike is independent of all other events and safety devices; it would qualify as an IPL. (Not all LOPA practitioners would count the dike as an IPL. They may also evaluate a second scenario where material is spilled to the dike instead of spreading over a wide area). | Yes |

The Layer of Protection Analysis, therefore, considers only that information shown in Table 2.

**Table 2. LOPA Form Entries**

| LOPA Entry | Description |
|---|---|
| Initiating Event | The initiating event (employee fails to close Valve B). |
| IPL | Level Switch D activates at 94% full, closing Valve E. (alternatively, #3 could be used as an IPL) |
| IPL | The SIS closes Valve F upon activation of Level Switch F. |

If the LOPA outcome shows that the maximum risk criteria have been met, no further safeguards to meet this requirement are needed. Numbers 1, 4 and 5 would be considered "have-to-have." numbers 2, 3 and 6 (if not counted in LOPA) would be considered "desirable-to-have."

## 5. Basis of Safety

The information labeled number 1 through 6 above pertains to one very simple scenario – the overfilling of a tank. Imagine that information multiplied by the number of different scenarios in a plant or unit, and you can easily see how one could be overwhelmed. On the operator level, we haven't even considered the task procedures he/she must know. At the maintenance level, there is no clear way to prioritize maintenance tasks – it can't all be critical to safety, can it? The Basis of Safety is designed to minimize, if not solve, these problems.

The Basis of Safety for a process should contain the following information (Shown in Figures 2 and 3):

- **Process Description** - a brief description of the process
- **Chemical Hazards** - a brief guide to the hazards of the main chemicals used in the process
- **Hazards Table** – Showing the Hazards, Causes and Worst-case Consequences
- **Safety Objectives** – the things which, if achieved, will allow the reasonably safe operation of the process.
- **Safeguards** (Have-to-haves only)
    - Preventive
    - Protective
    - Mitigative
- **Safe Operating Envelope** – the limits (e.g. temperature, pressure) within which the process is designed.
- **Appendix (optional)** - contains the "desirable-to-have" safeguards

## Figure 2. Sample Portion of a Basis of Safety Document

### Process Description

Epichlorohydrin is stored in a 10,000-gallon storage tank located in the flammable tank farm on the south side of the facility. The tank is rated for 60 psig, and constructed of 304 stainless steel. The storage tank sits on a concrete, diked pad, segregated from other storage tanks by dike walls. The storage tank is equipped with an automatic foam fire suppression system. Epichlorohydrin is unloaded from a tank truck by pump, with vapor balancing, into the storage tank. The tank pressure is maintained at approximately 12 psig with nitrogen. Epichlorohydrin is transferred to multiple synthesis vessels using a transfer pump and single flow meter, limiting the system to supplying a single vessel at a time.

### Hazardous Substances

| NFPA Rating | Health | Fire | Reactivity |
|---|---|---|---|
| **Epichlorohydrin** | **3** | **3** | **2** |

Epichlorohydrin is a corrosive liquid. It can cause corrosive injury to the eyes, skin, and respiratory tract, and can also cause sensitization – an allergic-like reaction that occurs after repeated exposures to the substance. Systemic effects from long-time exposure can include liver and kidney damage. Epichlorohydrin is listed as a potential human carcinogen by OSHA and is also listed by the International Agency for Research on Cancer.

At high temperatures, epichlorohydrin (Epi) can react violently and sometimes explosively with other chemicals, such as acids, alcohols, alkalis, amines and salts. Epi will form a two-phase mixture with water. Given time, an exothermic reaction will start at the interface between the epi and water layers, which can generate enough heat to create a pressure build-up in the vessel.

### Worst Case Scenarios

| Hazards | Causes | Consequences |
|---|---|---|
| 1. Storage tank failure due to nitrogen overpressure. | A. Failure of nitrogen control system. | Toxic release, vapor cloud explosion or flash fire, on/off site fatalities. |
| 2. Release through storage tank relief valve. | B. Failure of nitrogen control system. | Toxic release, vapor cloud explosion or flash fire, on/off site fatalities. |
| 3. Release through tank wagon pressure relief valve. | C. Failure of nitrogen control system on epi storage tank. | Toxic release, vapor cloud explosion or flash fire, on/off site fatalities. |
| 4. Piping leak. | D. Confinement (blocking in) of epi in piping segment. | Toxic exposure to on-site employees; potential pool fire/flash fire. |
| | E. Gasket failure | Toxic exposure to on-site employees; potential pool fire/flash fire. |
| | F. Piping Corrosion | Toxic release, pool fire, on/off site fatalities. |

**Figure 3. Safety Objectives, Safeguards and Safe Operating Envelope for Basis of Safety**

**Safety Objectives**

1. Inerting with nitrogen.
2. Avoid ignition sources.
3. Avoid loss of containment.

**Control measures for Basis of Safety**

***These control measures (or suitable alternatives) must be in place for the process to operate.***

**Prevention (including Process Control)**

- Temperature switch (TSH-773-018) on unloading line shuts down unloading pump on high temperature. 11O LAST LINE OF DEFENSE
- Temperature sensor TI-773-017 shuts down transfer pump on high temperature. 12P LAST LINE OF DEFENSE
- Low storage tank pressure alarm at -5 psig. (PI-773-002). 6I, 15S LAST LINE OF DEFENSE for 15S
- Power monitor (JSL-773-021) shuts down unloading pump on high or low power draw. 11O
- Recirculation line on transfer pump discharge dissipates heat buildup if pump is blocked in. 12P
- Level transmitter LT-773-094 alarms at 184". 7J

***Protection***

- Pressure relief valve PSV-077310/11 on epi storage tank, set at 60 psig. 1A, 1B, 5G, 5H, 6I SYSTEM RELIEF
- Relief regulator (PRV-077382) opens at 15 psig to relieve excess nitrogen pressure. 1A, 1B, 1C

**Safe Operating Envelope**

- Oxygen concentration in Epichlorohydrin storage tank less than 7.9 vol. % (60% of minimum oxygen concentration required for ignition).
- Storage tank pressure less than 60 psig.
- Epi storage tank level < 194" (90% of tank volume)

## 6. Uses for the Basis of Safety

The Basis of Safety (BoS) document contains the final results of all of the hazard studies, risk analyses, LOPAs, etc. that have been performed for the process. It contains only the most critical information for process safety. If, after all of the studies are complete, you were asked "Where are the results?" you could answer by handing them the Basis of Safety. If someone asks "How do you keep your process from causing a fire/explosion/toxic release?" you could hand them a Basis of Safety.

Site Directors, Unit Managers and others who need to understand the risks of the process and how they are controlled can review the Basis of Safety for each process on site. From the BoS, they can tell how bad the consequences might be, and can identify key risk control measures. They also can use it to quiz operators on process knowledge related to process safety.

Shift Managers and Operators must be aware of the potential consequences of an incident. However, they must also have a good understanding of the Safety Objectives and the safety systems used to achieve them. The BoS can help them to understand if a process upset could be the beginning of a catastrophic incident, and it can help determine if a process can continue to run if a sensor is malfunctioning.

Maintenance personnel, using the BoS, now have a smaller list of the highest-priority equipment to work from. Fewer jobs are competing for attention because fewer of the repairs will be to "have-to-have" equipment.

The likelihood of a major oversight when reviewing MOCs can be decreased if the BoS is reviewed. The reviewer can ensure that the safeguards in the Basis of Safety are not compromised by the change. One might even require escalation of an MOC for approval at higher levels of the organization if the BoS is being changed.

The BoS is an excellent training tool for many levels of the organization. Anyone who operates or is responsible for a process should be trained and tested on the BoS, and the BoS documents should be made available to the employees for quick reference

## 7. Last Line of Defense

The Last Line of Defense (LLoD) is a concept that arose from the Basis of Safety and LOPA. It provides a useful metric of process safety performance. Consider the LOPA form entries from Table 2:

A. The initiating event (employee fails to close Valve B).
B. Level Switch D activates at 94% full, closing Valve E. (alternatively, #3 could be used as an IPL)
C. The SIS closes Valve F upon activation of Level Switch F.

The Initiating Event and IPLs are listed in order of occurrence / activation. The employee should close Valve B when the tank is full. If that doesn't happen, the BPCS should close Valve E when the Level Switch D activates. If not, the SIS should stop the incident by closing Valve F. These three actions are sometimes called Lines of Defense. In the scenario described above there are three Lines of Defense, the last of which is the SIS. All three Lines of Defense must fail for the incident to occur.

If the SIS fails to function, the incident (overfilling the tank) will occur because there are no safeguards to stop the incident after it fails. It is, therefore, the Last Line of Defense. If the Last Lines of Defense for the various scenarios in your plant are identified, they can be used as a Key Performance Indicator. If an incident sequence is stopped by your Last Line of Defense, you came very close to actually having a potentially catastrophic incident. But for the last safeguard (LLoD), the incident would have occurred. Any preceding safeguards that should have stopped the sequence have failed. The entire event, including the initiating event, failure of the safeguards that precede the LLoD, and the activation of the LLoD, should be thoroughly investigated.

## 8. Process Safety-Critical Equipment

If the BoS lists the most critical safety systems in the plant, then the associated equipment (sensors, valves, etc.) should be included in a list of Process Safety-Critical Equipment. Calibration, maintenance, inspection, breakdowns, etc. on this equipment must be completed with the highest priority and utmost care. If deficiencies or other unexpected findings are discovered, they must be thoroughly inspected. Operations must know that if a piece of Process Safety-Critical Equipment fails, it must be addressed immediately. Safety systems not listed in LOPA would be considered less critical.

Plant personnel should be aware, however, that the BoS will not exhaustively or explicitly list some Process Safety-Critical Equipment. For example if the initiating event in a LOPA is a regulator failure, that regulator must be counted as Process Safety-Critical. Also, if the regulator failure will result in overpressure and rupture of a tank, then the tank should also be considered Process Safety-Critical. If we fail to identify these types of critical equipment, much of the equipment we depend on for containment of hazardous materials will be overlooked, as will equipment whose failure would initiate a sequence of events leading to a catastrophic loss.

## 9. Other Uses for the Basis of Safety Documentation

It is important that the information in the BoS remains fresh in employees' minds. If we can immerse them with constant reminders during their shift, and help them to recall information, they will be much more successful at running the plant safely. Some ways to achieve this include:

- Identify Process Safety-Critical Equipment – Give the components (valves, sensors, etc.) a unique appearance (color, shape, etc.) on operator display screens.
- Tag Alarms - If a LLoD activates, include in the alarm notification that it is a LLoD that activated.
- Record LLoD activation – If a LLoD activates, log it to a report that is then periodically viewed by unit management or engineering. This can help ensure that these events are thoroughly investigated.
- Physically Identify Equipment – Use a unique tag, paint color or other means to clearly mark Process Safety-Critical Equipment in the field. If an operator finds a device hanging by its wires, or a mechanic is called to trouble shoot an actuated valve, they should immediately know the criticality as Process Safety-Critical Equipment through identifying it in the field.
- Identify Process Safety-Critical Equipment on P&IDs. This emphasizes it to engineers who are marking up changes, and to Hazard Study Team members reviewing the drawings during a hazard study.

## 10. Conclusion

If information critical to avoiding catastrophic incidents is not retained after training, the likelihood of a major process incident increases. The Basis of Safety document can provide the most critical information to operators in a concise format that can be used both for training and as a reference. The list of potential incidents, the causes, and the most critical safeguards preventing these incidents can be presented in a logical order, enabling better retention due to both the organization of the document and the relatively small amount of information included. The document serves as a summary of all prior hazard studies / risk assessments.

The document can be leveraged to serve many purposes beyond operations. Senior managers, new engineers and others can use portions of the document to understand the major process risks at a facility. Each safeguard listed might be treated as "have-to-have," without which an alternative protective measure is required to continue operating. The document can also serve as a framework for identifying Process Safety-Critical Equipment, Last Lines of Defense, and to identify critical alarms, interlocks and devices in the field, on operator screens and drawings.

## 11. References

[1] Kaszniak, M. (2010), Oversights and omissions in process hazard analyses: Lessons learned from CSB investigations. Process Safety Progress, 29: 264–269. doi: 10.1002/prs.10373.

[2] Process Safety Management of Highly Hazardous Chemicals; US Occupational Safety and Health Administration; 29 CFR 1910.119(d).

[3] CCPS; Guidelines for Safety and Reliable Instrumented Protective Systems; 2009; Wiley-Interscience; Hoboken, NJ; Appendix B.

[4] CCPS; Guidelines for Developing Quantitative Safety Risk Criteria; 2009; John Wiley and Sons, Inc.; Hoboken, NJ.

[5] CCPS; Layer of Protection Analysis – Simplified Process Risk Assessment; American Institute of Chemical Engineers; 2001.

# 350026 The Need for a Unified Process Safety Map

**Abstract:**

Knowing Where You're Going -- The Need for a Unified Process Safety Map

**Mr. Teddy Bucher Mr. John T. Perez, P.E. Cognascents Consulting Group, Inc. 11777-A Katy Fwy Ste 438 Houston, TX 77079**

**teddy.bucher@cognascents.com**

**Keywords:** Process Hazard Analysis (PHA), Process Safety Management (PSM), Process Safety Information (PSI), effective communication, Independent Protection Layers (IPL), Hazard and Operability Study (HAZOP), hazard identification, Layer of Protection Analysis (LOPA), risk assessment, risk management.

Abstract

Since the passage of the OSHA Process Safety Mandate, the various methods of Process Hazard Analysis (PHA) have become commonplace. Process Safety Management (PSM) systems and the corresponding Process Safety Information (PSI) that feed into them have proliferated and permeated numerous aspects of production companies. As this has occurred, constituent groups within those companies have further developed the specific subsets of PSI relevant to them, yielding data sets that are increasingly useful and that allow them to operate more efficiently.

Yet with every PHA revalidation cycle, approaches in both preparation and execution frequently tend to concern themselves with the immediate - the presence of the direct inputs needed for the hazard analysis and the production of the direct outputs of the analysis. Viewing these separate components of PSM as distinct elements has a tendency to result in little consideration being given to the overall quality of the inputs, with the assumption that issues with the design of the input will be captured on the next revalidation cycle for that input. This approach of separate revalidation cycles for each element can be effective for the capture of design changes arising from major modifications to the process itself. However, it can also introduce the possibility for multiple changes to be made to a system in parallel, potentially rendering the input inadequate. It also overlooks the possibility that changes to one element of a PSM program could introduce new demands on another element previously thought to be adequate.

The authors posit that implementation of a unified process safety map that reflects the true interconnected nature of the various elements of a PSM program would minimize this potential for systemic error and allow for a more efficient PSM system. In this paper, the authors discuss the structure of an effective PSM program which utilizes a process safety map. This map would highlight how common PSM elements serve as both inputs to and outputs from the same processes through comprehensive communication of changes throughout the whole PSM program. In addition, the authors also provide examples of breakdowns in communication of PSI to illustrate the importance of a PSM program which continually improves as a whole, as a result of modifications to only one part of the program.

The target audience for this paper includes project managers, project engineers, EH&S managers, PSM coordinators, and operators; however, anyone involved with small or large capital projects may also benefit from this paper's content.

# 350100 How Is Your Process Safety Vision? - Try Pursuing "Perfect Process Safety"


**Abstract:**

Many companies have learned that only pursuing regulatory compliance is not enough to manage risks. Economic cycles, organizational changes, and aging assets and people all put great stress on companies and degrade their ability to improve and sustain safety performance. In occupational safety and HSE, some companies have adopted one form or another of "driving bad frequencies to zero" in order to keep their organizations focused on driving safety performance improvement. In process safety, why can't we do the equivalent - let's go after "Perfect Process Safety." This paper will describe a practical framework for pursuing perfect process safety:

- A culture based on proper ownership of process safety empowered by visionary leadership
- Risk-informed sensitivity that guides everything
- Effective, fit-for-purpose management systems
- PS practices embraced and followed with good operational discipline at ALL levels
- Learning from ALL sources – internal, external and outside industry group
- Well-formed/visible performance pyramid; metrics at every level that drive intended behaviors
- Goals and actual performance that improves

Effective process safety leadership throughout organizational levels and learning via management of a company's wisdom life cycle are two vital components of this pursuit strategy. This paper will define ways of measuring process safety leadership based on 15 key attributes of effective leaders and how to identify learning-disabled organizations.

# 348737 Using Metrics to Improve Emergency Management

**Abstract:**

**10[Th] Global Conference on Process Safety**

**Using Metrics to Improve Emergency Management**

**Rixio E. Medina**

**AcuTech Consulting Group**

**September 25, 2013**

As part of the process industry's efforts to manage process risk, facilities' emergency planning and response plans are essential. These plans have traditionally being developed by experienced fire chiefs or other emergency response personnel using professional and trade organizations' guidelines, local requirements and equipment manufacturers' recommendations. The OSHA Process Safety Management standard and EPA Risk Management Program regulation established minimum requirements for emergency planning and response for process safety. Unfortunately, the Chemical Safety and Hazard Investigation Board (CSB) continue to identify emergency planning and response gaps in some of the accidents they investigate. Industry's ultimate goal is to have emergency planning and response systems that address all credible accident scenarios they can potentially encounter and that the system is capable of delivering the intended results every time it is activated to minimize the consequences of events. To that end, quality assurance have been attempted by conducting period reviews of systems, equipment and resources using audits, checklists, emergency drills and exercises to identify the level of readiness.

The Center for Chemical Process Safety (CCPS) Risk-Based Process safety (RBPS) Guidelines (2007) provide additional insight on how to achieve an effective emergency management performance for PSM.

The author will examine the RBPS approach for measurement and metrics of emergency management and will offer advice to improve emergency management practices via metrics.

# Flame Propagation in Dust/Air Mixtures Under Reduced Pressure Conditions

**Hannes Kern**
**Montanuniversität Leoben**
**Franz-Josef Strasse 18, 8700 Leoben**
hannes.kern@unileoben.ac.at

**Gerald J. Wieser, Harald Raupenstrauch**
**Montanuniversität Leoben**
**Franz-Josef Strasse 18, 8700 Leoben**
gerald.wieser@stud.unileoben.ac.at, harald.raupenstrauch@unileoben.ac.at

Prepared for Presentation at
American Institute of Chemical Engineers
2014 Spring Meeting
10th Global Congress on Process Safety
New Orleans, LA
March 30 – April 2, 2014

UNPUBLISHED

# Flame Propagation in Dust/Air Mixtures Under Reduced Pressure Conditions

**Hannes Kern**
**Montanuniversität Leoben**
**Franz-Josef Strasse 18, 8700 Leoben**
hannes.kern@unileoben.ac.at

**Gerald J. Wieser, Harald Raupenstrauch**
**Montanuniversität Leoben**
**Franz-Josef Strasse 18, 8700 Leoben**
gerald.wieser@stud.unileoben.ac.at, harald.raupenstrauch@unileoben.ac.at

**Keywords:** dust explosion, flame propagation, non-atmospheric conditions

## Abstract

Industrial processes are often operated at non-atmospheric conditions. Nevertheless, the development of dust explosions and flame propagation under reduced pressure conditions has not received much attention. One explanation could be that standard laboratory equipment like the 20 l Siwek chamber does not allow investigations at very low pressures.

An experimental device was developed for the investigation of ignition and flame propagation under reduced pressure (0.25 bar to atmospheric pressure). Experiments have been carried out with lycopodium at dust concentrations of 100 g/m³, 200 g/m³ and 300 g/m³. Flame propagation was investigated using a video analysis system and the actual flame speed was measured by optical sensors.

Flame shape and velocity strongly differs from results at atmospheric conditions. Effects like buoyancy of hot gases during ignition and flame propagation are less strong than at atmospheric conditions. For the investigated dust concentrations, flame velocity reached speeds that are nearly an order of a magnitude higher than at ambient conditions. Results gained from these experiments are contributing to the setup of mathematical models for the description of gas explosions and ignition.

## 1. Introduction [1]

Dust explosion research still faces fundamental problems in the explanation and evaluation of basic parameters of dust explosions, such as minimum ignition energy and temperature, as well as those that define flame propagation, such as flame velocities. Much experimental and theoretical work has already been done but the sheer complexity of the problem still leaves many questions unanswered. Dust explosion testing, for example, still relies on methods that are only able to determine explosion effects under very specific circumstances. The deduction of physical or chemical mechanisms from such experiments is usually not possible. A more knowledge-based approach towards the fundamentals of ignition and combustion is needed for a better description of dust explosions. This paper should contribute to the understanding of dust explosion processes under conditions that deviate from standard ambient conditions. The results gained from these investigations should also lead to a better understanding of dust explosions in general.

## 2. Experimental Work

The experimental investigation of dust/air flames is rather difficult because dust/air mixtures are inherently turbulent. Any kind of mixing process that creates undefined turbulence levels strongly influences the results of the experiments. Different experimental approaches to determine the flame velocity of dust/air flames are known. For the present experiments, Lycopodium dust was used as a reference substance due to the well-investigated combustion characteristics and its physical parameters. Lyopodium shows a very narrow particle size distribution with a median value of around 30 µm. The shape of the lycopodium spores is nearly spherical with a fine structure that provides a big reactive surface.

### *2.1 Basic experimental Setup [2]*

In the development of dust explosions, turbulence is a key factor in addition to the specific characteristics of the dust itself. Therefore, an experimental setup had to be found that allowed the creation of a homogeneous dust cloud on the one hand and low turbulence conditions on the other hand. Apparatuses described by Krause et al. [3] are able to provide a good particle distribution by using a constant gas flow through a sinter plate under constant but comparatively high turbulence conditions. Other apparatuses use dust feed systems at the top of the testing device, which requires accepting an inferior particle distribution and the disadvantage of not being able to investigate dusts that tend to strongly agglomerate [4, 5, 6]. Pre-tests showed that the dispersion of lycopodium in air could be realized quite easily [7]. Therefore, a top feed arrangement for testing seemed adequate, especially in terms of low turbulence levels. Various dust concentrations are achievable with proper dust feeding systems. A top feed system also meets the requirements for investigations under reduced pressure conditions because no gas stream is required for dust dispersion.

Experiments in closed vessels often do not allow the visual observation of flame propagation and explosion development. For investigations of flame development, optical systems allowing the determination of flame propagation as well as flame geometry are obligatory. Experiments with pressure levels below atmospheric conditions required a completely sealed combustion section.

For further experiments a main explosion tube (length 2000 mm) made of Plexiglas® was chosen. A tube with an inner diameter of 140 mm and a wall thickness of 5 mm fulfilled the requirements for experiments at atmospheric as well as below atmospheric pressure conditions.

Basically the apparatus consists of a bottom section connected to a vacuum pump and housing a spark ignition system, the main combustion tube and a top section with the dust feeding device. Measurement and detection is conducted using an optical system with photodiodes and a video camera for the measurement of the flame shape (Figure 1).

The head element houses the dust input system and is equipped with a flame arrestor and an explosion flap. The dust input is carried out by a screw conveyor that can be regulated by the operating voltage. The input system allows the generation of dust concentrations from 50 g/m³ to 500 g/m³ (Lycopodium). The uniform distribution of the dust over the whole cross section of the tube was monitored by a light extinction of two lasers in a height of 100 mm and 900 mm above the igniton point. The particle distribution was also simulated by computational fluid dynamics and showed that particles are distributed evenly over the whole cross section 500 – 700 mm below the feed point.

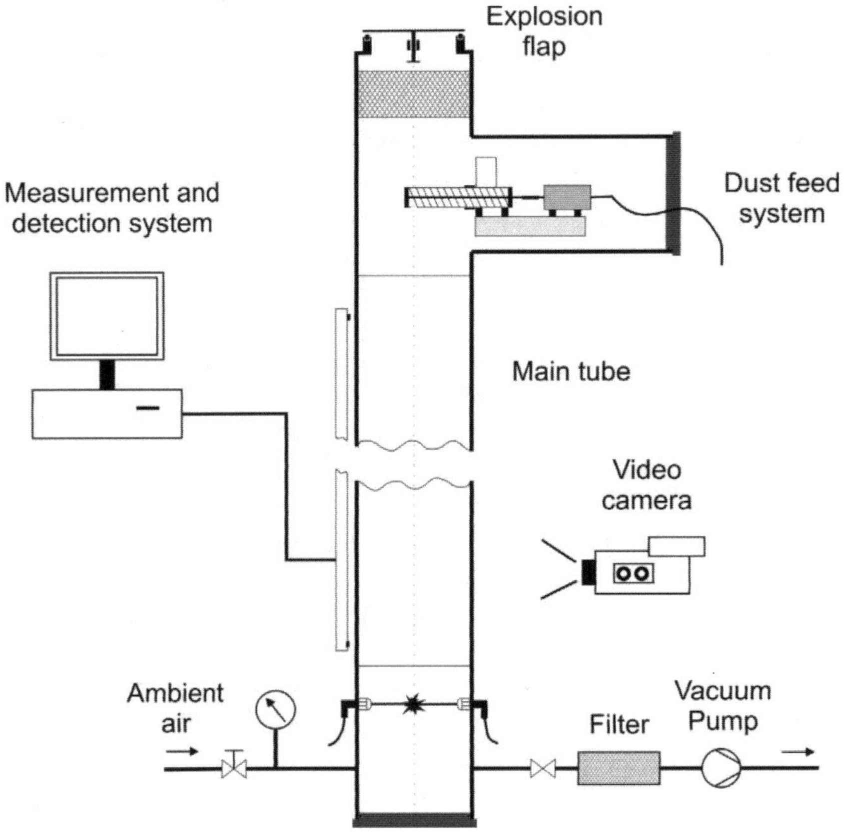


Figure 1: Testing assembly for the determination of flame velocity below atmospheric pressure conditions

The vacuum pump used for the creation of the proper pressure conditions allowed a minimum absolute pressure of around 90 mbar. The total volume of the assembly is around 40 litres. Due to Bartknecht's [3] findings that an ignition below 20 mbar is not likely and the fact that the spark igniter used provided considerably low ignition energies (~10 mJ), the pump characteristics proved to be satisfactory.

Since the experiments dealt with dust explosions that could cause injury to persons and significant damage to equipment, experimental safety was a key issue during the experiments. Main considerations were:

- control of flame spread
- pressure design
- relief systems for uncontrolled pressure build up
- fire protection

After a critical safety analysis, several modifications were made to the apparatus: flame arresters to keep the dust flame from spreading outside the combustion section and a pressure relief system to prevent the apparatus from uncontrolled overpressures.

## *2.2 Ignition System*

For the ignition of the dust/air mixture inside the combustion tube, a spark ignition system was used. The ignition source provided an electrical spark with an ignition energy of around 10 mJ, which was sufficient for the ignition of the dust/air mixture down to around 250-300 mbar. The circuit diagram of the ignition generator is demonstrated in Figure 2.

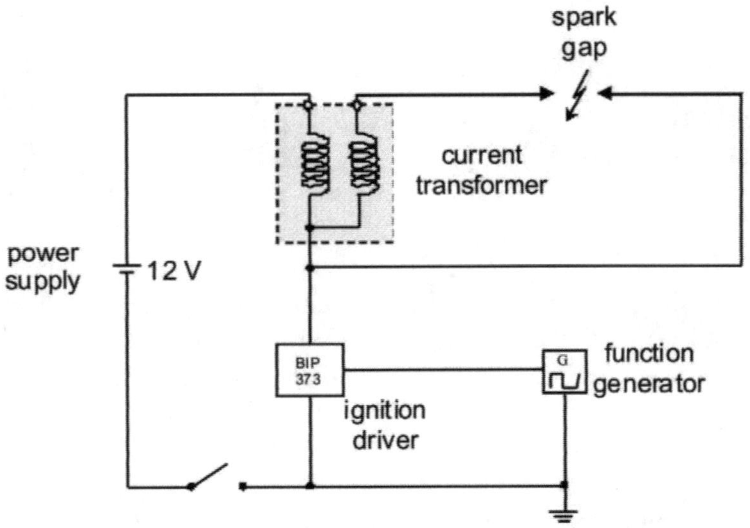


Figure 2: Circuit diagram of the spark generator used for the ignition of the investigated dust/air mixtures.

Two pointed steel-electrodes with a diameter of 2 mm were used as spark igniters. The electrodes were placed such as to form a gap of 6 mm right in the centre of the combustion tube. This electrode assembly is normally also used for dust explosion testing [8].

### *2.3 Results*

The main goal of the present work was to investigate flame propagation under non-atmospheric conditions. Investigations on phenomenological variations of the combustion process itself as well as the parameter of flame velocity were of special interest. The ignition process was also of certain interest, although in-depth research on its mechanisms was not possible with the chosen experimental setup.

#### 2.2.1 Flame shape

At ambient conditions the flame forms a parabolic shape due to wall effects [9]. During the experiments under reduced pressure conditions, a change in the appearance of the flame of the flame was noticed. The closed flame shape that is formed at ambient pressure becomes distorted with the reduction of pressure. At the given concentration, flames at 450 mbar and 250 mbar show nearly the same flame speed of around 7 m/s but appear visually as completely different.

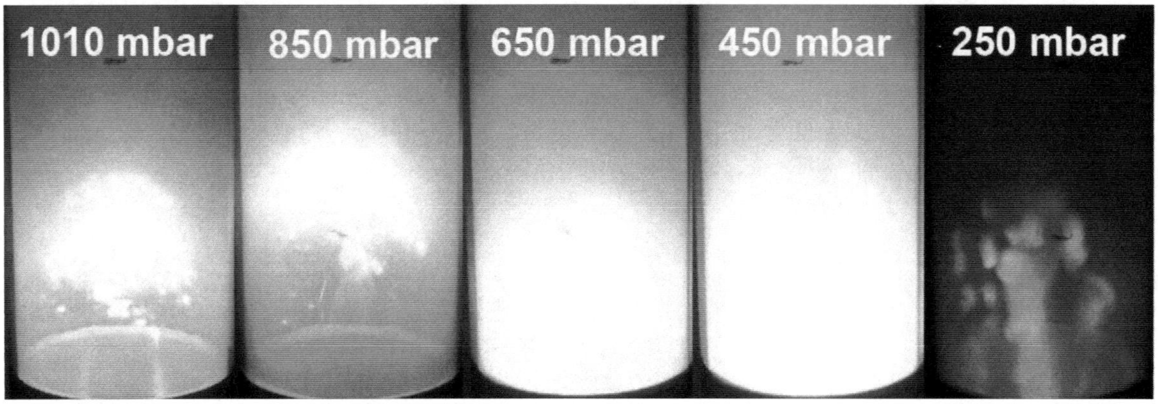


Figure 3: Development of the flame shape during the reduction of pressure (lycopodium/air at 100 g/m³)

At 250 mbar the flame does not form a uniform flame surface anymore and is dominated by single spot flames. In addition, the light intensity of the flames in the low pressure area is much lower than at ambient pressure, which may be caused by lower combustion temperatures.

#### 2.2.2 Flame speed

Investigations on flame speed were conducted in the apparatus described in figure 1. For igniting the dust/air mixture the spark generator described (IE ~10 mJ) was used. Flame speeds measured by Han et al. [5] lie at 0.5 m/s and below, but are not directly comparable with the results achieved during this experiments. The systems used works with a closed bottom end, which leads to a stronger acceleration of the flame due to a stronger increase of pressure behind the flame front.

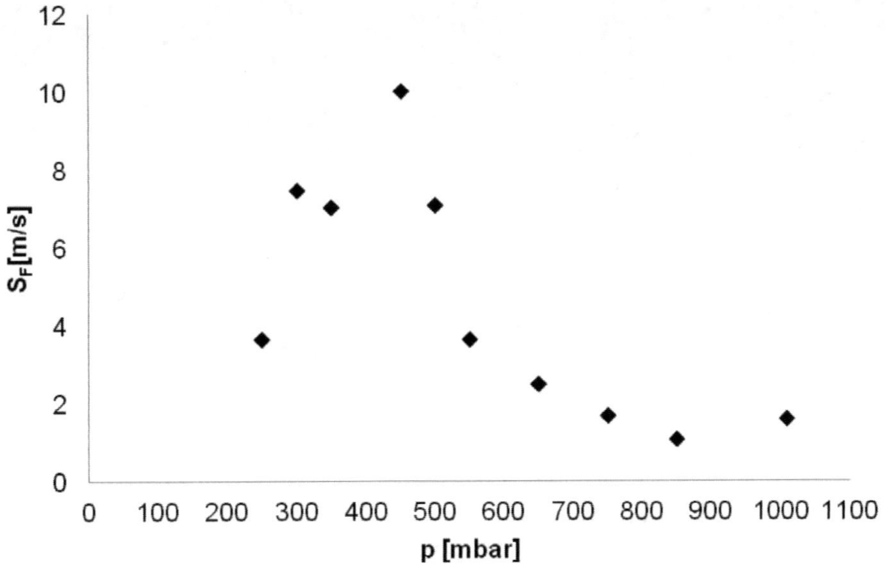


Figure 4: Pressure vs. flame velocity at 100 g/m³ lycopodium

For the experiments, the apparatus was evacuated down to the designated pressure level and ignited after around 30 s of dust feeding to assure homogeneous dust dispersion. Experiments were carried out at dust concentrations of 100 g/m³, 200 g/m³ and 300 g/m³. Figure 4 shows the results of the first set of experiments carried out at 100 g/m³.

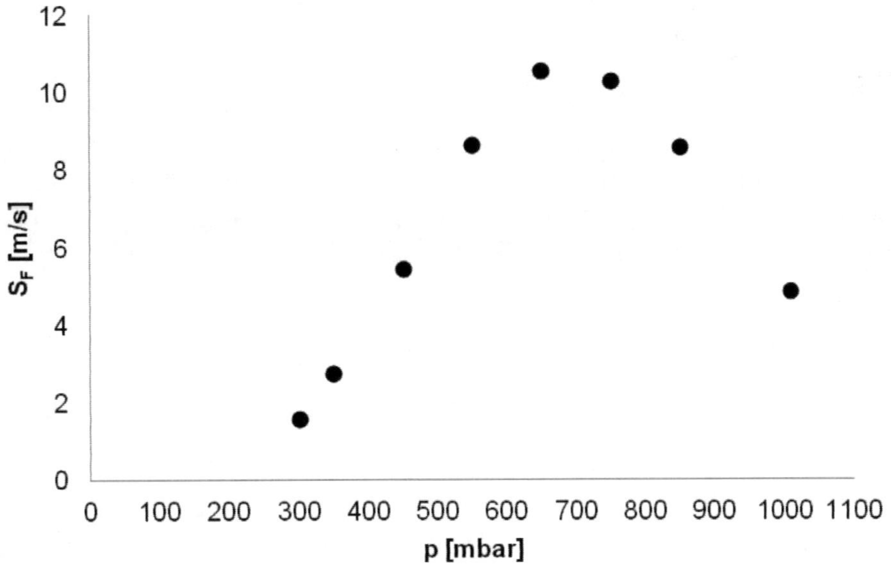


Figure 5: Pressure vs. flame velocity at 200 g/m³ lycopodium

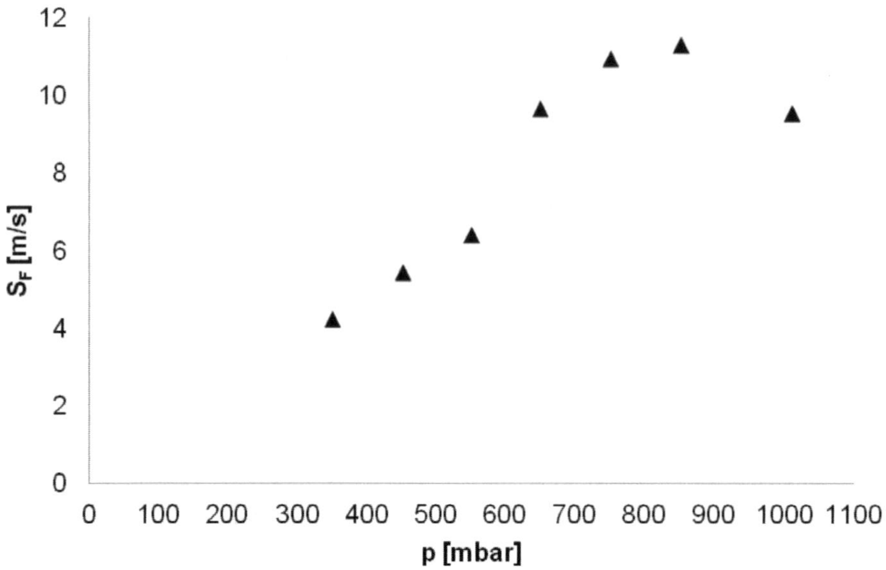


Figure 6: Pressure vs. flame velocity at 300 g/m³ lycopodium

A relatively stable flame speed was found between 1010 mbar and 750 mbar. Below 750 mbar the flame speed increases significantly reaching a maximum at 450 mbar. Further reduction of the pressure slows down flame propagation. With the spark generator used, ignition below a pressure of 250 mbar was impossible. As described above, the formation of a parabolic flame shape was not observed at very low pressures. The considerably high flame speed and the not very bright flame also led to a stronger variation of the values measured at reduced pressures.

Figure 5 shows flame speeds measured at a concentration of 200 g/m³ lycopodium in air. Corresponding to the previous set of experiments, the flame speed rises during the reduction of pressure. A section of constant flame speed was not found. Maximum flame speed was observed at around 650 mbar. Ignition became impossible at a pressure of 300 mbar. As can be seen in Figure 6, the maximum flame speed shifted further towards ambient pressure when the dust concentration was raised to 300 g/m³. The limit of ignitability shifted further and was reached at 350 mbar.

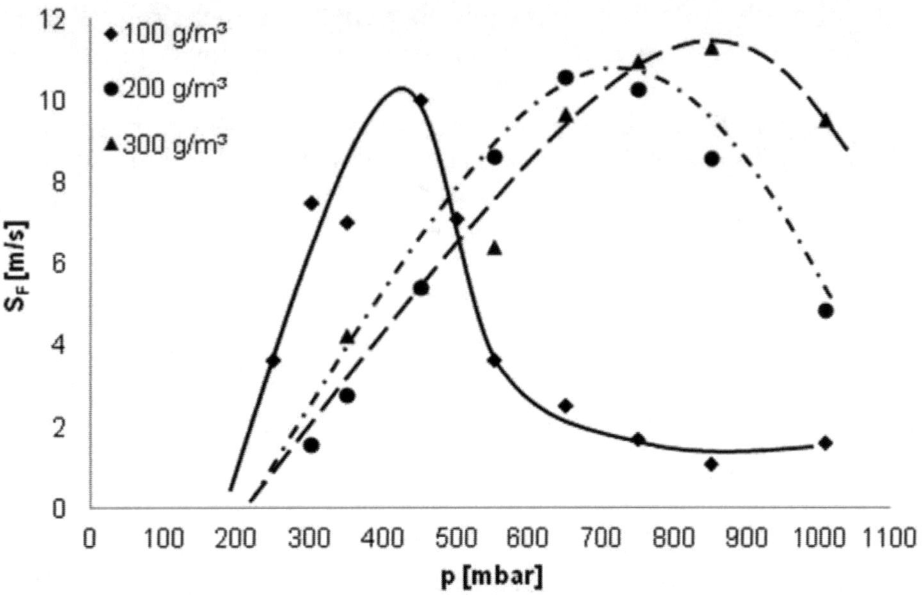


Figure 6: Pressure vs. flame velocity at 100 g/m³, 200 g/m³ and 300 g/m³ lycopodium

A comparison of all three datasets reveals that maximum flame velocities can be achieved at different pressures depending on the fuel concentration (Figure 6). The slope of the increasing as well as the decreasing flank of each curve also varies with the dust concentration. The curve at 100 g/m³ is more narrow than at the other two concentrations. The curves spread with increasing dust concentrations. Interestingly, the falling flanks of all three curves seem to join at a value of around 200 mbar when extrapolated. 1 m³ of air at 200 mbar contains the amount of oxygen that is necessary for the total combustion of 20-25 g of lycopodium. The lower explosion level of lycopodium lies right in this area. Therefore, the result seems consequent.

## 3 Conclusion

Flame velocity is not a very common parameter for the evaluation of combustible dust/air mixtures, but the experimental settings chosen in this work allowed investigations of flame spread under reduced pressure conditions. In contrast to present testing equipment, the equipment used for this study allowed a closer look at the combustion process itself. It was possible to investigate the influencing parameters on and the conditions at the ignition process in a more detailed way than possible with the 20 l Siwek chamber or the MIKE 3 apparatus.

The experiments below atmospheric conditions revealed effects that had not been predictable in the first place. Flame speeds generally increased during pressure reduction up to a certain peak value and decreased after passing that maximum speed. Flame speeds measured at low pressure conditions reached an order of magnitude higher than at ambient conditions for the investigated dust concentrations of 100 g/m³, 200 g/m³ and 300 g/m³. The velocity maximum shifted towards atmospheric pressure with increasing dust concentrations. Also the flame shape and appearance differs strongly from atmospheric conditions. Within the measurement and detection system there is still room for improvement but the results presented in this paper revealed some new perspectives in the research of flame propagation. Especially the speed measurement system has to be improved to be able to provide better results especially in the lower pressure regions.

The results may be helpful for the design of systems working below atmospheric pressure, but further research on the mechanisms and influencing parameters of this phenomenon have to be done. From the knowledge thus gained, theories of combustion processes under atmospheric conditions can be improved.

# 4 References

[1] Kern, H., Explosible Dust/Air Mixtures - Investigations on flame propagation under non atmospheric conditions, PhD Thesis, Montanuniversität Leoben, Leoben, 2013.

[2] Wieser, G., Untersuchung der Flammenfortpflanzung in Staub/Luft-Gemischen unter reduzierten Druckbedingungen, Master Thesis, Montanuniversität Leoben, Leoben, 2013.

[3] Steen, H., Handbuch des Explosionsschutzes, WILEY-VCH, Weinheim, 2000.

[4] Proust, C., A few fundamental aspects about ignition and flame propagation in dust clouds, Journal of Loss Prevention in the Process Industries 19 (2006), 104–120.

[5] Han, O. S., Yashima, M., Matsuda, T., Matsui, H., Miyake, A., Ogawa, T., Behavior of flames propagating trough lycopodium dust clouds in a vertical duct, Journal of Loss Prevention in the Process Industries 2000 (13), 449–457.

[6] Palmer, K. N., Tonkin, P. S., Coal dust explosions in a large-scale vertical tube apparatus, Combustion and Flame 1971, 17, 159–170.

[7] Kern, H., Held, K., Raupenstrauch, H., Investigations on the influence of the oxygen concentration on the flame propagation in lycopodium/air mixtures, ACHEMA Congress 2012, Frankfurt 2012

[8] Bartknecht, W., Zwahlen, G., Explosionsschutz, Springer, Berlin [etc.], op. 1993.

[9] Eckhoff, R. K., Dust explosions in the process industries, Butterworth-Heinemann, Oxford [u.a.], 1997.

# Using Incident Risk Analysis to Learn from Near Misses

**Sharon Tinker**
**ExxonMobil Research and Engineering Company**
**4500 Bayway Drive**
**Baytown, Texas 77520**
**sharon.k.tinker@exxonmobil.com**

**Kelly Keim**
**ExxonMobil Research and Engineering Company**
**4500 Bayway Drive**
**Baytown, Texas 77520**
**kelly.k.keim@exxonmobil.com**

Prepared for Presentation at
American Institute of Chemical Engineers
2014 Spring Meeting
10th Global Congress on Process Safety
New Orleans, LA
March 30 – April 2, 2014

UNPUBLISHED

# Using Incident Risk Analysis to Learn from Near Misses

**Sharon Tinker**
**ExxonMobil Research and Engineering Company**
**4500 Bayway Drive**
**Baytown, Texas 77520**
**sharon.k.tinker@exxonmobil.com**

**Other Author(s)**
**Affiliation(s)**

**Keywords:** incident risk analysis, incident investigation

## Abstract

Ineffective or missing layers of protection or protective barriers can result in incidents with consequences that range from insignificant to catastrophic. We usually refer to incidents with minor or insignificant consequences as a near miss. Investigation findings from these near misses can be as valuable to organizational learning and incident prevention as the findings from serious accident investigations. The challenge is to determine which near misses will provide valuable lessons and justify a full investigation. This paper will present a process to identify which near misses to investigate by analyzing the risk associated with each near miss.

The risk analysis considers the actual consequences of the incident, as well as the potential consequences of the incident. The more severe the actual and potential consequences, the higher the risk associated with the incident. Effective protective barriers that prevented the incident from escalating to the potential consequences are also identified and considered. The absence of effective barriers indicates higher risk, while numerous effective barriers reduce the risk. Near misses that are analyzed as high risk receive a full investigation, similar to catastrophic accidents, to capture the invaluable lessons learned.

## 1. Introduction

ExxonMobil Chemical[1] wanted to apply the safety pyramid concept to process safety, similar to evaluating personnel safety performance, in the early 2000s. Large fires and explosions were visible incidents and should be the top layers of the process safety pyramid. But how could the company capture other process safety incidents and near misses to populate the lower levels of the pyramid?

[1] References to "ExxonMobil", "ExxonMobil Chemical", "ExxonMobil Research and Engineering", "we" and "our" and any references to countries and other geographic areas are used for convenience and may refer to one or more of Exxon Mobil Corporation and its affiliates.

There was also a need to create a leading metric to monitor process safety performance. At the time, there were no process safety performance measures used widely in the industry. The low number of catastrophic process safety incidents did not provide a meaningful indication or trend of process safety performance.

ExxonMobil Chemical Company developed the Process Safety Measurement System (PSMS) in 2001 as a risk-based method to analyze the risk of each process safety incident – whether it had actual consequences or was a near miss. The risk analysis considered actual consequences, potential consequences, and barriers that prevented the potential consequences for the incident. The analysis was named the Incident Risk Analysis Tool or IRAT.

The incident risk, as indicated by the IRAT score, has been used for the following in the last several years:

- Determine investigation requirements
- Identify risks not found by the Process Hazard Analysis (PHA) process
- Analyze incidents for common causes to identify areas for improvement

IRAT was implemented at ExxonMobil's chemical plants first, and then use was expanded to the ExxonMobil refineries in 2005. The implementation of industry lagging and leading metrics for process safety by API RP-754 has provided some standardization of process safety metrics, but has not replaced the use of IRAT within ExxonMobil to evaluate incident risk. IRAT is still used to focus on the lower levels of the process safety pyramid.

This paper will briefly explain the application of the IRAT, demonstrate how the creation of the leading metric has been applied to increase focus on process safety risk in ExxonMobil, and compare the difference in incident evaluation between RP-754 Process Safety Events (PSEs) and our incident risk analysis.

## 2. Incident Risk Assessment Using IRAT

Chuck Fellman, Scott Ostrowski, and Don Williams, from ExxonMobil Chemical, have previously presented a paper titled "A Process Safety Measurement System" at the Mary Kay O'Connor Process Safety Center's International Symposium in 2005. It demonstrated the four-step process of IRAT scoring. This is the same information that I will present first to explain the application of the Incident Risk Assessment Tool.

## *2.1 Determine the Actual Consequences*

The first step is to determine the actual consequence for the incident or near miss. The actual consequence is evaluated against the levels in the corporate risk matrix and considers the impact on safety and health, the environment, the public, and direct cost financial costs. Each level is assigned a multiplier for the risk analysis calculation.

An IRAT Worksheet provides guidance for determining the IRAT score. An accurate description is written for the incident so that anyone not present will be able to understand what happened. This description is documented at the bottom of the Actual Consequence Column in the IRAT Worksheet in Figure 1. The IRAT Worksheet provides guidance to determine the level of actual consequence. If there are multiple consequences, the highest level is chosen for the analysis.

The example incident shown in Figure 1 is the activation of a safety instrumented system which prevents the incident from escalating. The highest level actual consequence is the public impact from neighbor's complaints, which is an Actual Consequence Level III. The multiplier for Actual Consequence Level III is 5.

**Figure 1: IRAT Actual Consequence Level**

| Actual Consequence Level |
|---|
| ❑ **LEVEL I** **Multiplier Value = 20**<br>– Fatality<br>– Significant financial loss<br>– Significant public disruption<br>– Significant adverse environmental impact |
| ❑ **LEVEL II** **Multiplier Value = 10**<br>– Serious injury / illness<br>– Serious financial loss<br>– Serious public disruption<br>– Serious adverse environmental impact |
| ✓ **LEVEL III** **Multiplier Value = 5**<br>– Moderate injury / illness<br>– Moderate financial loss<br>– Moderate public disruption<br>– Moderate adverse environmental impact |
| ❑ **LEVEL IV** **Multiplier Value = 2**<br>– Minor injury / illness<br>– Minor financial loss<br>– Minor public disruption<br>– Minor adverse environmental impact<br>– Near miss |
| Comments **(Describe actual consequence)**:<br>*The PE-1 reactor trip system vented the reactor due to high pressure, preventing a potential decomposition in the reactor. The vented cloud dissipated without any impact. Two neighbors called to complain about the noise created by the vent.* |

## *2.2 Determine Potential Consequence Level*

The second step is to determine the potential consequence level by developing potential consequence scenarios. These scenarios include the failure of barriers that did work or were available to prevent the incident from escalating to the potential consequence. The potential consequence scenario must be credible and explain the progression of the incident from the point at which the incident actually ended to the potential consequence. If there are multiple potential consequence scenarios with different consequences, each should be considered and the potential consequence scenario with the highest resulting IRAT score should be selected to determine the investigation level and for reporting.

The potential consequence scenario is documented at the bottom of the Potential Consequence Level column in the IRAT Worksheet as shown in Figure 2. The IRAT Worksheet provides the same guidance from the corporate risk matrix for Potential Consequence Level.

**Figure 2: IRAT Potential Consequence Level**

| **Potential Consequence Level** |
|---|
| ✓ LEVEL I **Multiplier Value = 20**<br>– Fatality<br>– Significant financial impact<br>– Significant public disruption<br>– Significant adverse environmental impact |
| ❑ LEVEL II **Multiplier Value = 10**<br>– Serious injury / illness<br>– Serious financial loss<br>– Serious public disruption<br>– Serious adverse environmental impact |
| ❑ LEVEL III **Multiplier Value = 5**<br>– Moderate injury / illness<br>– Moderate financial loss<br>– Moderate public disruption<br>– Moderate adverse environmental impact |
| ❑ LEVEL IV **Multiplier Value = 2**<br>– Minor injury / illness<br>– Minor financial loss<br>– Minor public disruption<br>– Minor adverse environmental impact<br>– Near miss |
| Comments: **(Describe assumed potential scenario & its consequence)**<br>*The PE-1 reactor experienced a high pressure excursion and the trip system did not vent the reactor. The reactor pressure continued to increase and initiated an ethylene decomposition reaction. The decomp reaction caused the reactor pressure to exceed design pressure resulting in a reactor rupture. The blast wall around the reactor failed when the hydrocarbons released from the reactor ignited in a VCE and caused fatal burn injuries to three maintenance technicians working on the PE-2 recycle gas compressor.* |

The potential consequence for the example incident is shown in Figure 2. It includes failure of the safety instrumented system and failure of a blast wall resulting in a fatality of a worker in the area. The highest potential consequence level is the fatality, which is a Potential Consequence Level I. The multiplier for Potential Consequence Level I is 20.

### 2.3 *Determine Barriers*

The third step is to determine barriers to the potential consequence. The type of barriers that can be counted are human intervention, safety instrumented systems (SIS), safety relief valves, automatic response of the process control computer, and post-release mitigation. These are shown on the IRAT Worksheet in Figure 3. The barriers should be identified in the potential consequence scenario description. The multiplier value for the IRAT calculation is determined by the number of barriers.

**Figure 3: IRAT Barriers to Potential Consequence**

**Barriers to Potential Consequence**

***Note: To be counted, the barriers must have or could have helped prevent, alert, or mitigate the actual incident from becoming the potential incident scenario. They should be independent from the initiating event and from each other.***

❑ **Human intervention** (supported by appropriate alarms, procedures, training, response PPE, etc.) (1 barrier)

- Alarms to alert (including gas detectors)
- Emergency Block Valves
- Manual Overspeed Trip
- Emergency shut-down systems (Manual activation)
- Manually activated depressuring systems
- Operator response to process computer alarm / data

✓ **Safety Instrumented Systems** (SIS) (1 barrier)

- Safety Critical shut-down / protective systems (**Automatic** response action required)

❑ **Safety relief valves and rupture discs** (1 barrier)

❑ **Process control computer**, automatic response action.

❑ **Post-release mitigation** (1 barrier)

- Fireproofing, properly designed & maintained
- Firewater, firefighting equipment, & fire team
- HAZMAT / spill control team (Trained & equipped)
- Safe Havens
- Closed disposal, spill containment system, secondary containment system (dike, process sewer, flare, etc.)
- Emergency medical response

✓ **Personal Protective Systems** (1 barrier)

- Personal Protective Equipment (PPE)
- Equipment guards & interlocks

**Other Independent Barriers** (list):

❑ ______________________________

| Total barriers | Multiplier Value |
|---|---|
| 0 | 20 |
| 1 | 10 |
| 2 | 5 |
| 3 + | 2 |

Barriers must meet the following guidelines to be counted in the IRAT calculation:

- Independent of the initiating event of the incident
- Independent of other barriers
- Reliable, with an availability of greater than 90%

In the example, the safety instrumented function that vented the reactor and stopped the incident from progressing is counted as a barrier. The blast wall that surrounds the reactor and would have directed the fire and blast wave away from people in the potential consequence scenario is also counted as a Personal Protective Systems barrier, even though it was not needed in the actual incident it was available and would have prevented the potential consequences. The process control computer could not respond fast enough to prevent the reaction from progressing to decomposition, so would not be counted as a barrier. This results in two barriers and a Multiplier Value of 5.

### *2.4 Calculate the Score to Determine Risk Level*

The final step is multiply the multiplier values to calculate the IRAT score and determine the incident risk level. The risk level categories were developed based on historical incidents. The example incident has multipliers of 5, 20, and 5 resulting in an IRAT score of 500, as shown in Figure 4. This would be a Medium Risk incident.

**Figure 4: IRAT Calculation and Risk Level**

**Actual Multiplier Value __5__ X Potential Multiplier Value__20__ X Barriers Multiplier Value __5__ = __500__ IRAT Score**

**Lower Risk = 399 and below Medium Risk = 400-799 Higher Risk = 800 and above**

### *2.5 Use of the Incident Risk Data*

Incidents with severe actual consequences are highly visible and are investigated to satisfy the demands of stakeholders at the site and outside of the site. The IRAT score provides guidance on which near misses are significant and should be fully investigated to identify learning opportunities. This is a risk-based approach to assigning limited resources to near miss investigation. As noted in the book *Incidents That Define Process Safety* by Atherton and Gil:

> "Sadly, there are many cases where the second accident occurs before the investigation of the first, or preceding near miss, has taken place. Not every event construed as an accident, or near miss, can be investigated to the extent that a major fatal accident is, although the process for doing so should be the same. This is why assessing the true potential of a near miss or minor accident is vital to ensuring that safety critical issues are identified and effort can be placed to resolve these in a timely fashion."

The immediate use of the IRAT score is to determine the level of investigation. Higher Risk and Medium Risk incidents require an investigation team and use of the company's root cause analysis tool to identify causal factors and related management system gaps. The example incident is a Tier 3 PSE and has two barriers, but the potential consequences increase the risk to a Medium Risk and indicate an investigation with full root cause analysis is justified.

Another benefit of IRAT is the increased awareness of potential consequences and barriers in the first-line supervisors and technicians. In particular, the discussion of effective barriers has helped the operating teams understand the purpose of multiple barriers and the importance of their effectiveness in preventing the Level I Potential Consequences from becoming Level I Actual Consequences. This has created a risk screening mind-set of identifying potential consequences and what is preventing those consequences.

The Medium Risk and Higher Risk incidents are also evaluated to determine if a risk has been found that was not identified by the site's PHA work processes. The incident data used to determine the IRAT score and investigation findings are entered in the corporate incident reporting database. This allows site-wide and company-wide analysis of incidents for common causes by risk level.

## 3. Comparing Incident Risk Analysis with Process Safety Events

ExxonMobil has completed an annual analysis of process safety events in Refining and Chemicals since about 2007. One aspect of the analysis is the health of process safety incident reporting and whether the lower risk, low visibility, incidents are being captured.

**Figure 5: 2012 Process Safety Pyramid by Incident Risk**

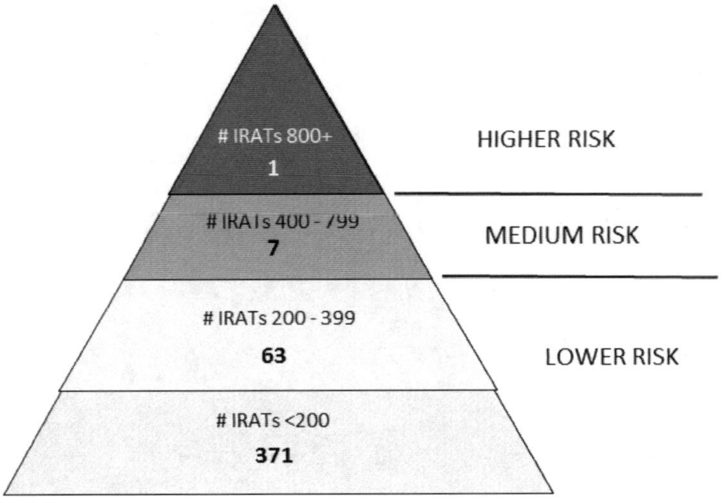


The shape of the process safety pyramid for 2012 based on process safety IRAT scoring is shown in Figure 5. The pyramid shows very healthy reporting with seven Medium Risk incidents for every Higher Risk incident and 434 Lower Risk incidents for every Higher Risk incident. This is similar to the shape of the pyramid in 2011. Both 2011 and 2012 had a much broader base than 2010 and earlier, indicating improvement in miss reporting from earlier years.

Stable and healthy incident reporting is required to use the incident risk scores as a leading indicator of process safety health. Previously, upward trends in cumulative IRAT score were an indication of increased reporting of near misses and lower risk incidents. With stable incident reporting, an upward trend in cumulative IRAT scores could indicate deterioration in management system effectiveness and an increase in site risk. This would indicate further analysis of investigation findings and management system gaps is required.

**Figure 6: 2012 Process Safety Pyramid by PSE Tier**

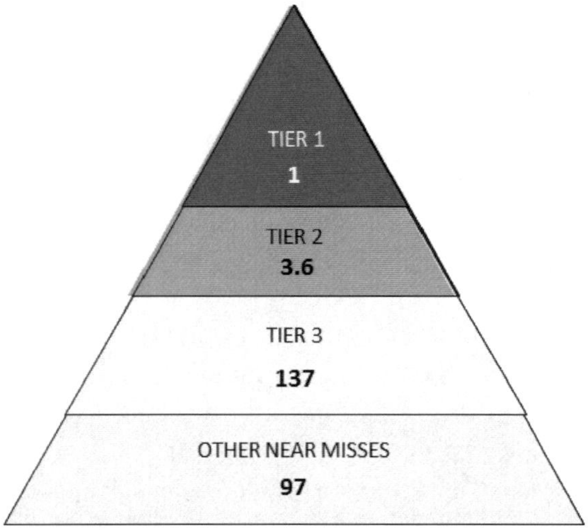


The data analysis team also looked at the same incidents by Process Safety Event tier. The process safety pyramid in Figure 6 shows the ratio of incidents by PSE Tier. There were 3.6 Tier 2 PSEs for every Tier 1 PSE. There were 234 near misses for every Tier 1 PSE and 65 near misses for every Tier 2 PSE. Over 40% of the near misses did not meet the definition of a Tier 3 Process Safety Event.

A comparison between the tiered process safety events and incident risk analysis shows that about half of the higher risk incidents were also Tier 1 Process Safety Events, but 29% of the higher risk incidents were Tier 3 Process Safety Events or other near misses, as shown in Figure 7. Figure 8 shows the contribution of near misses to risk is even higher for the medium risk incidents, with 81% of them being from near misses.

**Figure 7: 2012 Higher Risk Incidents**

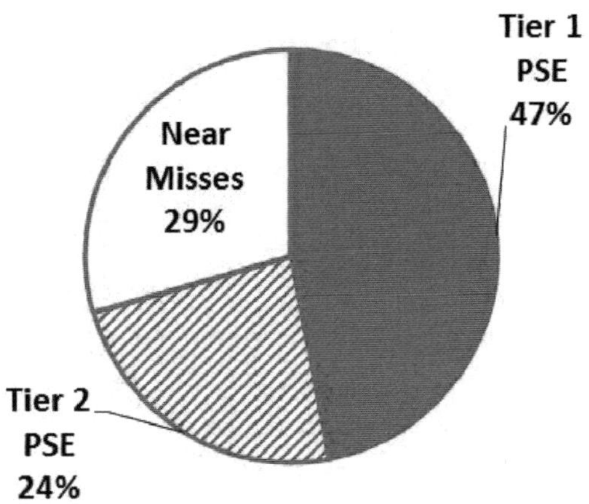


**Figure 8: 2012 Medium Risk Incidents**

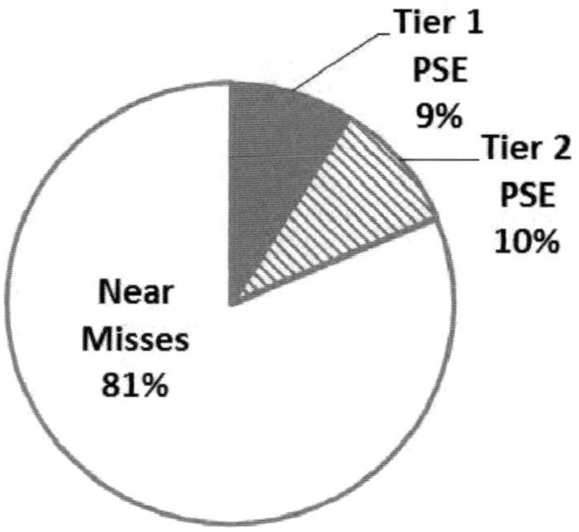


This demonstrates the value of the IRAT tool for identifying risk by using the combination of potential consequences and barriers in addition to the actual consequences that are captured by the APR RP-754 lagging indicators. Investigating only the Tier 1 and Tier 2 PSEs would be missing the opportunities to learn and improve from the higher risk and medium risk near misses and potentially allow repeat incidents that could have severe consequences.

**Figure 9: Center Cut of Incidents for Common Cause Analysis in PSE Pyramid**

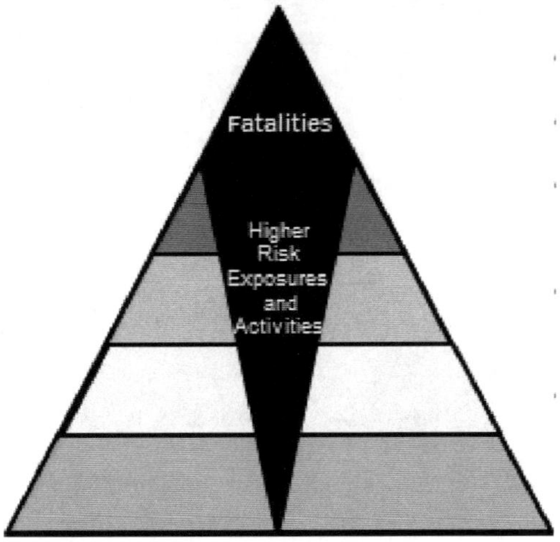


Using the incident risk to determine which incidents are included in the common cause incident analysis focuses the analysis resources on the subset of incidents that provide the most learning opportunity, as shown by the black diamond in the center of the process safety pyramid. This is extending the risk-based approach to improving process safety.

## 4. Conclusion

ExxonMobil has developed an incident risk assessment process that has been fully deployed at the chemicals and refining manufacturing sites for several years. The immediate, short-term benefits of the tool were to provide a structured method to identify higher risk and medium risk near miss incidents that justified investigation to determine root causes and corrective actions to prevent repeat incidents that could result in more serious consequences. Use of the IRAT also increased the awareness of potential consequences and barriers and the understanding of risk in the people most exposed to the risk – the technicians and first-line supervisors.

A quote from Trevor Kletz, a well-known authority on chemical process safety, on reporting and investigating near misses in the Chemical Safety Board Video *Anatomy of a Disaster* captures the importance of identifying the near misses that should be investigated:

> "Well, it is widely recognized now that you should investigate near misses, or really near accidents that are called near misses, as thoroughly as you investigate accidents. For it is only a matter of luck whether a leak of flammable liquid or gas catches fire or not."

The longer term benefit of IRAT is the development of a leading process safety metric to expand the process safety pyramid. The increased and stable reporting of all incidents, including near misses, and use of IRAT to evaluate the risk has produced a useful leading indicator to evaluate process safety health. We are beginning to evaluate trends in cumulative IRAT scores and average IRAT score by incident across the company and at individual sites to identify focus areas for improvement.

The comparison between incident risk and tiered Process Safety Events demonstrates that risk can be found before it results in severe consequences. This provides additional risk identification outside of the PHA processes. By focusing on identified risk before the risk progresses to an incident with serious consequences, the company is using IRAT to be proactive at improving management system performance and reducing risk, This is always better than being reactive after a serious incident.

## 5. References


[1] Fellman, C. H., Ostrowski, S. W., and Williams, D.D., "A Process Safety Measurement System," *Mary Kay O'Connor Process Safety Center 8$^{th}$ Annual Symposium Proceedings*, 2005

[2] Atherton, John and Gil, Fredric. *Incidents That Define Process Safety.* Center for Chemical Process Safety, New York, NY, 2008. page 210.

[3] U. S. Chemical Safety Board. *Anatomy of a Disaster.* 2008. accessed at www.csb.gov.

# 350205 Use of Kpi's for Process Safety

**Abstract:**

## INTRODUCTION

In recent years, the concept of Key Performance Indicators (KPI) and their application to Process Safety has taken prominence following incidents like the ones at BP Texas City Refinery in the USA and Buncefield Oil terminal in the UK.

KPIs could be leading or lagging indicators and their proper usage could help prevent incidents involving hazardous materials and manage Process Risk.

## TEXT

The intent of this paper is to briefly walk though the concepts of KPIs, define and explain Leading and Lagging indicators, explanation of the Four Tiers for Process Safety Events (PSE) as identified in API RP 754, usage of Tier 1 and 2 PSE KPIs to asses Company performance, and Tier 3 and 4 PSE KPI's, to monitor weaknesses in Independent Protection Layers (IPL) with the help of Pro-active and Reactive KPIs and application of these KPIs during the lifecycle management of Safety systems.

An example from the Process industry will be used to explain the concepts and how the KPIs could be identified and optimally used to reduce and manage the Process Safety Events in a Process plant. In particular, for Safety Instrumented Systems (SIS) designed based on IEC61511 / ISA84.01, the paper will indicate how such KPIs help measure the 'real' performance at any given time, rather than theoretical performance based on reliability calculations during the design phase, at both the Plant and Corporate level.

## CONCLUSION

Use of Key Performance Indicators (KPI) to reduce all levels of Process Safety Events (PSE) is gaining a lot of importance globally. It is understood that this will lead to reduction in Process related accidents and reduce harm to Personnel and the Environment. This paper hopes to re-iterate this message.

# 349873 Pssr: The Easiest and Sometimes Forgotten PSM Element


**Abstract:**

**PSSR: The easiest and sometimes forgotten PSM Element**

Pre-Startup Safety Reviews (PSSR) is to Process Safety what Personal Protective Equipment is to Personal Safety. Often, the PSSR is the last line of defense to ensure that a change or other major plant activity has not introduced an unintended or unmitigated risk. The principles behind the PSSR process are straightforward. The actual PSSR effort is basic and logical. And everyone says they do them and that they are easy. Is there data to support these statements? In this presentation, not only will PSSR best practices be reviewed but common PSSR weaknesses and improvement opportunities will be exposed along with suggested remedies and solutions.

**Kinetic identification and risk assessment based on non-linear fitting of calorimetric data**


***Charles Guinand***
*ISIC-GSCP*
*EPFL*
*Lausanne, Switzerland*

*HES-SO, Ecole d'ingénieurs et d'architectes de Fribourg*
*Institut ChemTech*
*Fribourg, Switzerland,*
*charles.guinand@epfl.ch*

***Michal Dabros***
*HES-SO, Ecole d'ingénieurs et d'architectes de Fribourg*
*Institut ChemTech*
*Fribourg, Switzerland,*
*michal.dabros@hefr.ch*

***Bertrand Roduit***
*AKTS AG*
*Sierre, Switzerland*
*b.roduit@akts.com*

***Thierry Meyer***
*ISIC-GSCP*
*EPFL*
*Lausanne, Switzerland*
*thierry.meyer@epfl.ch*

***Francis Stoessel***
*Swissi Process Safety GmbH*
*Basel, Switzerland*
*francis.stoessel@swissips.com*

Prepared for Presentation at
American Institute of Chemical Engineers
2014 Spring Meeting
10th Global Congress on Process Safety
New Orleans, LA
March 30 – April 2, 2014

UNPUBLISHED

**Kinetic identification and risk assessment based on non-linear fitting of calorimetric data**

C. Guinand[1, 2], M. Dabros[1], B. Roduit[3], T. Meyer[2], F. Stoessel[4]

[1]HES-SO Haute école spécialisée de Suisse occidentale, Ecole d'ingénieurs et d'architectes de Fribourg, Institut ChemTech, 1705 Fribourg, Switzerland

[2]Ecole Polytechnique Fédérale de Lausanne, 1015 Lausanne, Switzerland

[3]AKTS Inc., www.akts.com, TECHNOArk 1, 3960 Sierre, Switzerland

[4]Swissi Process Safety GmbH, Schwarzwaldallee, 4058 Basel, Switzerland

## Abstract

Chemical processes safety requires a thorough knowledge of the kinetic and thermal parameters regarding the chemical reactions involved. The main goal of this work was to develop a method in order to extract the kinetic parameters using different calorimetric technics, experimental design and optimization tools. In the first stage, the esterification of acetic anhydride by methanol was used as a case study for the kinetic identification. In the second stage, a risk assessment was performed using the identified reaction model. Thus, the global kinetic parameters of this reaction, as well as its activation energy and reaction orders were characterized.

**Keywords:** Calorimetry; Reaction kinetic; Design of Experiment; Thermal Safety; Risk assessment, Optimization

## 1. Introduction

In most cases, chemical reaction incidents are caused by loss of control, resulting in runaway reactions caused by process deviations in different units. These incidents could be foreseen and avoided, or at least their frequency decreased, with an appropriate process development and risk assessment tool [1].

Reaction characterization is aimed at providing a better understanding of reaction pathways and thermal behavior [2]. In order to achieve this goal efficiently, the system dynamics can be reproduced by a mathematical model whose parameters are estimated based on measured data. Such models are based on ordinary differential equations describing, on one hand the reaction kinetic law (reaction kinetic model) and on the other hand, the measurement model.

After an initial estimation, the measurement model does not represent the overall reality, but only the one in the measurement conditions. For this reason, several experiments at different scales have to be done to maximize the disturbances and explore the overall reaction kinetics [3].

Nevertheless, even with this method, the investigation can be very long due to the increasingly complex models, involving a large number of parameters and needing a huge amount of measurements. Therefore, a new approach to solve these problems has to be developed.

This approach can be summarized in two points: the first one is to plan the measurements in an optimal way to cover the experimental space [4-6]. The second is to develop a numerical method, able to extract the reaction kinetic parameters, for a given model structure, from the measurements available [7].

## 2. Theoretical part

### 2.1. *Reaction kinetic model [1, 8-11]*

Assuming a reaction *j* taking place in a homogeneous solution, the reaction rate *r* is given by the kinetic law:

$$r_j = k_j \cdot \prod_{i=1}^{I} C_i^{m_i} \qquad \text{[Eq. 1]}$$

The rate constant *k* is described by the Arrhenius law:

$$k_j = k_{j,0} \cdot \exp\left(-\frac{E_{a,j}}{RT_r}\right) \qquad \text{[Eq. 2]}$$

The concentration *C* of a component *i* is calculated in function of the number of mole *N* in the reaction mass $m_r$:

$$C_i = \frac{N_i}{m_r} \qquad \text{[Eq. 3]}$$

The transformation rate *R* of a component *i* is:

$$R_i = \frac{dC_i}{dt} = \sum_{j=1}^{J} v_{ij} r_j \qquad \text{[Eq. 4]}$$

where the sum is made over all J compounds. The thermal behavior is directly linked to the reaction kinetics; indeed any reaction implies a production or consumption of energy which can be detected as heat *q* by calorimetric techniques and is described as:

$$q_{rx,j} = m_r \left( r_j \cdot \left(-\Delta_r H_j\right) \cdot \sum_{i=1}^{I} v_{ij} M_i \right) \qquad \text{[Eq. 5]}$$

The reaction heat flow depends on many factors as for example the reaction rate *r* [Eq. 1], its enthalpy $\Delta_r H$ and the reaction mass $m_r$. The molar mass $M_i$ is used to transform the reaction enthalpy in [J·mol$^{-1}$] into the specific heat of reaction given in the units of [J·g$^{-1}$].

### 2.2. *Optimization Non-linear fitting [12]*

The system considered here is a least squares optimization problem. Its objective consists of adjusting the parameters $\theta$ of a model in order to fit in the best way a data set, e.g using a least squares approach defined as:

$$F(\theta)=\frac{1}{2}\sum_{p=1}^{P}\left(f_p(\theta)\right)^2 \qquad \text{[Eq. 6]}$$

Where $F$ is the objective function and $f_i$: $\mathbb{R}^n \longrightarrow \mathbb{R}$, $p=1,\ldots,P$ the residuals. $p$ are the studied functions (e.g. the measurement simulation) with $p \geq n$ and $n$ the number of parameters in the model.

It is assumed that it exists a vector of parameters $\hat{\theta}$ which allows the model to describe the response $y$ as:

$$y_i = Model\left(\hat{\theta},t_i\right)+\varepsilon_i \qquad \text{[Eq. 7]}$$

where $\varepsilon_i$ is the measurement error compared to the model.

For any choice of $\theta$, the residuals can be calculated as:

$$f_i(\theta)=y_i - Model(\theta,t_i) \qquad \text{[Eq. 8]}$$

For least squares fits, the optimal vector $\hat{\theta}$ is considered as the minimizer of the sum of residual squared of the model:

$$\hat{\theta} = \arg\min_{\theta}\{F(\theta)\} \qquad \text{[Eq. 9]}$$

In general, such problems are very hard to solve due to non-linearity; however, some methods are presented by K. Madsen et al.[7]. The one retained for this work is the secant version of the Levenberg-Marquardt (L-M) algorithm [7, 13, 14]. This method is used to optimize the model parameters (in this case, the kinetic parameters) without the need for analytical derivative solutions, in order to quickly find the optimal minimum using gradient methods. Indeed, they are numerically estimated.

# 3. Experimental part

## *3.1. Studied reaction*

The chemical reaction chosen to apply the method proposed above is the exothermic esterification reaction of acetic anhydride and methanol [15]. This reaction is widely used as a test for safety-oriented studies and as such, is adequate for the modeling and simulation of dynamic processes during runaway scenarios.

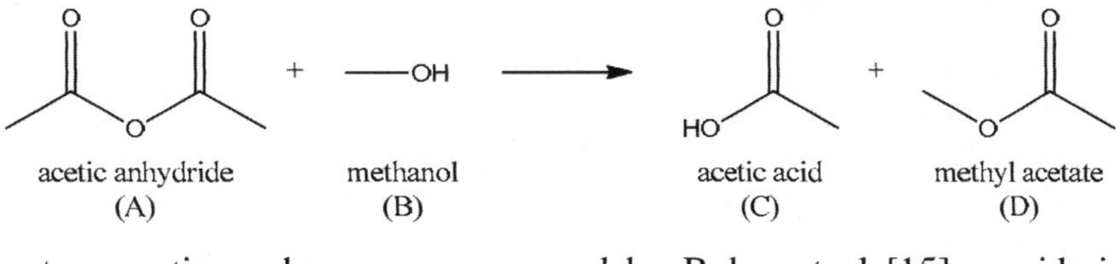


A two-step reaction scheme was proposed by Bohm et al [15] considering the autocatalytic effect of the acetic acid on the kinetic of this esterification as:

$$A + B \rightarrow C + D$$
$$A + B + C \rightarrow 2C + D$$

This reaction scheme will be used as hypothesis to define the different kinetic parameters for the reaction model.

### *3.2. Determination of the work space*

Two non-isothermal measurements at different heating rates between 25 and 300°C have been carried out at 0.1 and 4°C/min respectively in Setaram C80® and Mettler Toledo® DSC. Considering the esterification reaction, the temperature range should be below the boiling point, but nevertheless sufficiently high to have an appropriate experiment duration and with a high range of experimental temperatures. On the basis of these measurements, the working temperature range was determined in order to keep the kilo scale measurements safe and clean. Moreover no significant secondary decomposition reaction was detected in the temperature range of the measurement.

Other factors are relevant at the kilogram scale, for example the number of moles fed into the reactor ($n_{i,fd}$) and the time to feed ($t_{fd}$). Consequently, in order to determine all the different ranges, a rough optimization of the reaction model parameters was performed on the preliminary measurements. Then, using the physical properties (boiling points, heat capacities) and the kinetic parameters, it was possible to predict the process behavior at some different temperatures (Table 1).

**Table 1.** Variable ranges for the Design of Experiment

| | **Temperature ($T_{r,set}$)** | **Time to feed ($t_{fd,B}$)** | **Number of added moles ($n_{feed,B}$)** |
|---|---|---|---|
| Units | [°C] | [min] | [mol] |
| Min | 40 | 60 | 0.7 |
| Max | 60.3 | 120 | 1.5 |

### *3.3. Design of Experiment (DoE)*

An experimental design, based on the Wootton, Sergent and Phan-Tan-Luu's (WSP) algorithm and a Space Filling Design method [16] was used to choose six experimental points, as shown in Table 2. Each of the three factors shown in Table 1 appears at a minimum of three levels – a criterion which was needed to extract the value of the activation energy.

### *3.4. RC measurements*

The RC experiments are performed under isothermal fed-batch conditions in a Mettler Toledo® RC1e where the methanol (B) is fed at different rates. The acetic anhydride (A) is initially present in the reactor, up to 1.5 moles, and the acetic acid (C) and methyl acetate (D) are produced during the process. The results are shown in Table 2.

**Table 2. DoE study:** Space Filling Design results resulting from WSP algorithm[16] performed on 10000 points in the candidate set randomly generated with minDist = 0.8500 and Cov = 0.0020. Results: Feed rate, time of experiment Obtained heat ($Q_r$), average heat capacity ($c_{p,average}$) and reaction enthalpy ($\Delta_r H$) for the measurements at RC scale.

| | DoE plan | | | Results | | | |
|---|---|---|---|---|---|---|---|
| Exp. | $T_{r,set}$ | $n_{feed,B}$ | $t_{fd}$ | $m_{fd}$ | $c_{p,average}$ | $Q_r$ | $\Delta_r H$ |
| | [°C] | [mol] | [s] | [g·min$^{-1}$] | [J·K$^{-1}$·g$^{-1}$] | [KJ] | [kJ·mol$^{-1}$] |
| 1 | 50.15 | 1.09 | 5425 | 0.39 | 2.08 | -61.30 | -56.17 |
| 2 | 60.30 | 0.69 | 3843 | 0.35 | 2.05 | -36.26 | -52.65 |
| 3 | 41.35 | 0.69 | 7247 | 0.18 | 2.09 | -43.59 | -63.58 |
| 4 | 40.00 | 1.49 | 7034 | 0.41 | 2.05 | -76.38 | -51.13 |
| 5 | 60.04 | 1.48 | 3776 | 0.76 | 2.10 | -84.10 | -56.65 |
| 6 | 60.03 | 0.69 | 7200 | 0.19 | 2.10 | -40.78 | -58.69 |
| **Average** | | | | | **2.08** | | **-56.5 ± 4.7** |

The determined mean reaction enthalpy corresponds to -56.5 ± 4.7 [kJ·mol$^{-1}$], a lower value than the one found in literature, 67.3 ± 3.0 [kJ·mol$^{-1}$] [17] as well as 65.99 [kJ·mol$^{-1}$] based on online tabulated standard enthalpies of formation [18].

This deviation could be explained by a too high working temperature, causing heat losses by evaporation considering a boiling point of 56°C and 64°C for methyl acetate and methanol respectively. Nevertheless, these data are a good input to create the data basis for the optimization and to create a reaction model.

## 4. Results and discussion

### *4.1. Model*

Using the non-linear optimization based on the Levenberg-Marquardt algorithm on the obtained calorimetric data (DSC, RC) and the reaction scheme proposed by Bohm[15], it was possible to determine the kinetic parameter values (Table 3).

**Table 3.** Obtained kinetic parameters for the esterification of acetic anhydride with methanol considering an autocatalytic effect of the acetic acid (product).

| **Reaction 1** | | | | |
|---|---|---|---|---|
| | $k_0$ | $E_a$ | $\Delta_r H$ | Orders |
| Units | [g$^{mA+mB-1}$·s$^{-1}$·mol$^{-(mA+mB-1)}$] | [J·mol$^{-1}$] | [J·g$^{-1}$] | [-] |
| Value | 4.462e8 | 71678 | -662.3 | $m_A$=0.96<br>$m_B$=0.71 |
| **Reaction 2** | | | | |
| Units | [g$^{(mA+mB+mC-1)}$·s$^{-1}$·mol$^{-(mA+mB+mC-1)}$] | [J·mol$^{-1}$] | [J·g$^{-1}$] | [-] |
| Value | 3.293e7 | 51815 | -260.1 | $m_A$=1.16<br>$m_B$=0.6<br>$m_C$=0.52 |

### *4.2. Validation*

In order to validate the model, the simulation was confronted to two different measurements at the milligram (DSC) and kilogram (RC1) scales, as shown in Figure 1.

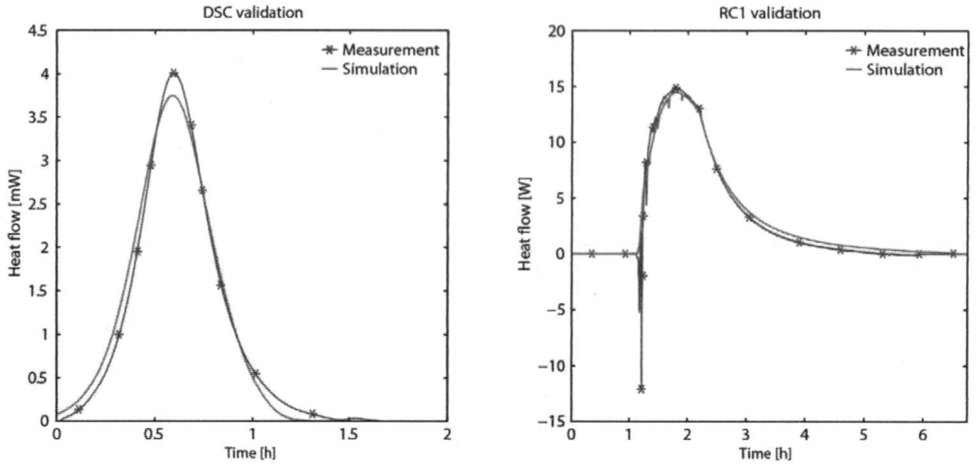


**Figure 1** Validation: comparison of the measurements (DSC at 1.5K/min: left; RC1 at 62°C: right) versus their respective simulations using the optimized reaction kinetic model.

It can be observed that the model fits the RC measurement somewhat better than the DSC data. This can be explained by a large number of measurements at this scale and focused on a narrow temperature range. In other words, the DSC leads to an extrapolation (large temperature range) where RC1 measurement are closer to interpolation. Hence, the optimization is more disturbed by these latter which leads to an optimum displaced toward the kilogram scale. Other facts as micro-mixing or difficulties in the baseline definition may have caused the slight misfit observed in the DSC validation.

Another point to keep in mind is the fact that a model is valid only in the range where it was defined. In this case, the goal is to perform a scale-up of this reaction; consequently, a good fit at the kilogram scale points out the right direction.

### *4.3. Risk assessment*

In order to get a safe process, several points regarding the thermal safety should be addressed. The best way to assess the risk, is the cooling failure scenario [1].

The main variable to consider in a process risk assessment is the maximum temperature that can be reached in the reaction vessel as a result of a cooling failure (Maximum Temperature of the Synthesis Reaction, MTSR). Would this temperature be sufficiently high to initiate secondary reactions? Would the consequences be worse? What would the level of temperature be after such events? Is it the worst moment during the process for cooling failure to happen?

### 4.4. *Batch process*

Based on the reaction kinetic model, the thermal behavior of the autocatalytic esterification can be evaluated. Thus, the worst case can be assessed depending on the kind of reactor and cooling system used. A simulation with a standard reactor of $2.5m^3$ was used to demonstrate the thermal risk assessment procedure.

#### 4.4.1. Cooling failure scenario

The scenario in case of loss of control from a starting temperature of 25°C would lead to a MTSR of 236°C ($\Delta T_{ad}$ = 211°C). Such a temperature would lead to loss of containment with spillage of flammable compounds and possibly a secondary explosion. Moreover, the Time to Maximum Rate in case of adiabatic conditions is very short ($TMR_{ad}$ = 1.36h). The two limit temperatures defining the thermal stability of the reaction mixture were determined using simulation for 24h and 8h are respectively -8°C ($T_{D24}$) and 4°C ($T_{D8}$). Thus, the lack of control of the reactor temperature results in a thermal explosion.

Another point to consider is the boiling point of the reaction which is about 60°C at the end of the process. This temperature is reached during the runaway, and the energy available at this temperature is sufficient to result in a pressure increase, leading eventually to a burst of the reactor. Consequently the Maximum Technical Temperature (MTT) should be based on the boiling point of the components involved.

All these points lead to a final conclusion: in order to work under safe conditions, the reaction cannot be performed as a batch without emergency measures. It requires a reliable and powerful cooling system or should preferably be performed in the fed batch mode allowing a better control of the reaction rate.

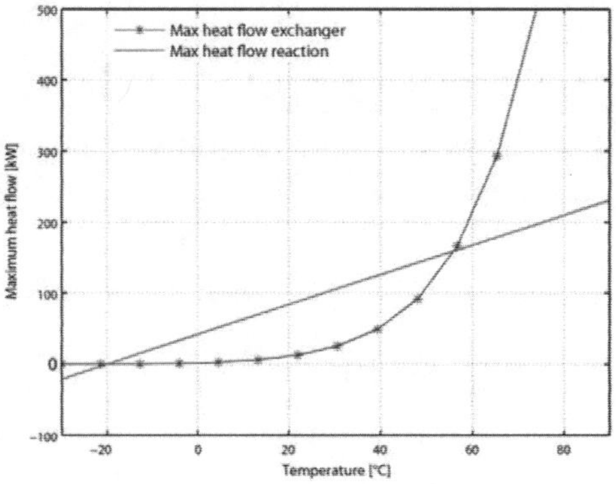


**Figure 2** Semenov diagram: comparison of the maximum heat release of the autocatalytic esterification with the heat removal of the cooling system, considering standard reactor of $2.5m^3$ (Area = 4.6 $m^2$; $T_c$ = -20 °C; U = 500 $W \cdot K^{-1} \cdot m^{-1}$) with an equimolar mixture (15kmol of A and B).

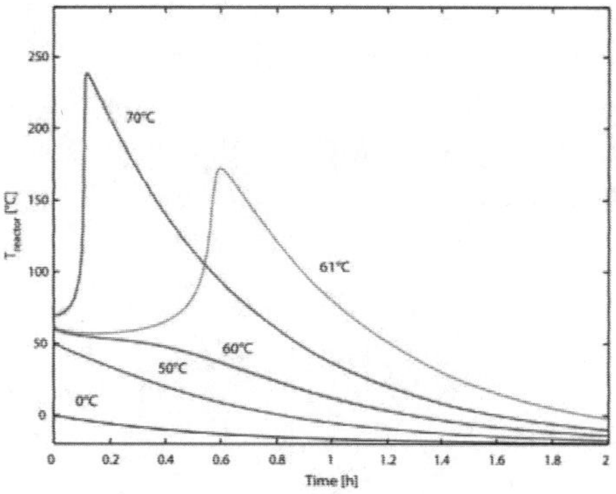


**Figure 3** Influence of the initial temperature of the process for a constant temperature of the coolant ($T_c$ = -20 °C)

#### *4.4.2. Semenov diagram and assessment of cooling capacity*

This criterion allows the discrimination of stable operation conditions from a runaway situation. The kinetic model was used to evaluate the maximum heat flow leading to the Semenov diagram (Figure 2) which can be used to define a working area for this reaction.

Nevertheless, it should be kept in mind that a too high temperature can produce small deviations resulting in excess heat production and leading to a runaway. In case of an autocatalytic system, a small deviation in temperature will produce more product, thereby accelerating the main reaction and leading to a quicker runaway.

Moreover, the kinetic model can be used to determine the influence of the initial temperature under isoperibolic control as shown in Figure 3.

The results of these simulations show that a too high initial temperature can lead to a runaway. In such conditions for the autocatalytic esterification, the Semenov Diagram predicts a runaway region around 60°C which is confirmed here.

Moreover, this temperature is higher than the boiling point of the final reaction mixture. This latter can be used as a barrier only if the process is designed to work under such conditions (reflux). Consequently, the risk of a pressure increase leading to an explosion is not negligible in the present case.

#### *4.4.3. How to control this reaction*

Different types of temperature control are available for batch processes such as adiabatic, isoperibolic or polytropic. In this project, only the polytropic control will be discussed, whereby the reaction temperature is first brought to a setpoint which is maintained by varying the jacket temperature.

As shown above (Figure 3), the initial temperature is a primordial factor. Hence, if this latter is poorly designed, it can lead to a loss of control and runaway reaction. Nevertheless, in order to explore the process behavior, knowledge of the dynamic behavior of an industrial reactor is necessary. Consequently, to define this last point, the different time constants for the heating and cooling were determined at different scales [19].

The time constants show how the cooling system responds to a deviation in temperature. Indeed, two factors need to be taken in account: the heat exchange variables (U, A, $c_p$ and mass) and the controller (usually PID). In practice, even if the controller is designed to deliver a fast response, the heat carrier response will be limited due to the thermal inertia of the system at the industrial scale, therefore, the adjustment is not instantaneous.

Consequently, the reaction temperature must be chosen appropriately to obtain an economically reasonable process time, meaning as short as possible and sufficiently low to control the heat release rate. In this case, a low temperature would be preferred in order to keep the process safe. Unfortunately, maintaining a low temperature is not economically viable. Moreover, in order to keep isothermal conditions, the jacket temperature has to respond really quickly. Consequently, depending on the plant available, this kind of conditions may be difficult to achieve in practice. If this is the case, a possible way to manage this kind of reaction is to carry out a step control at the beginning: start at a low temperature and then let the reaction progresses adiabatically until the temperature reaches a switch value ($T_s$), at which time the cooling system is started (Figure 4).

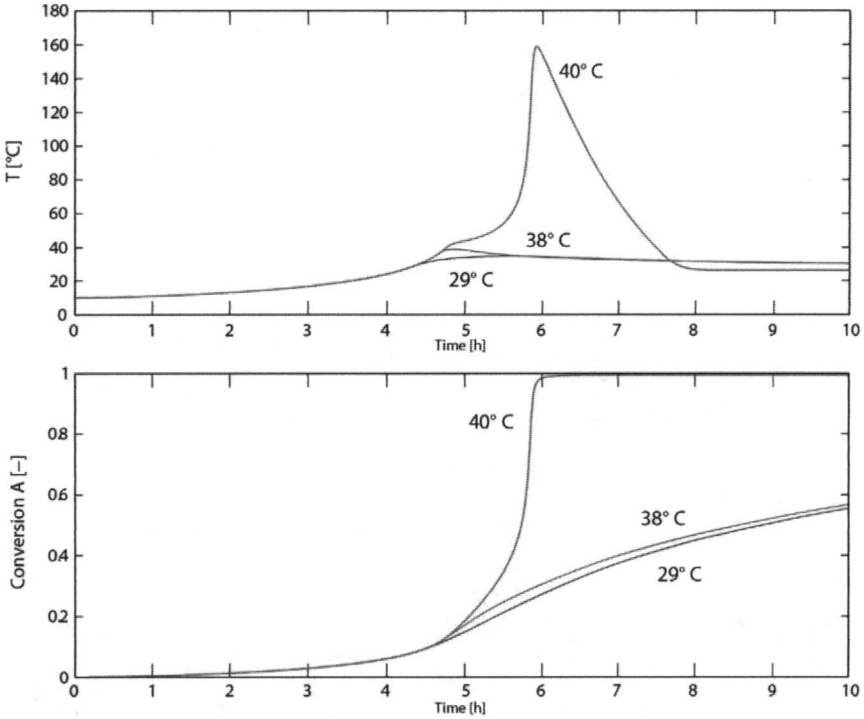


**Figure 4** Reactor temperature as a function of time and respective conversion for different switching temperatures of the cooling system from a starting temperature at 10°C.

As it can be observed, even with this method, the conversion remains poor for batch processes under control. Another critical point remaining is the maximum Temperature reached in case of Cooling Failure ($T_{cf}$). Indeed, as mentioned previously, the boiling temperature of the final reaction mixture is low (around 60°C). Consequently, in order to avoid a rise of pressure, the $T_{cf}$ should remain under the boiling point. Unfortunately, this condition can never be achieved in case of batch processes; only the reactor temperature can be adjusted to keep the temperature under control.

This assessment remains global and does not take in account supplementary equipment such as burst disk or condenser. Nevertheless, it was shown that for this kind of reaction, batch processes are not the most suitable. Moreover, the worst moment for a cooling failure is, in the case of this process, present from the start, as the accumulation is maximal, making this mode not safe at all for such autocatalytic reaction.

Considering all these points, in order to find a viable a safe solution, semi-batch process will be explored from a thermal safety point of view.

## 4.5. *Semi-Batch process*

### *4.5.1. How to control this reaction?*

In semi-batch operation, many strategies can be used to improve the process safety: the temperature control strategy, the feed control strategy, and also the choice of reactant(s) to be initially added and, respectively, to be fed [1]. The most stable reactant is generally chosen to be initially added. Consequently, due to its higher boiling point, the acetic anhydride was chosen. Thus, the methanol is used as the fed reactant in order to limit the accumulation and evaporation. Moreover, in order to maintain isothermal conditions, the use of polytropic control is preferred.

The feed may be controlled in different ways such as:

- Constant feed rate: should be sufficiently low to avoid or at least to minimize reactant accumulation.
- Working under constraints based on physical considerations (volume at the end of the operation, flow rate,...) and safety considerations (the system must remain safe under normal operating condition,...) [20].

Nevertheless, the choice of the reactor temperature and feed strategy remains of primary importance during the thermal safety assessment. The next section will discuss a semi-batch process working under constraints based on safety considerations.

### *4.5.2. How to determine an optimal feed profile working under constraints*

As mentioned previously in the batch process section, to maintain isothermal conditions, the heat exchanger system should be able to remove the heat release rate of the reaction at any moment, especially at the maximum. In order to control the reaction, the temperature that would be reached in case of a cooling failure $T_{cf}$ can be used as constraint: if the calculated $T_{cf}$ is greater than the maximum safe temperature ($T_{max}$), the feed is stopped until the content reaches a stable point anew [1]. The $T_{max}$ is usually defined by the process itself or on the data basis of the reaction mixture (boiling point, decomposition temperature...).

The constraint is applied as follows: the simulation is performed under normal conditions and then every 60 seconds, the simulation is performed under cooling failure conditions (adiabatic). If the $T_{cf}$ is higher than the set point $T_{max}$, the feed is stopped. The result is the evaluation of the best feed profile satisfying the constraint as shown in Figure 5.

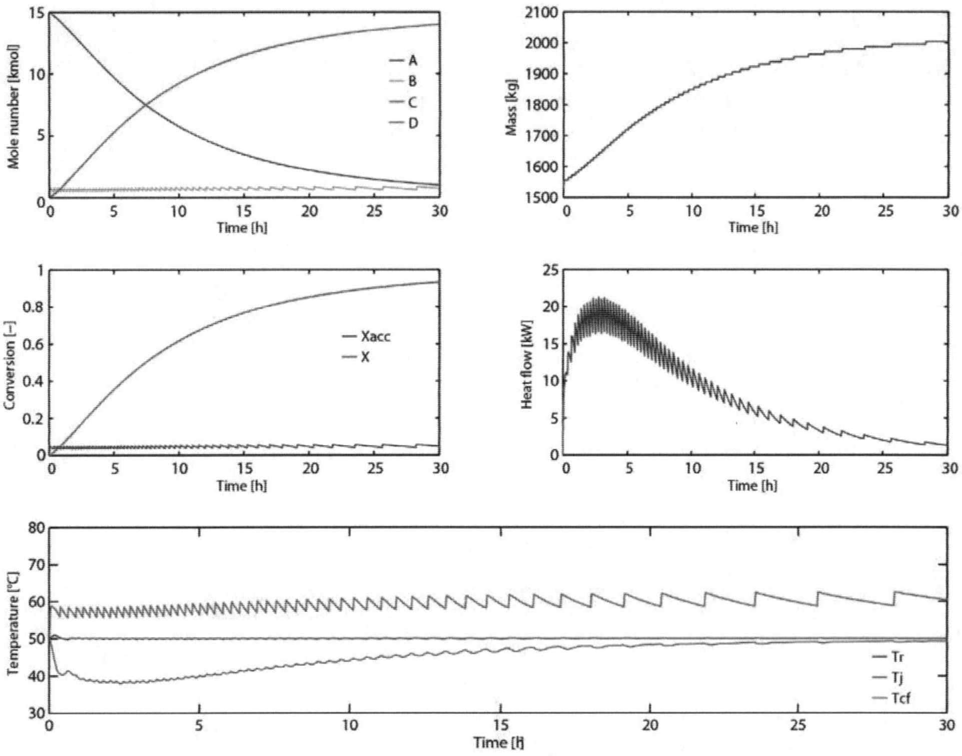


**Figure 5** Concentration, mass, conversion (of A, and accumulation), heat flow profiles obtained for a semi-batch process working under $T_{cf}$ constraint

With this strategy, the feed profile is optimized in order to ensure that $T_{max}$ may not be surpassed even in case of failure and considering the autocatalytic reaction.

## 5. Conclusion

A first insight into a way of investigating reaction kinetics using non-linear model fitting has been presented. It was possible with limited experimental data to determine the reaction's kinetics and to predict its behavior under different operation conditions.

Industrial reactors dynamics were simulated, in order to optimize the thermal safety of a semi-batch reactor. Furthermore, the application of constraints has led to an optimized feed profile keeping the process thermally safe.

In order to validate the optimal feed profile, the next step in this study involves performing an experiment at industrial scale. To achieve this purpose, it is necessary to apply the scale-down approach presented by Zufferey [19]; a method used to simulate industrial reactor at laboratory scale.

In the future, an interesting study might involve the extension of this investigation procedure to other kind of reaction, leading to a generalized method to get reaction model and kinetics from calorimetric measurements.

## 6. Symbols

| | | |
|---|---|---|
| $A$ | Area | ($m^2$) |
| $C$ | Concentration | ($mol \cdot kg^{-1}$) |
| $c_p$ | Specific heat capacity | ($J \cdot kg^{-1} \cdot K^{-1}$) |
| $Cov$ | Covariance | (-) |
| $DoE$ | Design of Experiment | |
| $E_a$ | Activation energy | ($J \cdot mol^{-1}$) |
| $F$ | Objective function | (-) |
| $k$ | rate constant | ($\propto$) |
| $k_0$ | Frequency factor | ($\propto$) |
| $M$ | Molar mass | ($kg \cdot mol^{-1}$) |
| $m$ | Mass | (kg) |
| $MinDist$ | Minimal distance between the points | (-) |
| $MTSR$ | Max. Temperature of Synthesis Reaction | (K) |
| $MTT$ | Max. Temperature for Technical reasons | (K) |
| $N$ | Number of moles | (mol) |
| $n$ | Number of parameters in the model | (-) |
| $p$ | Studied function | (-) |
| $PID$ | Proportional-Integral-Derivative | (-) |
| $Q$ | Thermal energy | (J) |
| $q$ | Heat flow | (W) |
| $r$ | Reaction rate | ($mol \cdot kg^{-1} \cdot s^{-1}$) |
| $R$ | Transformation rate | ($mol \cdot kg^{-1} \cdot s^{-1}$) |
| $TMR_{ad}$ | Time to maximum rate adiabatic | (s) |
| $T$ | Temperature | (K) |
| $t$ | Time | (s) |
| $U$ | Overall heat transfer coef. | ($W \cdot m^{-2} \cdot K^{-1}$) |
| $WSP$ | Wootton, Sergent, Phan-Tan-Luu's algorithm | |
| $X$ | Conversion | (-) |
| $\Delta_r H$ | Enthalpy of reaction | ($J \cdot mol^{-1}$) |
| $\Delta T_{ad}$ | Adiabatic temperature rise | (K) |

***Greek***

| | |
|---|---|
| $\varepsilon$ | Error |
| $\theta$ | Parameters |
| $\nu$ | Stoichiometric coefficient |

***Subscript***

| | |
|---|---|
| $acc$ | Accumulation |
| $cf$ | Cooling failure |
| $f$ | Final |
| $fd$ | Feed |
| $i$ | Component |
| $j$ | Number of the reaction |
| $r$ | Reactor |
| $rx$ | Reaction |
| $s$ | Switch |
| $set$ | Set point |

***Superscript***

$m$ Partial reactant order

## 7. References


[1] Stoessel, Francis. *Thermal Safety of Chemical Processes: Risk Assessment and Process Design*. Wiley. Weinheim. 2008.

[2] Blackmond, Donna G. "Reaction Progress Kinetic Analysis: A Powerful Methodology for Mechanistic Studies of Complex Catalytic Reactions," *Angewandte Chemie International Edition,* 44, 28, 2005

[3] Stegmaier, Johannes, Skanda, Dominik, et al. "Robust Optimal Design of Experiments for Model Discrimination Using an Interactive Software Tool," *PLoS ONE,* 8, 2, 2013

[4] Zaglauer, Susanne "Bayesian design of experiments for nonlinear dynamic system identification," Proceedings of the 5th International ICST Conference on Simulation Tools and Techniques, Desenzano del Garda, Italy, Institute for Computer Sciences, Social-Informatics and Telecommunications Engineering, 2012.

[5] Lazic, Z.R. *Design of Experiments in Chemical Engineering*. Wiley. 2004.

[6] Tinsson, W. *Plans d'expérience: constructions et analyses statistiques*. Springer. 2010.

[7] Madsen, K., Nielsen, N. B., et al. *Methods for non-linear least squares problems (2nd ed.)*, IMM, 2004.

[8] Ingham, J., Dunn, I.J., et al. *Chemical Engineering Dynamics: An Introduction to Modelling and Computer Simulation*. Wiley. 2008.

[9] Levenspiel, O. *Chemical Reaction Engineering*. 1999.

[10] Fogler, H.S. *Elements of chemical reaction engineering*. Prentice Hall PTR. 2006.

[11] Capellos, C. and Bielski, B.H.J. *Kinetic systems: mathematical description of chemical kinetics in solution*. R. E. Krieger Pub. Co. 1980.

[12] AKTS-*Thermokinetics Software, Reaction Calorimetry Version*, http://www.akts.com/reaction-calorimetry.html

[13] Nocedal, J.A. and Wright, S.J. *Numerical optimization*. Springer-Verlag New York. 1999.

[14] Wojciechowski, B. and Rice, N. *Experimental Methods in Kinetic Studies*. Elsevier Science. 2003.

[15] Bohm, Stefan, Hessel, Günther, et al. "Auto-catalytic effect of acetic acid on the kinetics of the methanol/acetic anhydride esterification," *Institute of Safety Research*, 2005

[16] Santiago, J., Claeys-Bruno, M., et al. "Construction of space-filling designs using WSP algorithm for high dimensional spaces," *Chemometrics and Intelligent Laboratory Systems,* 113, 2012

[17] Duh, Yih-Shing, Hsu, Chang-Chia, et al. "Applications of reaction calorimetry in reaction kinetics and thermal hazard evaluation," *Thermochimica acta,* 285, 1, 1996

[18] "National Institute of Standards and Technology (NIST)", http://webbook.nist.gov/chemistry (accessed February 2014).

[19] Zufferey, Benoît. *Scale-down approach: chemical process optimisation using reaction calorimetry for the experimental simulation of industrial reactors dynamics*. Swiss Federal Institute of Technology, Lausanne. 2006.

[20] Ubrich, O. *Improving safety and productivity of isothermal semi-batch reactors by modulating the feed rate*. Swiss Federal Institute of Technology, Lausanne. 2000.

# 349619 Integration of Learned Knowledge into the Technical Information System

**Abstract:**

How to integrate acquired knowledge of experienced employees into the technical information systems to drive continuous process safety improvement through training and risk assessment. Creation of mentoring programs and networking will minimize organization knowledge loss when key personnel are lost due to retirement, organization changes, or asset divestitures. This will create interaction locally, throughout the organization, and industry. It will enhance employee training and development, understanding of hazard and risk identification and mitigation systems, improve process technical understanding, and reduce the potential for a catastrophic event.

# 349712 Control Systems Integrity Review - Selective Application of Controls Systems HAZOP (CHAZOP) Study

**Abstract:**

A Control Systems Hazard and Operability Study (CHAZOP) provides a means of systematically reviewing the design and operation of a control and safety system to identify potential hazards and/or operability problems. It focuses on how a control or safety system may deviate from the design intent, and draws on the expertise of the team members and their past experience with the design and/or operations of similar facilities.

The CHAZOP technique stemmed from the evolution of control and protection system designs, from single-element analog devices a couple of decades ago, to the increased application of electronics, logic solvers with redundant processors and voting capabilities, and feedback loops to identify various sensor and end-device failures, including covert. Although the control and protection systems were evolving, many of the HAZOP Study applications and Teams did not always involve participants that could properly weave these new control and protection system features into the HAZOP Study scenarios. To address this gap, CHAZOP was developed to help ensure that the safety evaluation objectives of the HAZOP were achieved.

For high-quality HAZOP Studies that are supported by individuals with the correct "skill sets" and thoroughly address instrumentation issues (to the extent necessary for the conservative scenarios evaluated for the HAZOP Study), it can be argued that a complete CHAZOP isn't necessary. CHAZOP can also be a significant effort that is difficult to justify to management, if a HAZOP Study has been performed. However, there can still be value in reviewing some of the key characteristics associated with the Integrated Control and Safety Systems (ICSS) that are applied at modern process facilities. With the HAZOP Study Team (and any necessary supplemental instrumentation specialists), there can often be value in a general review of network architecture, workstations, switching networks, key communications/power cable routing, data transfer strategies, human-machine interface issues, back-up strategies, and power sources/reliability. This Control Systems Integrity Review can be implemented in a focused manner to review single failures and potential vulnerabilities.

The objective of this paper is to review appropriate applications of CHAZOP and to define a streamlined approach for Control Systems Integrity Reviews that can benefit many capital and instrumentation retrofit projects in process facilities. Tips will be provided to maximize the effectiveness of the Team's efforts when performing the Control Systems Integrity Review.

equion

Energía para la vida

# Are We in Control of Our Safety Critical Equipment in Drilling Operations?

Claudio Castaneda– April/2014

equion
Energía para la vida

## From hazard to major accident

10TH GLOBAL CONGRESS ON PROCESS SAFETY

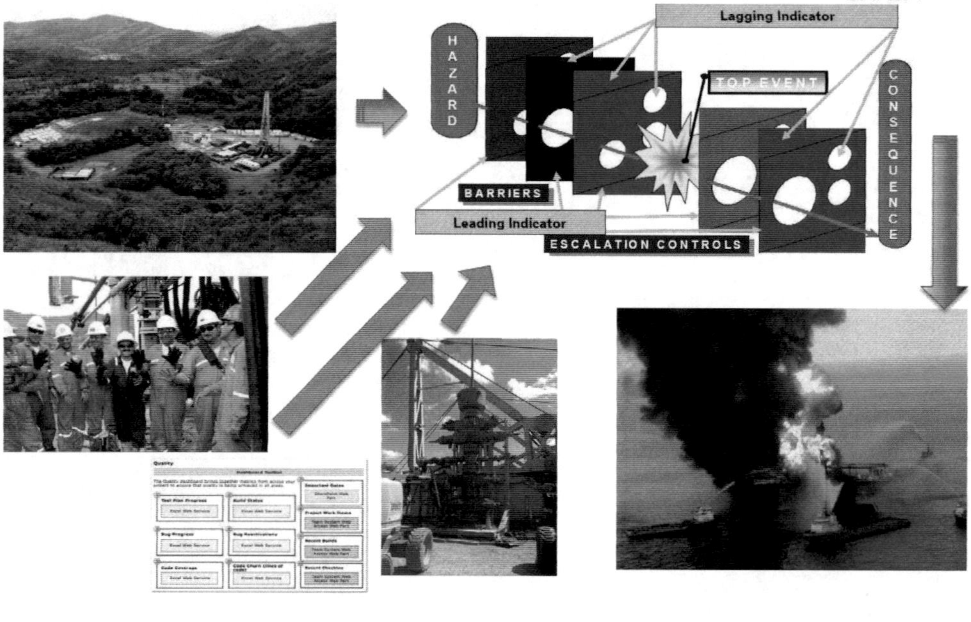

equion
Energía para la vida

# Well barriers for Major Hazards risk

10TH GLOBAL CONGRESS ON PROCESS SAFETY

HARD → SOFT

| PLANT | PROCESS | PEOPLE |
|---|---|---|
| Safety Critical Equipment – SCE | GPS | Engineering authorities - TA |
| Protective systems Integrity of design, drilling, completion & well intervention | LEP´s & DWOP | Clear R & Rs |
| Assets integrity & reliability | IM Strategies | Competency assessment & training |
| Equipment dossiers | Risk Management | Contractor´s key positions |
| | MoC | |
| | QA-QC | |
| | Contractor´s Management Incident Investigation | |
| | OPS-ERP practices | |

PERFORMANCE

equion
Energía para la vida

# Land rig map of SCEs

10TH GLOBAL CONGRESS ON PROCESS SAFETY

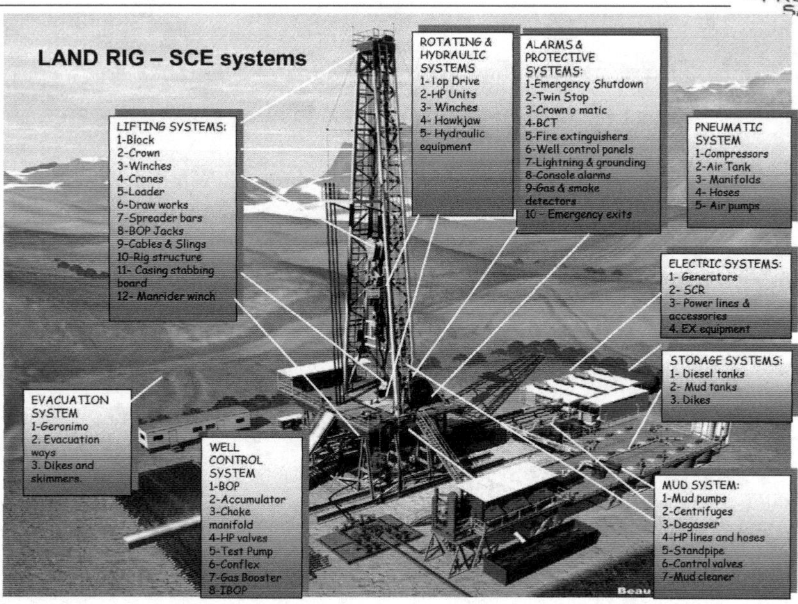

## HOW ARE WE DOING?

10TH GLOBAL CONGRESS ON PROCESS SAFETY

### LOWLIGHTS

- Asset integrity normally delegated to rig contractor.
- Backlog on PM tasks
- SCE concept and application barely applied.
- Criticallity of equipment based on operations rather than barriers for major incident risks
- Equipment modifications or alteration not following MoC
- Incident investigation- RCFA??

### HIGHLIGHTS

- Drilling Contractors enforcing asset integrity – R.A.M. -
- Raising Awareness on key people competence and certification.
- OEM recommendations mainly followed
- Incorporating lessons learnt from industry
- Renewal / upgrade of rig fleets

## SCE into PS model

10TH GLOBAL CONGRESS ON PROCESS SAFETY

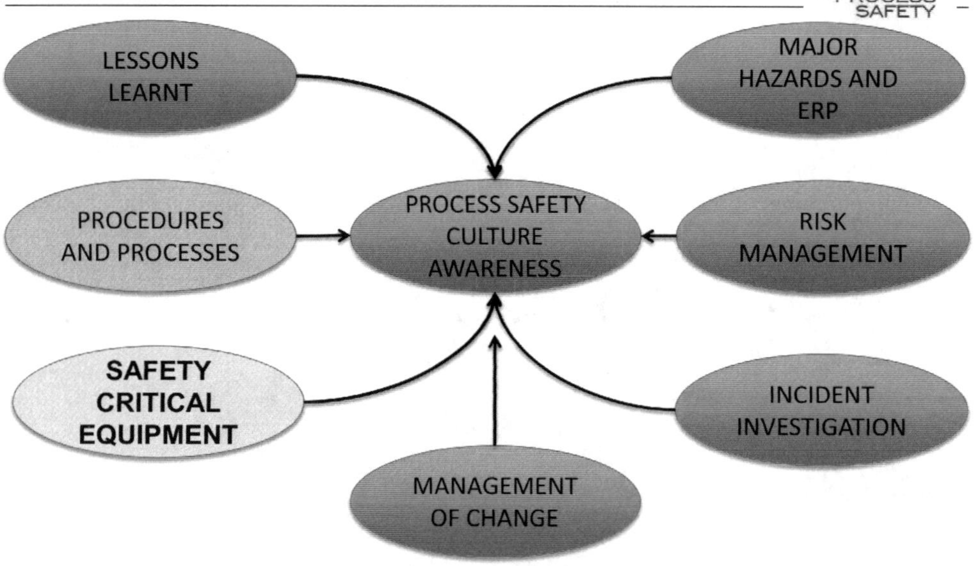

equion
Energía para la vida

# What´s next for SCEs !!

10TH GLOBAL CONGRESS ON PROCESS SAFETY

1. Define performance standards for rigs/platforms' SCEs addressing functionality, availability, reliability and failure modes, survival if designed for and interaction with other systems to operate.

2. Identify critical maintenance and test activities for SCEs and duly incorporate them into Contractor´s asset management systems.

3. Set key performance indicators – KPI – for maintenance and reliabilty related activities that lead performance and help to focus on compliance of PM tasks.

4. Define test frequencies and acceptance parameters for protective systems associated with SCEs.

5. Enforce SCE failure reporting, root cause analysis and sharing lessons

equion
Energía para la vida

10TH GLOBAL CONGRESS ON PROCESS SAFETY

# QUESTIONS ???

# Development of low-charring nano-composites to aid in fundamental understanding of nano-composite flame-retardancy

**Logan Hatanaka**
**Mary Kay O'Connor Process Safety Center**
**Artie McFerrin Department of Chemical Engineering, Texas A&M University**
**College Station, TX 77843**
**Lchatanaka@tamu.com**

**Sonny Sachdeva**
**Mary Kay O'Connor Process Safety Center, Texas A&M University,**

**Agustin Diaz**
**Department of Chemistry, Texas A&M University**

**Qinsheng Wang**
**Department of Chemical Engineering, Oklahoma State University**

**M. Sam Mannan**
**Mary Kay O'Connor Process Safety Center, Texas A&M University**

Prepared for Presentation at
American Institute of Chemical Engineers
2014 Spring Meeting
10th Global Congress on Process Safety
New Orleans, LA
March 30 – April 2, 2014

UNPUBLISHED

# Development of low-charring nano-composites to aid in fundamental understanding of nano-composite flame-retardancy

**Logan Hatanaka[ab], Sonny Sachdeva[ab], Agustin Diaz[c], Qingsheng Wang[de], M. Sam Mannan[abf] and Zhengdong Cheng[abf]**

[a] *Artie McFerrin Department of Chemical Engineering, Texas A&M University, College Station, TX 77843-3122, USA.*
[b] *Mary Kay O'Connor Process Safety Center, Artie McFerrin Department of Chemical Engineering, Texas A&M University, College Station, TX 77843-3122, USA.*
[c] *Department of Chemistry, Texas A&M University, College Station, TX 77843-3125, USA*
[d] *Department of Fire Protection & Safety, Oklahoma State University, Stillwater, OK 74078-4082, USA*
[e] *Department of Chemical Engineering, Oklahoma State University, Stillwater, OK 74078-5021, USA*
[f] *Material Science and Engineering, Texas A&M University, College Station, TX 77843-3003, USA*

**Keywords: Fire, Retardant, Nano-composites, Polymer, Loss Prevention**

## Abstract

For decades, nano-sheets comprised of various different materials have been utilized to construct nano-composites with enhanced properties including high thermal stability, flammability reduction, and modified mechanical properties. Although research in this field has been actively pursued in the past, there is little understanding as to why these materials exhibit enhanced behavior compared to their pure polymeric counterparts.

Two major mechanisms for enhanced flame-retardancy have been proposed: (1) a physical barrier effect and (2) a catalytic charring effect. As the material burns, a physical barrier of carbonaceous char and stacked nanoparticles forms at the material surface, acting as a physical barrier for radiant, conductive, and convective heat transfer as well as a physical barrier for mass transfer of the fuel to the burning front. Secondly, the presence of nanoparticles can act as a catalyst for charring, increasing the char yield of the material and effectively altering the physical barrier that forms. By utilizing a nano-composite system with low charring it is possible to observe the physical effects caused by the presence of nano-sheets in the polymer matrix. As charring agents are added to the composite, the enhancement of flame-retardancy based on catalyzed charring can be observed.

Furthermore, the research into the field of thermally stable and flame-retardant nano-composites has only focused on a relatively narrow view of polymers and nanoparticles, so any advancement into new materials is useful to further understand how these systems work. The current research project incorporates α-zirconium phosphate, a material capable of forming atom-thick, nanoscale-width nano-sheets, and poly (methyl methacrylate), a widely used polymer in commercial application. Aside from enhanced thermal stability, these materials have altered glass-transition temperatures, excellent optical transparency with UV scattering, and a relatively low char yield, making them an excellent candidate to study the physical barrier effects and catalytic charring effects of polymer nano-composites.

## 1. Introduction to Flame-Retardant Nano-composites

Fires can affect anyone, whether at home or working in an industrial plant. Although it is important to understand the hazards of fires in order to prevent them, once a fire occurs it is necessary to protect property and life from further damage. For this purpose, flame retardant materials are critical in slowing the spread and severity of damage due to a fire. However, not everything about common flame-retardants is as good as they seem. A recent study by Duke University studied the types of materials being used as flame-retardant additives in commercial couches. With a sample size of 101 couches, the study found that 52% of couches contained tris (1,3-dichloroisopropyl) phosphate (TDCPP, commonly called "Tris")[1], a mutagen in animals[2,3] and a neurotoxin in mice cell[4].

The need for non-toxic and effective flame-retardant materials has promoted the research of technologies that aim to replace the antiquated materials of the past. Specifically, the last two decades have seen major advancements into polymeric nano-composites as flame-retardant materials. These materials are composed of a nano-particle dispersed within a polymer matrix and have the potential to replace flame retardant technologies of the past with materials that are more effective and safer to use.

## 2. Flame Retardant Mechanisms

Polymeric nano-composites have shown great potential as flame-retardant materials with high thermal stability, but a detailed understanding of the mechanisms is not currently available. However, it has been shown that two main mechanisms contribute to the flame-retardant effects found in these types of polymeric nano-composites.

### *2.1 The Catalytic Charring Effect*

The first of the two main flame-retardant effects present when nano-composites burn is due to the formation of char. Upon heating, a linear (unbranched) polymer will go from a glassy to rubbery state near the polymer's glass transition temperature. If the polymer has crystalline domains, a melting transition will then occur at the polymer's melting temperature, or if the polymer is entirely amorphous, the rubbery state will slowly become more fluidic until the polymer is a liquid. Past this temperature, the polymer will begin to degrade and break apart in a process known as pyrolysis[5].

In non-charring polymers, pyrolysis creates lighter compounds from the larger original polymer chains that then vaporize and diffuse away from the material surface where they burn in the gas phase. This creates a region of pyrolysis at the surface of the material near the flame followed by a region of heated polymer melt and solid polymer away from the material surface[5]. However, if the temperature of the fire is high enough, more and more polymer will decompose over time, reducing the amount of solid and liquid polymer until essentially none is left.

In charring polymers, pyrolysis also forms lighter compounds from the original polymer chains. However, instead of diffusing away from the surface and then burning, these lighter components have the ability to cross-link the existing polymer chains. In this case, the flame impinges on the cross linked polymer, forming a char layer that covers the unburned cross linked material with an unreacted polymer melt and solid below it[5]. In order to burn, vaporized material must travel through this char layer to ignite in the gas phase, slowing the rate of burning and heat release in the polymer material. In addition, charring of the polymer leads to incomplete combustion of the polymer, reducing the overall heat released, and increasing flame-retardancy performance[5]. Some examples of non-charring polymers include poly (methyl methacrylate) (PMMA), polypropylene (PP), and ethylene-vinyl acetate (EVA), while an example of a charring polymer includes most cross-linked epoxies.

Up to this point, charring has only been considered for a pure polymer, but it has an important effect on flame-retardancy of polymeric nano-composites as well. The presence of nano-filler in the polymer can act as a charring agent similar to how Lewis acids and Friedel-Crafts alkylation reagents can increase the char yield of a normally non-charring polymer, increasing the overall flame-retardancy and heat resistance of the material[6-9]. For example, with a non-charring polymer filled with nano-filler, the addition of a carefully chosen nanoparticle could act as a catalytic charring agent to increase the over char-yield of a nano-composite, increasing flame-retardancy.

### *2.2) Physical Barrier Effect*

The second effect that plays a role in the flame-retardancy of a nano-composite is the physical barrier effect. As a nano-composite burns, nanoparticles begin to agglomerate at the surface of the material near the flame, forming a physical barrier towards gas transport from the material to the flame. This layer also acts as an insulating layer between the flame and the unburned polymer below the physical barrier and a radiant heat barrier to prevent re-radiation of heat to the flame. The physical barrier formed consists of intumescent, stacked nano-sheets with small floccules of carbonaceous char trapped between plates[10-20]. These small floccules together make up the physical barrier[21].

### *2.3) Synergistic Effects*

The physical barrier effect and the catalytic charring effect are coupled. The catalytic charring of the polymer effects the composition of the physical barrier and similarly, the physical barrier affects the conditions in which char formation occurs. Often in nano-composites, these two effects work together synergistically to produce flame-retardant properties but the interaction between these two effects is not well understood.

Cinausero *et al.* (2011) showed how these two effects act synergistically by first adding an alumina polyphosphate (AP) charring agent to PMMA and polystyrene (PS), then observing how the addition of silica and alumina nano-filler affected the thermal stability and flame-retardancy of PMMA[22]. The results showed that the limiting oxygen index (LOI) was increased for PMMA with AP char agents added when compared to only PMMA. To then observe the synergistic effects, nano-silica, organo-modified nano-silica, nano-alumina, and organo-modified nano-alumina were added to the polymer. The result was an increased LOI for all nano-fillers except organo-modified nano-alumina. Furthermore, the addition of AP to PMMA decreased the heat release rate (HRR), and further addition of the previously mentioned nano-filler reduced the HRR even more, showing synergistic effects between the charring agents and the nano-fillers. In addition to testing the synergistic effects of PMMA with AP and varying nano-filler, the same was done for PS. Similarly to PMMA, the addition of AP increased the catalyzed charring effect and increased the LOI. However, in the case of PS, only the addition of organo-modified silica increased the LOI further, while the other three nano-fillers (nano-alumina, organo-modified nano-alumina, and nano-silica) reduced the LOI and decreased flame-retardant performance. The cone calorimetry tests showed that PS and AP had a slightly reduced HRR, but the addition of any of the four nano-fillers further decreased the HRR, increasing the flame-retardant performance of the material. In this study, it was determined that a reaction occurs between the nanoparticle and charring agent, forming a new crystalline material which acts as a good physical barrier with catalytic charring abilities[22].

Laachachi (2007) has a similar study that utilizes titanium oxide and aluminum oxide as the physical barrier forming nanoparticle with a charring agent to provide charring effects to show synergistic effects[23]. Alone, aluminum oxide only provides a physical barrier effect, but when combined with charring agents, synergistic flame-retardant effects are also observed. This reinforces the idea that two separate mechanisms are at play, a physical barrier and a catalytic charring effect within the material. However, in order to fully understand how these effects work, each effect must be observed independently and synergistically[23].

## 3. Current Research

In order to further advance the flame-retardant properties of polymeric nano-composites, it is necessary to gain a deeper understanding of the flame-retardant mechanisms that make these materials effective. Observing the physical barrier effect separately from the catalytic charring effects and then together in varying degrees allows a complete understanding of the interactions between flame-retardant mechanisms in nano-composites. This knowledge would allow nano-composites to be optimized with a fine-tuned mix of physical barrier and catalytic charring effects. To start this research project, a physical barrier dominated nano-composite must be produced.

### *3.1 Synthesis of Physical Barrier Dominated Nano-composites*

Initial experiments have successfully prepared nano-composites containing alpha-zirconium phosphate (ZrP) as the nano-sheet filler and PMMA as the polymer matrix. ZrP is useful in nano-composite research because it can readily form individual sheets of atomic-scale thickness and a controllable diameter [24-27].

ZrP can undergo intercalation and exfoliation under the correct chemical conditions. If an amine such as propylamine or tetrabutylammonium is present and at a controlled ratio of ZrP:amine, the ZrP becomes exfoliated and individual nano-plates of ZrP are formed[24-27]. Once individual ZrP nano-sheets are created, they can then be dispersed into a polymer to form a nano-composite. By mixing a dispersion of exfoliated ZrP with a dissolved polymer solution, a dispersion of ZrP and polymer in solvent can be established. The solvent is then dried to produce a

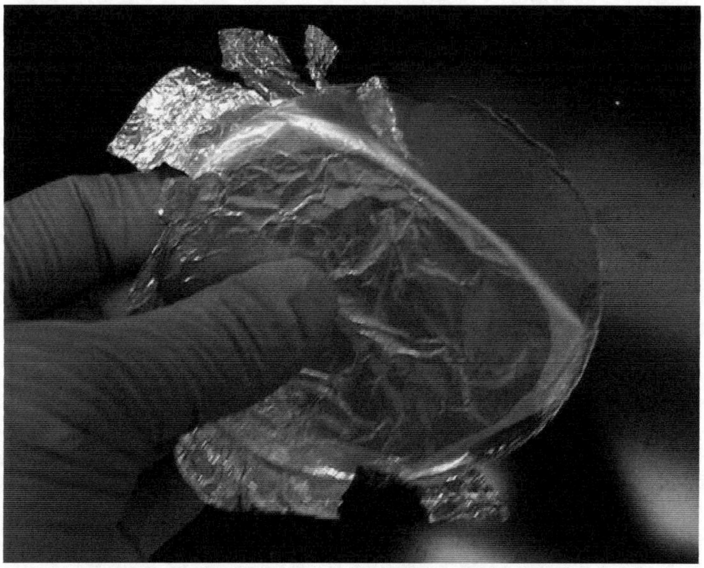

Figure 1: 5 wt% ZrP in PMMA nano-composite being removed from foil substrate.

ZrP/polymer nano-composite film.

This general method was used to produce ZrP/PMMA nano-composites. Figure 1 shows a 5 wt% ZrP in PMMA nanocomposite produced using this method. In the figure, a newly cured nanocomposite is peeled from a foil substrate, revealing an optically transparent ZrP/PMMA nanocomposite. Initial materials tests on a 1 wt% ZrP in PMMA composite have shown that intercalated and exfoliated nanoparticles are present, as seen in the X-Ray Diffraction (XRD) data in Figure 2. The agglomerated plate-to-plate distance of ZrP normally produces a peak at 11.6° which is absent in Figure 2, indicating that there is not a significant amount of agglomerated nanoparticles. In future work, scanning electron microscopy (SEM) and transmission electron microscopy (TEM) imaging will be used to confirm this. Instead, a peak at lower a 2Θ value indicates intercalated nanoparticles are likely present.

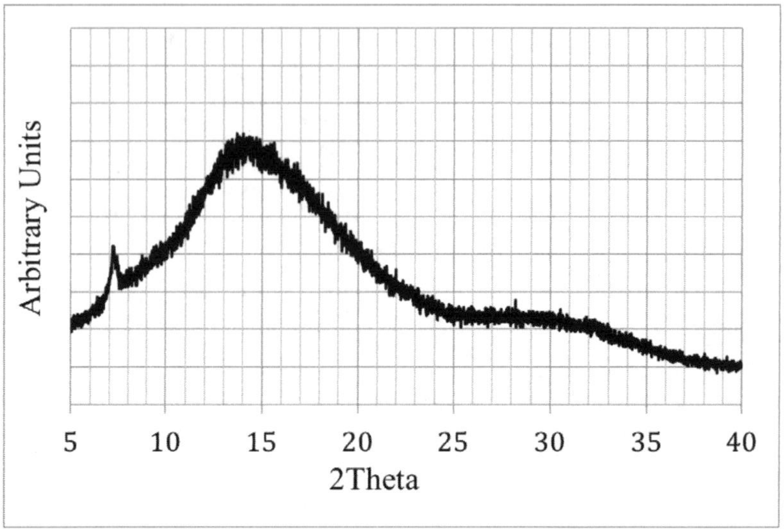


Figure 2: XRD results for a 1 wt% ZrP/PMMA nano-composite.

Initial thermogravimetric analysis (TGA) of ZrP/PMMA composites show that they most likely have a physical barrier dominated mechanism of heat resistance, as shown in Figure 3. Every composite shows the same 3 mass loss peaks, showing that the mechanism for degradation has not changed significantly. Additional cone calorimetry tests will be conducted to confirm if catalytic effects are also contributing to heat resistance during decomposition, but this is unlikely since the shape of each TGA curve is similar to PMMA.

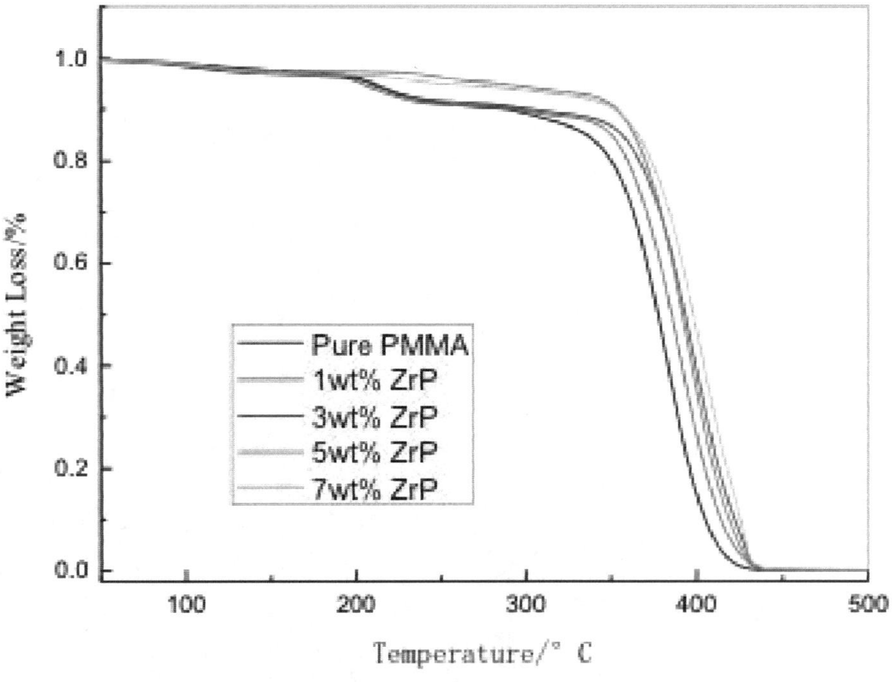


Figure 3: Initial TGA of various ZrP nano-composites.

### *3.2 Synthesis of Catalytic Char Dominated Nano-Composites*

To view the opposite side of the spectrum, a material must be produced that is dominated by the catalytic charring effect. This nano-composite will be produced similarly to the first nano-composite, using the same polymer as the first composite for consistency. However, this charring nano-composite will utilize an entirely different nanoparticle that induces charring but is spherical rather than layered. This change in geometry is meant to hinder floccule formation while inducing polymer charring. Future research will utilize phosphotungstic acid, a noted charring nanoparticle, in PMMA in the same weight fraction as the nano-sheet in the physical barrier-dominated material[28].

### *3.3 Synthesis of Synergistic Nano-Composites*

At this point, a physical barrier-dominated material and a catalytic char-dominated material has been established, but no composite exists to exhibit synergistic effects. Conveniently, the production of homogenous catalysts consisting of PTA and silica (a physical barrier forming nano-particle) can be through a sol-gel synthesis process[29]. Silica/PTA composites have been produced with up to 80 wt% PTA on silica, allowing a wide range of PTA to silica weight ratios to be examined. This could potentially uncover a wide range of synergistic effects with varying degrees of the physical barrier effect to the catalytic charring effect. It allows the chemical composition of the nano-filler to be finely tuned so that when it is introduced into the polymer, it gives precise control over the physical barrier and catalytic char behavior.

## 4. Conclusions:

Flame retardant nano-composites have shown great promise in providing non-toxic and effecive loss prevention materials. However, in order to increase the effectiveness of flame retardant nano-composites, the exact nature of the flame retardant mechanisms within these materials is necessary. Since the two mechanisms for nano-composite flame retardancy behave differently together than they do independently, it is important to observe the physical barrier effect and the catalytic charring effect independently and synergistically. Initial research has produced a ZrP/PMMA nanocomposite system which appears to be physical barrier dominated with little charring. This is the first step in developing nanocomposites which exhibit a complete range of nanocomposite flame-retardancy mechanisms. The next step is to produce nano-composites with catalytic char dominated behavior along with nano-composites with varying ratios of the physical barrier effect to the catalytic charring effect to observe the full range of flame retardant mechanisms.

## Acknowledgements:

This material is based upon work supported by the National Science Foundation under Grant No. CBET1336162.

## 5. References

1. Stapleton, H.M., S. Sharma, G. Getzinger, P.L. Ferguson, M. Gabriel, T.F. Webster, and A. Blum, *Novel and High Volume Use Flame Retardants in US Couches Reflective of the 2005 PentaBDE Phase Out*. Environ Sci & Technol, 2012. **46**(24): p. 13432-13439.
2. Blum, A. and B.N. Ames, *Flame-retardant additives as possible cancer hazards*. Science, 1977. **195**(4273): p. 17-23.
3. Gold, M.D., A. Blum, and B.N. Ames, *Another flame-retardant, tris- (1,3-dichloro-2-propyl)-phosphate, and its expected metabolites are mutagens*. Science, 1978. **200**(4343): p. 785-787.

4. Dishaw, L.V., C.M. Powers, I.T. Ryde, S.C. Roberts, F.J. Seidler, T.A. Slotkin, and H.M. Stapleton, *Is the PentaBDE replacement, tris (1,3-dichloropropyl) phosphate (TDCPP), a developmental neuro- toxicant? Studies in PC12 cells.* Toxicol. Appl. Pharmacol. , 2011. **256**(3): p. 281-289.
5. Pochiraju, K.V., G.P. Tandon, G.A. Schoeppner, and A. Morgan, *Additives, Nanocomposites, and Barrier Coatings*, in *Long-Term Durability of Polymeric Matrix Composites*. 2012, Springer: US. p. 39-75.
6. Cai, Y., Y. Hu, L. Song, S. Xuan, Y. Zhang, Z. Chen, and W. Fan, *Catalyzing carbonization function of ferric chloride based on acrylonitrile-butadiene-styrene copolymer/organophilic montmorillonite nanocomposites.* Polymer Degradation and Stability, 2007. **92**(3): p. 490-496.
7. Jang, J., J. Kim, and J.-Y. Bae, *Effects of Lewis acid-type transition metal chloride additives on the thermal degradation of ABS.* Polymer Degradation and Stability 1999. **88**(2): p. 324-332.
8. Jang, J., J. Kim, and J.-Y. Bae, *Synergistic effect of ferric chloride and silicon mixtures on the thermal stabilization enhancement of ABS.* Polymer Degradation and Stability, 2005. **90**(3): p. 508-514.
9. Wang, Z., D.D. Jiang, C.A. Wilkie, and J.W. Gilman, *Further studies on fire retardant polystyrene by Friedel-Crafts chemistry.* Polymer Degradation and Stability 1999. **66**(3): p. 373-378.
10. Bras, M.L., G. Camino, S. Bourigot, and R. Delobel, *Fire Retardancy of Polymers: The use of intumesence.* 1998: Cambridge: Royal Society of Chemistry Information Services.
11. Bras, M.L., C.A. Wilkie, and S. Bourbigot, *Fire Retardancy of Polymers: New Applications of Mineral Fillers. .* 2005: Cambridge: Royal Society of Chemistry.
12. Fina, A., D. Tabuani, A. Frache, and G. Camino, *Polypropylene-polyhedral oligomeric silsesquioxanes (POSS) nanocomposites.* Polymer, 2005. **46**(19): p. 7855-7866.
13. Gilman, J.W., *Flammability and thermal stability studies of polymer layered-silicate (clay) nanocomposites.* Applied Clay Science 1999. **15**(1-2): p. 31-49.
14. Schartel, B., P. Potschke, U. Knoll, and M. Abdel-Goad, *Fire behaviour of polyamide 6/multiwall carbon nanotube nanocomposites.* European Polymer Journal 2005. **41**(5): p. 1061-1070.
15. Tidjani, A., *Polypropylene-graft-maleic anhydride-nanocomposites: II - fire behaviour of nanocomposites produced under nitrogen and in air.* Polymer Degradation and Stability 2005. **87**(1): p. 43-49.
16. Torre, L., E. Frulloni, J.M. Kenny, C. Manferti, and G. Camino, *Processing and characterization of epoxy–anhydride-based intercalated nanocomposites.* Journal of Applied Polymer Science, 2003. **90**(9): p. 2532-2539.
17. Zanetti, M., G. Camino, and R. Mulhaupt, *Combustion behaviour of EVA/fluorohectorite nanocomposites.* Polymer Degradation and Stability 2001. **74**(3): p. 413-417.
18. Zanetti, M., G. Camino, P. Reichert, and R. Mülhaupt, *Thermal Behaviour of Poly(propylene) Layered Silicate Nanocomposites.* Macromolecular Rapid Communications 2001. **22**(3): p. 176-180.
19. Zanetti, M., T. Kashiwagi, L. Falqui, and G. Camino, *Cone Calorimeter Combustion and Gasification Studies of Polymer Layered Silicate Nanocomposites.* Chemistry of Materials, 2002. **14**(2): p. 881-887.
20. Zhao, C., H. Qin, F. Gong, M. Feng, S. Zhang, and M. Yang, *Mechanical, thermal and flammability properties of polyethylene/clay nanocomposites.* Polymer Degradation and Stability, 2005. **87**(1): p. 183-189.
21. Gilman, J.W. *Flame Retardant Polymer Nanocomposites.* in *Pollution Prevention through Nanotechnology Conference.* 2007. Arlington, VA.
22. Cinausero, N., N. Azema, J.M. Lopez-Cuesta, M. Cochez, and M. Ferriol, *Synergistic effect between hydrophobic oxide nanoparticles and ammonium polyphosphate on fire properties of poly(methyl methacrylate) and polystyrene. .* Polymer Degradation and Stability, 2011. **96**(8): p. 1445-1454.
23. Laachachi, A., M. Cochez, E. Leroy, M. Ferriol, and J.M. Lopez-Cuesta, *Fire retardant systems in poly(methyl methacrylate): Interactions between metal oxide nanoparticles and phosphinates.* Polymer Degradation and Stability 2007. **92**(1): p. 61-69.
24. Alberti, G., M. Casciola, and U. Costantino, *Inorganic ion exchange pellicles obtained by delamination of alpha-zirconium phosphate crystals.* Journal of Colloid and Interface Science, 1985. **107**: p. 256-263.
25. Maclachlan, D.J. and K.R. Morgan, *Phosphorus-31 solid-state NMR studies of the structure of amine-intercalated .alpha.-zirconium phosphate. 2. Titration of .alpha.-zirconium phosphate with n-propylamine and n-butylamine.* J. Phys. Chem. , 1992. **96**: p. 3458-3464.
26. Sun, L., J.Y. O'Reilly, D. Kong, J.Y. Su, W.J. Boo, H.J. Sue, and A. Clearfield, *The effect of guest molecular architecture and host crystallinity upon the mechanism of the intercalation reaction.* Journal of Colloid and Interface Science, 2009. **333**(2): p. 503–509.
27. Clearfield, A. and Z. Wang, *Organically pillared microporous zirconium phosphonates.* Front cover Journal of the Chemical Society, Dalton Transactions,, 2002. **15**: p. 2937-2947.
28. Bourbigot, S. and S. Duquesne, *Fire retardant polymers: recent developments and opportunities.* Journal of Materials Chemistry, 2007. **17**: p. 2283-2300.
29. Kukovecza, Á., Z. Balogia, Z. Kónyaa, M. Tobab, P. Lentzc, S.-I. Niwab, F. Mizukamib, Á. Molnárd, J.B. Nagyc, and I. Kiricsia, *Synthesis, characterisation and catalytic applications of sol–gel derived silica–phosphotungstic acid composites.* Applied Catalysis A: General, 2002. **228**(1-2): p. 83-94.

# 349935 Best Practices in IPL Integrity Management - Case Study in Kuwait Oil Company


**Abstract:**

**16th Process Plant Safety Symposium (PPSS) – Validation and Maintenance of IPLs**

This paper presents Management of Independent Protection Layers at Kuwait Oil Company (KOC), Kuwait. Several methods and practices were initially followed by various project teams involved in SIL Assessment / Verification / Validation over several years since the SIL compliance was introduced. However, little interaction existed across projects and Operation & Maintenance personnel. As a result, valuable experience and observations from the actual plant operation were never fed back and were not integrated into the LOPA methodology. The conditional modifiers remained un-changed for almost a decade.

In order to revalidate the IPLs, a task was carried out to systematically compile, analyze and prepare in easily understandable form, failure modes and effects over the past seven (7) years from five major process plants. Mechanical, Electrical and Instrument failures were separately listed with their cause-consequence relations. The consequences are rationalized against the Company's LOPA criteria, in order to be able to sort them in the order of maximum to minimum consequence. For the high consequence events, the initiating events that have caused them were compared with the cause-consequence pairs listed in the Company standard. It was observed that for many of the high consequence events, the initiating events that were assumed were not the real ones as per the 7-year history. A detailed report that followed presented this analysis to the stake-holders for further consideration, to be reflected in a revised Company standard and also to make this method of revalidation of IPLs a Company practice.
An independent "Alarm Management" exercise issued a Company Alarm Management Philosophy document (APD). After implementing the Alarm Management per the new APD, several of the existing Process alarms were either removed, or modified to give them lower priority. The credits allowed by the LOPA standard now need revision to be in line with the APD. This issue was highlighted and soon, the LOPA standard will be revised to reflect the same.

The Flare & Blow-down systems in several of the company's process plants were provided with SIL approved SDVs under one major project. However, the credit for this SIF was not being reflected in most LOPA workshops. This issue was highlighted and was subsequently corrected.

The access security for a few SIFs was agreed to be given to the senior process operators, thus increasing the risk profile, as against the original assumption that only the Operations Manager will be approving changes. Hence these SIFs no longer deserve credits in LOPA and needed suitable modification in the LOPA standard.

The 'type' of equipment and thus their technology and reliability has changed significantly. For example, the Tank Vapour compression was changed to electric motor driven centrifugal / screw type from previous gas-engine driven reciprocating type, thus decreasing vibrations in the tanks and piping systems. Similar changes in other process units either decreased or increased the initiating events but these changes were not reflected in the frequencies and cause-consequence relationships of the initiating events.

All the above issues significantly impact the accuracy of the LOPA exercises and thus directly impact the Risk reduction in the process plants under consideration. Sustained efforts over several years resulted in noticeable change in perceptions of Operation and Maintenance personnel and the 7-year failure data has significantly contributed to rationalizing the Company's LOPA criteria to make it more meaningful for both existing and future process facilities.

# 349800 Multiple Perspectives on the Role of Safety Leadership in Major Hazard Organisations

**Abstract:**

There is widespread agreement between industry, regulators, and academics that leadership is a key component of a safe organisation. This view is widely supported by findings from almost all major incident inquiries and investigations, which have demonstrated that managerial failure is at least as important as technical failure and operator error. Managers play an important role because their decisions have a fundamental impact on the resources invested in safety and more broadly on how work is organised and managed. Despite the increased focus on managers', and particularly senior managers', roles and responsibilities both in legislation and in major incident enquiries, comparatively little research has been conducted on the types of behaviours and leadership styles that are effective in managing safety.

This paper will bring together different sources of evidence on the role of leadership particularly in major hazard organisations. First, based on a review of academic literature and an analysis of major incidents, the requirements for effective safety leadership will be presented. Specifically, a review of published evidence over the last ten years was carried out in order to capture current knowledge in the area of leadership. This was complemented by an analysis of accident reports and published papers for sixteen major incidents in order to identify any recurrent themes in the causation of these incidents, particularly those that represented leadership failures or have leadership implications for the management of safety. Second, an empirical study was conducted on the perceptions of senior managers regarding the roles and responsibilities of Board members in terms of safety leadership across three different major hazard organisations. The paper will discuss the implications of the findings from these different sources for the roles and responsibilities of managers, and what constitutes effective safety leadership in major hazard organisations.

Please note that this paper will draw on the third author's publication 'A review of the literature on effective leadership behaviours for safety', Research Report 952, Health and Safety Executive: Sudbury Books.

# Using Explicit Finite Element Analysis to Simulate the Dynamic Response and Predict the Structural Damage Associated with a Real-Life Process Equipment Failure Due to an Internal Detonation

**Phillip E. Prueter**
**The Equity Engineering Group Inc., Shaker Heights, Ohio**
peprueter@equityeng.com

Prepared for Presentation at
American Institute of Chemical Engineers
2014 Spring Meeting
10th Global Congress on Process Safety
New Orleans, LA
March 30 – April 2, 2014

UNPUBLISHED

# Using Explicit Finite Element Analysis to Simulate the Dynamic Response and Predict the Structural Damage Associated with a Real-Life Process Equipment Failure Due to an Internal Detonation

**Phillip E. Prueter**
**The Equity Engineering Group Inc., Shaker Heights, Ohio**
**peprueter@equityeng.com**

**Keywords:** Explicit Finite Element Analysis, Blast Loading, Accidental Explosions, Internal Detonations, Structural Response, Heat Exchanger

## Abstract

Simulating the realistic blast loading associated with an internal detonation occurring within a pressure vessel or heat exchanger is extremely challenging. Unlike evaluation of external blast loading on structures due to far-field explosions, where typical overpressure-time histories can be reasonably defined based on empirical data, investigating confined detonations presents additional complications such as accounting for internal blast wave reflection that can yield multiple blast waves and considering potential pressure build up due to the accumulation and temperature rise of gaseous explosion by-products. The subsequent impulsive peak reflected overpressure from confined detonations acting on a structure can be extremely high due to the often times, close proximity of the blast source to the vessel wall or pressure boundary. This establishes the possibility of significant structural damage for process equipment subjected to an internal detonation, even for relatively modest amounts of concentrated explosive products.

Performing dynamic computational simulations of internal detonations is valuable in assessing the structural response and possible failure modes of critical process equipment such as process piping, pressure vessels, or heat exchangers. Furthermore, assessing the potential damage from an internal detonation within such a structure and predicting ensuing damage to surrounding equipment if the containment vessel catastrophically fails can provide valuable information about protecting structures from explosions and implementing designs that promote blast damage mitigation and process safety. Additionally, determining a damage threshold for a given amount of concentrated chemical explosives can guide process engineers in mitigating the potential for accidental detonations that pose a significant risk to the structural integrity of pressure containing equipment.

This paper discusses the underlying theory of blast analysis and examines the practical application of non-linear, finite element based, explicit computational techniques for simulating the load acting on a structure due to internal and external blasts. The investigation of a recent, real-life industry failure of a heat exchanger due to a suspected internal detonation is discussed. This particular failure resulted in gross rupture of the pressure boundary and significant damage to nearby equipment. Explicit, three-dimensional blast analysis is performed on the heat exchanger in question, and an internal detonation is simulated to reasonably replicate the considerable damage actually observed in the field. This analysis permits the determination of an approximate amount of concentrated product that caused the accidental explosion; that is, the plausible equivalent amount of explosives is back-calculated based on the predicted damage to the finite element model of the equipment in question. Computational iterations of varying charge amounts are performed and the predicted amount of permanent damage is documented so sensitivity to the hypothesized charge amount can be quantified from a structural response standpoint. Furthermore, explicit blast analysis of nearby equipment is performed to predict the magnitude of plastic deformation actually observed due to external blast loading.

## 1. Introduction

In general, an explosion in air creates a blast wave as a result of the atmosphere surrounding the explosion being pushed back, and the resulting pressure of the compressed air at the blast wavefront decays as it moves away from the blast source. As discussed in references [1-6], a typical blast wave, observed at a location removed from the center of a chemical explosion, reaches a peak value of overpressure and begins to decay exponentially (as shown in Figure 1), eventually decreasing below atmospheric pressure. For conventional chemical explosions, the overpressure profile is divided into the positive pressure portion (positive phase) and negative pressure portion (negative phase). Furthermore, the amount of time it takes the peak overpressure to decay below atmospheric pressure is known as the positive phase duration and is generally, closely related to the overall damage capabilities of a given blast wave. Similarly, the amount of time it takes for the negative pressure to return to atmospheric pressure is the negative phase duration [3]. The area under the curve during the positive phase duration is the positive blast wave impulse. In general, peak overpressure observed at a location removed from the blast source decreases as the distance from the detonation center increases.

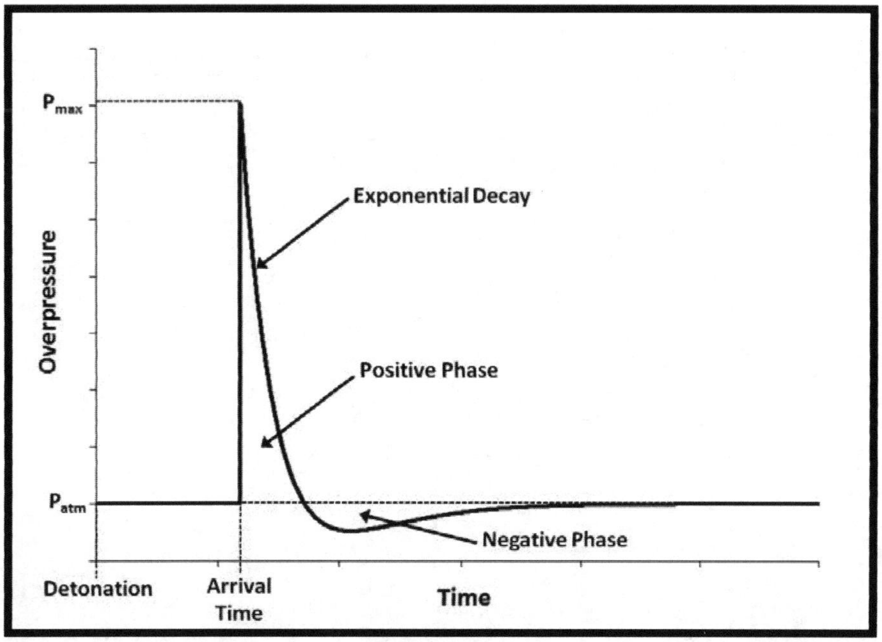


**Figure 1: Typical overpressure amplitude for a chemical explosion [7].**

This paper presents the underlying methodology for employing explicit finite element analysis to investigate a recent, real-life process industry failure of a vertical heat exchanger that experienced an unexpected internal detonation. The primary computational tool used to generate realistic blast wave profiles in this study is based on the Conventional Weapons Effects Blast Loading Model or CONWEP [8].

In this study, computational results for both the heat exchanger (subjected to internal blast loading) and surrounding equipment (subjected to external blast loading) are in good agreement with the measured plastic deformations and failure modes that were actually observed in the field. Commentary on the observed failure modes, supplemented by explicit computational finite element results, is provided. Additionally, an advanced finite element failure criterion is employed where portions of the computational model are removed from the simulation once a certain plastic strain threshold is reached. This approach facilitates simulation of the gross heat exchanger pressure boundary failure actually observed in this case. The explicit, three dimensional finite element based analyses discussed in this study reasonably predict the structural response and damage characteristics corresponding to a recent, real-life industry failure.

## 2. Blast Wave Reflection Effects

A complication associated with computational blast modeling is accounting for overpressure wave reflection effects. In general, the peak blast pressure that a structure is subjected to is greater than the peak incident overpressure and is known as the reflected pressure. This pressure is a function of the peak incident overpressure and the angle of incidence; that is, the angle between the blast wave loading direction and the surface normal of a given point on the loading surface.

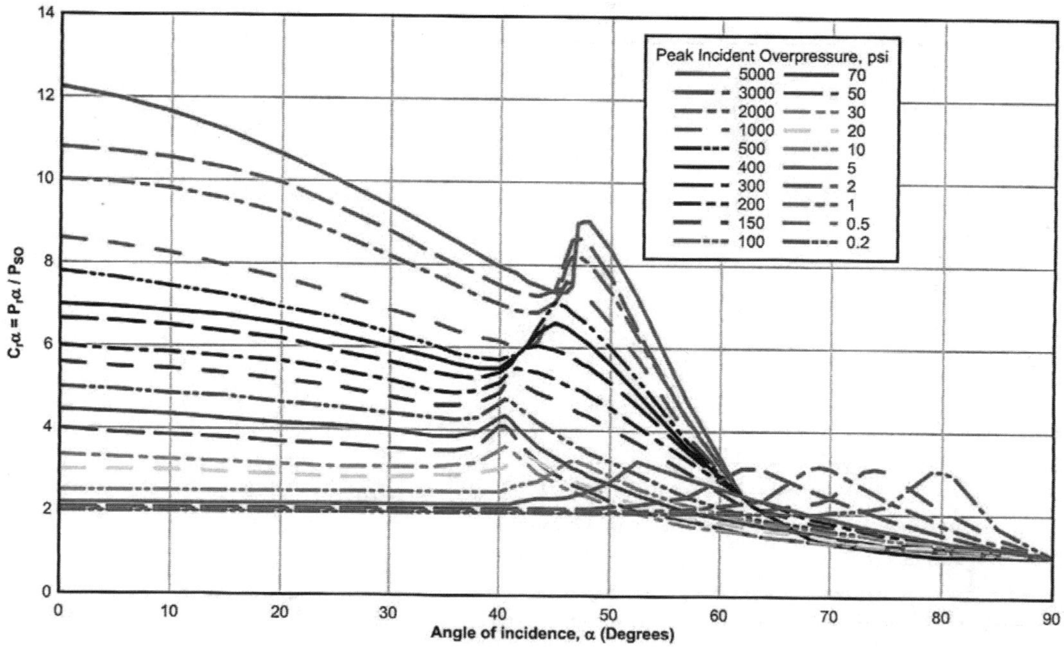


**Figure 2: Reflection Coefficient vs. Angle of Incidence for Varying Peak Overpressures [3].**

Figure 2 shows how the reflection coefficient changes with angle of incidence and peak incident overpressure. It can be observed that for extremely high overpressures, the actual reflected pressure that a structure experiences can be as high as 10 or 12 times the incident blast pressure alone. Neglecting incident wave reflection in a computational analysis can produce non-conservative results, particularly for high peak overpressures. On the contrary, assuming the maximum peak reflected pressure acts on an entire structure is potentially overly conservative. To this end, structural response due to impulsive loading is often times, sensitive to capturing the correct incident wave reflection. Accurately simulating realistic overpressure amplitudes acting on a structure for a given explosive scenario is challenging because the peak overpressure, positive phase duration, amount of explosives, and distance from the explosion source all affect the overpressure amplitude. Furthermore, accounting for reflection effects on a loading surface adds further complication to the blast loading model.

For internal detonations, accurately predicting blast wave reflection is extremely difficult because of the effects of confinement. As discussed in [9], an internal detonation within a containment vessel can be characterized by two distinct types of loading. First, immediately following a detonation, the structure is subjected to a transient impulsive pressure load. If this initial pressure wave does not cause catastrophic failure or rupture of the pressure boundary, a secondary, long-term quasi-static pressure build-up occurs in the vessel due to the evolution and heating of gaseous detonation by-products. Additionally, as discussed in [10], resonance effects between the vibration of the vessel and the reflected internal shock wave is a potential damage mechanism. In the analyses discussed herein, only the initial impulsive pressure wave is considered as a damage mechanism because it is believed that this type of loading caused gross failure of the pressure boundary for the actual equipment failure discussed. Furthermore, the interaction between any internal components and a propagating blast wave are not explicitly considered due to the complexity of the heat exchanger geometry and the computational cost of analyzing fluid-structure interaction in detail, which would require computational fluid dynamics (CFD) techniques.

## 3. CONWEP Blast Loading Model

The CONWEP blast loading model is based on a collection of conventional weapons effects calculations of TM 5-855-1 established by the United States Military [8]. The main advantage of using CONWEP is that realistic overpressure amplitudes and other blast wave parameters are calculated based on a defined charge amount, and the appropriate impulsive loading is automatically employed in the explicit finite element analysis. Additionally, overpressure amplitude decay and reflection formulations are incorporated into the model. The CONWEP blast loading model does not require modeling the fluid medium to account for overpressure reflection (eliminating the need for coupled structural-acoustic analysis and offering a significant improvement in computational efficiency, especially for complex simulations). For a more in-depth overview of the underlying theory behind the CONWEP blast loading model, refer to Reference [11]. An example of a computational study that employs CONWEP through explicit finite element analysis to investigate the response of storage tanks subject to external blast loading is provided in [12].

In order to develop the blast parameters required to generate realistic overpressure curves for conventional explosives, the equivalent TNT method is most commonly used. TNT equivalency of an explosive is simply the ratio of the mass of TNT to the mass of explosive such that both yield equal pressure and impulse [13]. As discussed in [3], the ratio of energy output of a chemical explosive material relative to a given amount of TNT is calculated following

$$W_{effective} = \left[\frac{H^d_{exp}}{H^d_{TNT}}\right] W_{exp}$$

where $W_{effective}$ is the effective charge weight, $W_{exp}$ is the weight of explosive in question, $H^d_{exp}$ is the heat of detonation of a given explosive, and $H^d_{TNT}$ is the heat of detonation of TNT. In order to compare shock wave parameters to available published data, a scaled distance from a given explosion source to a structure may be used. As discussed in [14], the most common form of blast scaling is the Hopkinson-Cranz or cube-root scaling method following

$$Z = R/W^{(1/3)}$$

where $Z$ is the scaled distance, $R$ is distance from the explosion source to the loading point, and $W$ is the weight of TNT. This formulation asserts that self-similar blast waves are produced at identical scaled distances when two different sized explosive charges (of similar geometry and chemical composition) are detonated in the same atmosphere. Using available empirical data [3], this scaled distance from the explosion source to a loading surface can be used to determine specific blast parameters such as peak reflected overpressure, arrival time, and positive phase impulse. The CONWEP blast loading model discussed herein is based on a TNT explosion (assumed surface blast in this study).

## 4. Recent Industry Failure Case Study

The results of an explicit finite element analysis investigating a recent industry failure of a vertical heat exchanger are summarized below. It is believed that this specific failure was the result of an accidental internal detonation within the heat exchanger. The overall damage to the exchanger involved gross failure/rupture of the shell-side and tube-side pressure boundaries and the fragmentation of numerous tubes and other internal components. Furthermore, after the detonation event, an extensive debris field was present in the vicinity of the exchanger and noticeable distortion to the bottom tubesheet and other internal components was observed. In addition, there was extensive damage and plastic deformation of nearby equipment, particularly the support skirt of an adjacent tower.

A schematic of the vertical exchanger is shown in Figure 3. The approximate diameter, of the exchanger is 4 feet with an overall length of approximately 15 feet. The exchanger tubesheets are approximately 4 inches thick, and there are roughly 1,000 tubes. The shell-side of the exchanger contains a flanged and flued expansion joint. The exchanger is supported vertically near the center where chair supports attach to surrounding structural steel.

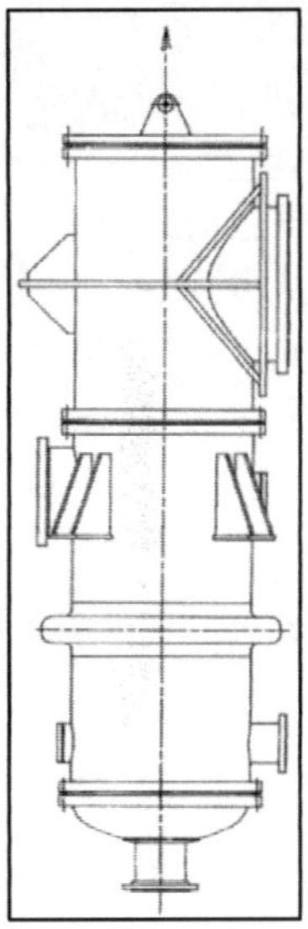

**Figure 3: Schematic of the Vertical Heat Exchanger.**

The goal of the analysis presented herein is to gain a better understanding of the detonation that occurred by looking at the structural response and failure modes of the heat exchanger and the surrounding equipment. In order to establish a reasonable approximation of the equivalent charge amount that led to the damage observed in the field, computational iterations are first performed for the adjacent tower and support skirt. A 3D finite element model of the tower and support skirt is constructed out of shell elements and subjected to external blast loading corresponding to varying charge amounts at the appropriate distance away from the tower (actual distance between the vertical exchanger and the tower). The overpressure amplitudes applied to the outside of the tower are calculated using CONWEP and account for reflection effects.

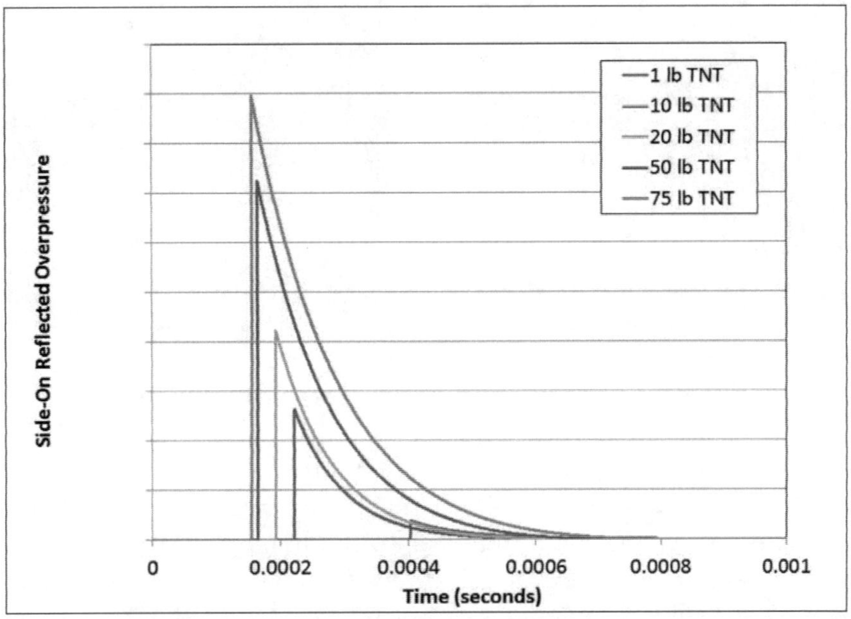


**Figure 4: Peak Reflected Overpressures at the Tower Support Skirt for Varying Equivalent Charge Amounts.**

The material model employed in the analysis is elastic-plastic and reflects quasi-static monotonic behavior (i.e. strain rate dependent material properties are not explicitly considered due to lack of available dynamic test data for the material in questions). Figure 4 shows the peak reflected overpressure amplitudes for varying equivalent amounts of TNT that would be observed at the base of the tower (note that specific overpressure values are intentionally omitted on the vertical axis of this figure). Based on these iterations, an equivalent charge amount is determined that approximates the amount of plastic deformation in the adjacent tower skirt that was measured in the field. Figure 5 shows contours of plastic deformations and equivalent plastic strains (based on this approximate charge amount) predicted in the tower skirt from explicit, non-linear FEA (again, specific contour values are intentionally omitted). This equivalent charge amount is then used to simulate an internal detonation from within the vertical heat exchanger shown in Figure 3, again using 3D explicit FEA techniques.

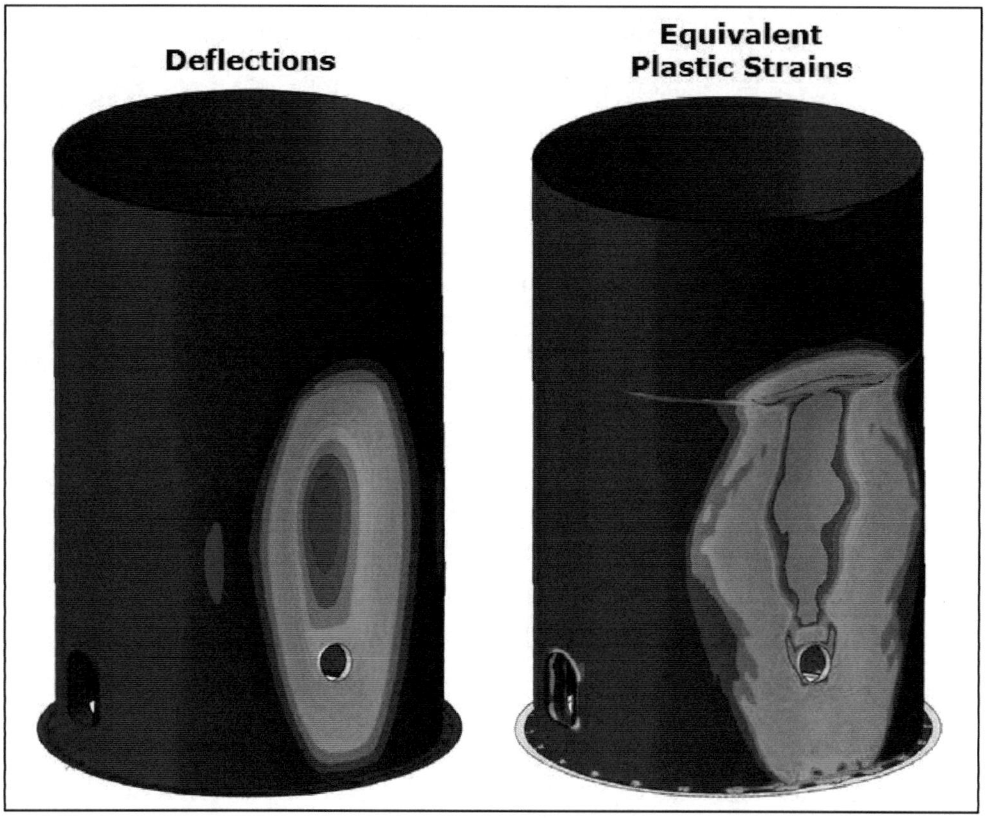


**Figure 5: Permanent Deflections and Equivalent Plastic Strains in Adjacent Tower Skirt.**

The analyses discussed herein employ explicit FEA formulations, which in general, are well suited for analyzing highly transient and non-linear behavior and permit evaluation of models with highly discontinuous events with large deformations. Explicit FEA simulations are computationally expensive due to the small time increments generally needed to achieve convergence. Explicit dynamic simulations on complex, 3D models would generally not be feasible or practical without the use of parallel computing optimization. For example, the simulations presented herein were run across 16 cpus or more simultaneously in order to achieve practical solution times.

A 3D finite element model of the bottom portion of the vertical heat exchanger is constructed using solid elements. The 3D model is shown in Figure 6 and includes the flanged and flued expansion joint. It is noted that the model is truncated at the shell-to-channel interface to facilitate computational efficiency and the holes in the bottom tubesheet are included. The top of the finite element model is constrained using a kinematic coupling definition in a local cylindrical coordinate system to permit radial dilation and maintain planar behavior at the end face (this reasonably mimics the restraint of the un-modeled shell). The detonation source is placed at a location within the exchanger that most reasonably matches the maximum deflection measured in the bottom tubesheet and observed in the shell; that is, based on the permanent damage observed in the field, a postulated location of the detonation center is assumed in the FEA simulation.

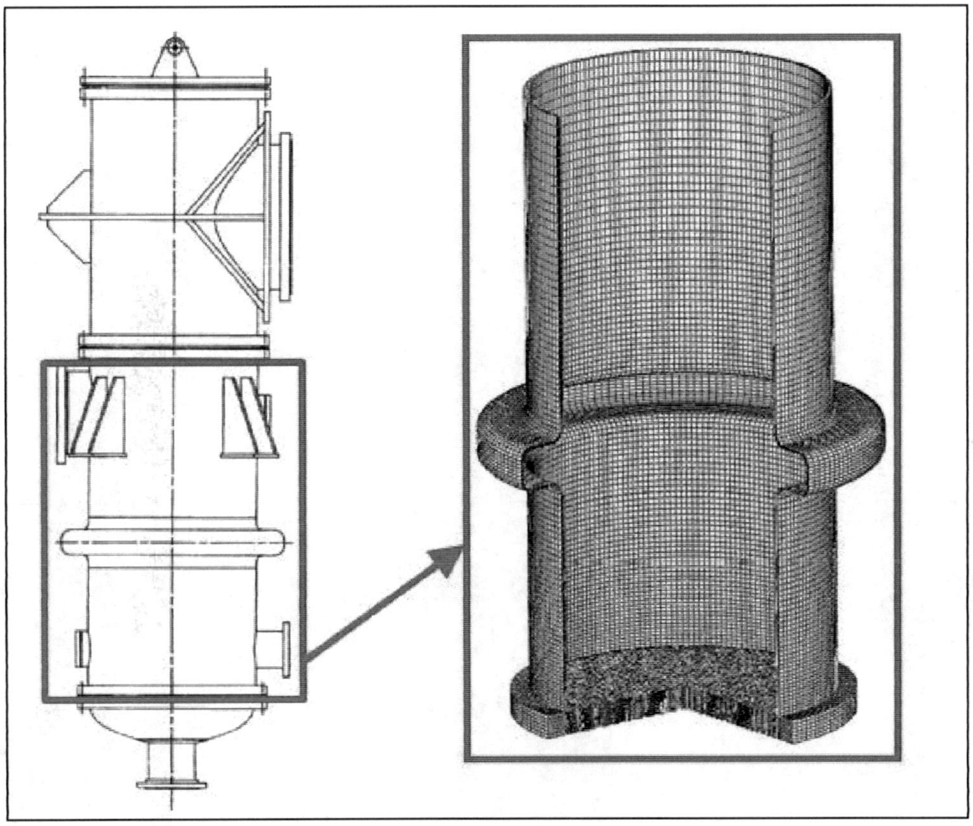

**Figure 6: Portion of the Heat Exchanger Modeled and Subjected to Internal Blast Loading (Cutaway View of FEA Model Shown).**

Explicit FEA is performed on the vertical exchanger where again, the CONWEP charge property is employed to generate the overpressure amplitudes applied to the inside surface of the exchanger. It is noted that in this case, a shear element failure criterion is implemented that is driven by plastic yielding where a given individual element loses stiffness and is removed from the model once it reaches a user-defined strain threshold. For the initial simulation presented herein, this threshold is defined as 100 percent equivalent plastic strain. This strain threshold is a somewhat arbitrary, but is used to bound critical behavior of the shell as discussed below.

In general, quasi-static material properties for carbon steel dictate that failure would occur between 25 and 30 percent strain. As discussed in Reference [15], increasing the strain rate in carbon steels generally results in a higher yield strength until eventually, the ultimate stress is approached without yielding. Essentially, higher rates of loading (impulsive blast loading results in very high loading rates) can cause steels to exhibit increased strain hardening and behave in a more brittle fashion (i.e. less ductility). In this case, because detailed dynamic material properties were not readily available for the material in question, some judgment was used in developing the explicit element failure criterion (with the intent being to conservatively bound failure). Figure 7 below shows the progression of damage in the pressure boundary of the vertical heat exchanger (contours of deflection are shown). It can be seen that a large portion of the shell fails (equivalent plastic strain threshold of 100 percent is exceeded) and is automatically removed from the model.

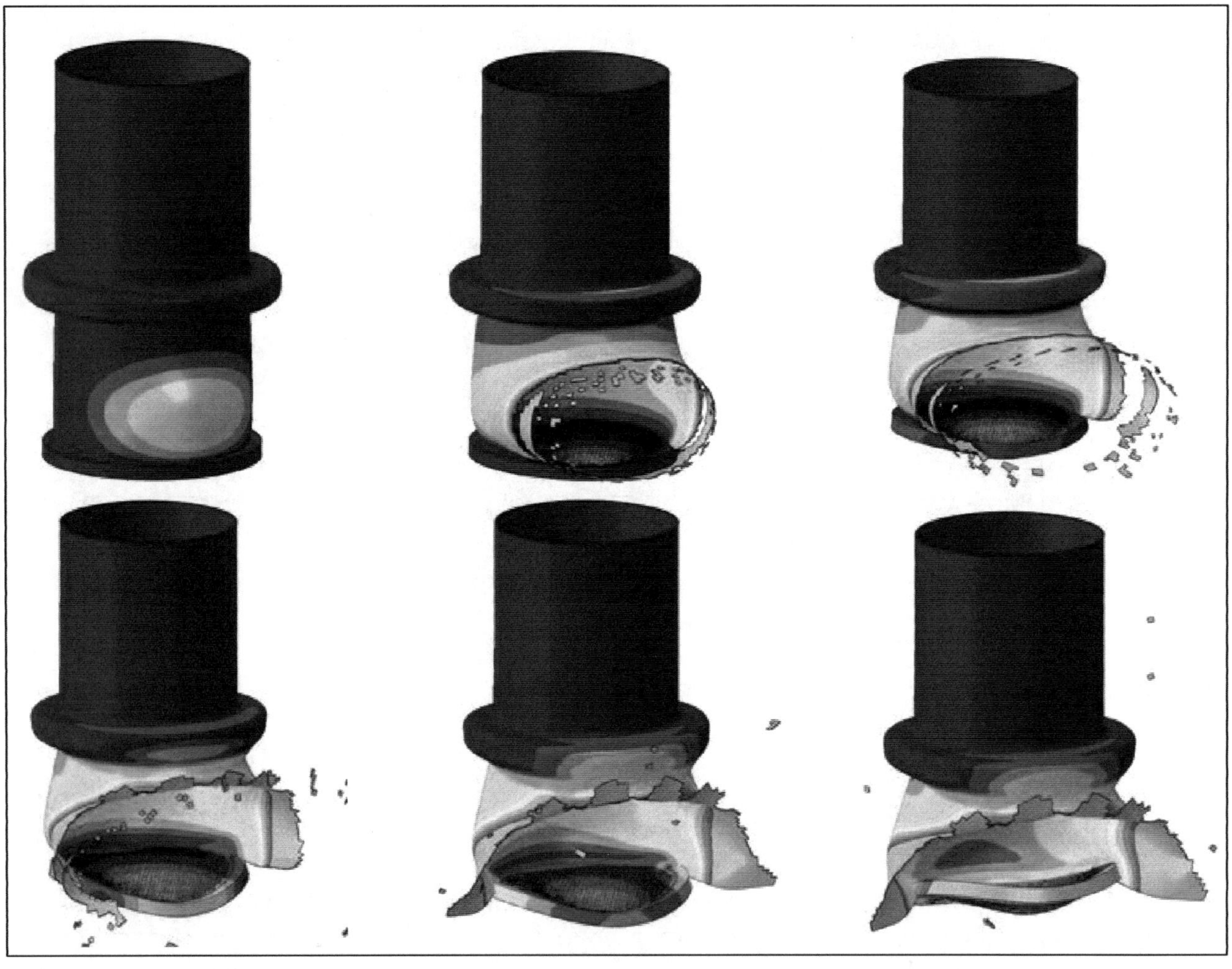

**Figure 7: Contours of Displacement Showing Damage Progression of the Shell Due to Internal Blast Loading (Element Failure Criterion Employed at 100 Percent Strain).**

Individual elements are also fragmented and thrown from exchanger in this simulation because all surrounding elements reached their user-defined strain threshold and were removed from the model; thus, there is no restraint surrounding these elements to hold them in place and the overpressure applied propels them outward from the blast center. It is noted that exchanger internals such as individual tubes, baffles plates, etc. are not explicitly modeled, and complex internal blast wave reflection from these components is neglected (as stated, investigating internal reflection effects due to complex geometry in detail would require the use of CFD analysis). In general, this simulation reasonably approximates the actual gross damage to the pressure boundary observed in the field; that is, the portion of the shell that was either fragmented or opened up following the presumed detonation event.

Sensitivity to the plastic strain threshold value used in the dynamic element failure criterion definition is investigated to examine the overall effect on structural response. Intuitively, lower user defined plastic strain thresholds predict more severe gross damage to the pressure boundary. Figure 8 shows contours of deflection for the exchanger shortly after the simulated internal detonation. In this case, the shear element failure criterion uses a strain threshold that is set to 30 percent equivalent plastic strain.

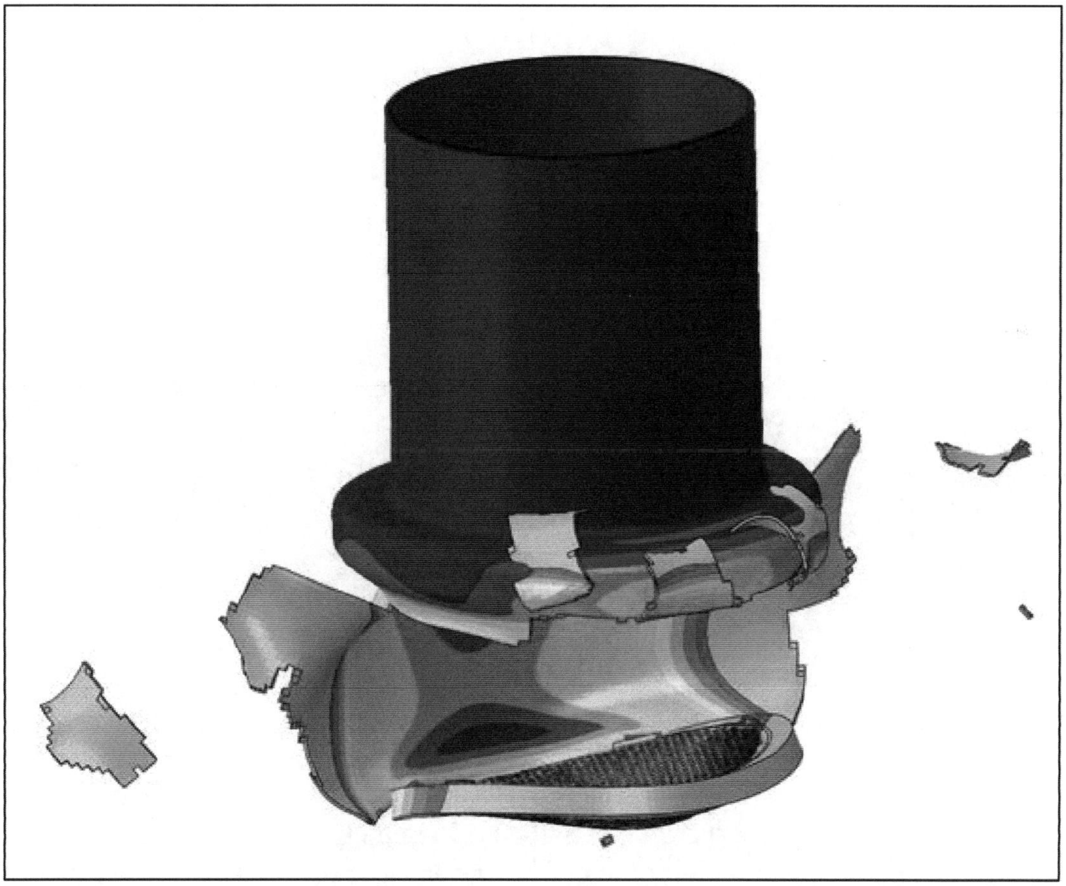

**Figure 8: Contours of Displacement Showing Damage of the Shell Due to Internal Blast Loading (Element Failure Criterion Employed at 30 Percent Strain).**

When compared to the damage progression depicted in Figure 7 (corresponding to a 100 percent equivalent plastic strain threshold), it can be observed that a greater portion of the shell is removed from the model. Furthermore, for this lower strain threshold, the overall damage extends further in the axial (element removal extends in the upward direction into the expansion joint) and circumferential directions, and plastic deformations are generally more severe in the shell that was not removed from the model.

In the explicit dynamic simulations discussed above, the stiffness of the tube bundle is neglected entirely. For this reason, the simulations over predict the amount of plastic deformation in the bottom tubesheet. Figure 9 shows the plastic deformation observed in the bottom tubesheet shortly after the simulated detonation. It can be observed that local to the assumed blast source within the exchanger, axial deflection in the tubesheet is predicted relative to the majority of the tubesheet area. This predicted deformation is consistent with the actual damage observed in the tubesheet in the field, although it is slightly higher than measured deformation. Additionally, there is some rigid body motion associated with the tubesheet after the simulated detonation because of the neglected stiffness of the tubesheet bundle from the model. After the removal of a large portion of the shell due to the assumed element failure criterion, there is little stiffness to hold the tubesheet in place. To this end, additional simulations are performed to investigate the effects of the tube bundle stiffness.

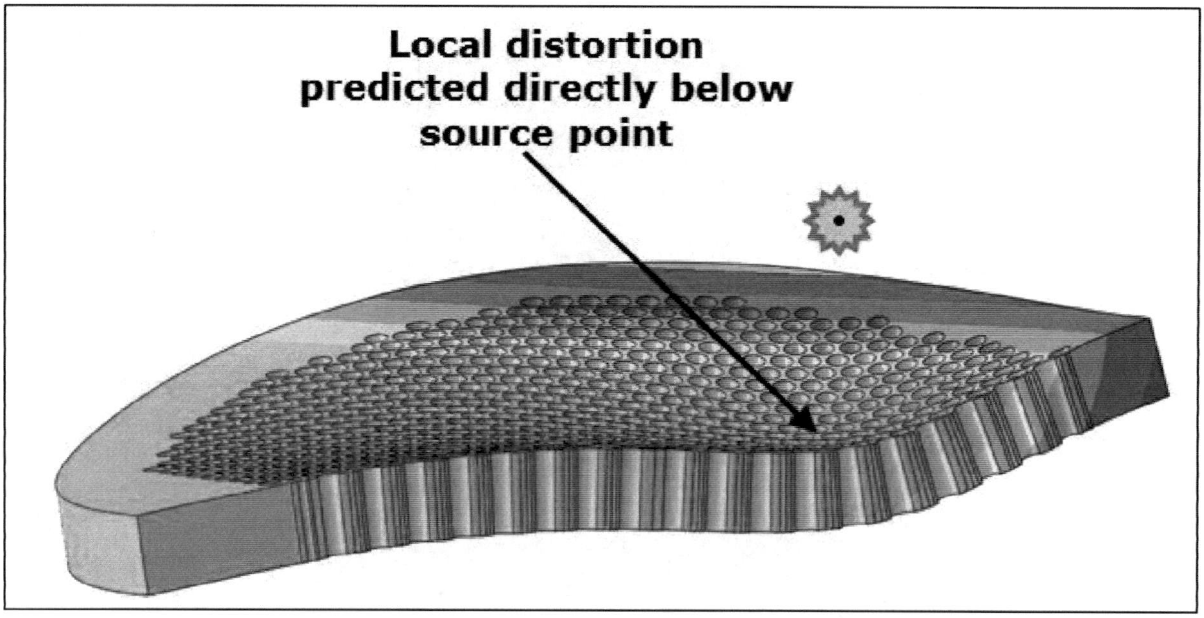


**Figure 9: Exaggerated Tubesheet Deflections Local to the Assumed Blast Source.**

In this case, explicitly modeling the tubes would be extremely computationally expensive and impractical due to required simulation times. In order to investigate the effect of the bundle on overall tubesheet deflections, equivalent beam elements are defined and coupled to each quadrant of the inside tubesheet face. This coupling definition permits the tubesheet to warp or bow while still distributing the stiffness of the bundle over each quadrant of the tubesheet face. Detailed discussion of this method of approximating the bundle stiffness is beyond the scope of this paper, but further explanation can be found in reference [16]. In this simulation, local tubesheet plastic deformations are under predicted (there is little local deformation of the tubesheet predicted directly below the assumed blast center). This occurs because after the detonation, the equivalent bundle stiffness is maintained and does not account for any damage/fragmentation (and ensuing removal) of individual tubes.

These results indicate that the amount of plastic deformation in the tubesheet actually observed in the field is in between the predicted amount of deformation in the model with the bundle completely neglected and the model with the equivalent bundle stiffness incorporated. The most rigorous approximation of tubesheet deformation would entail explicit modeling of the tubes and incorporation of progressive damage (and element deletion) such that the tubes in the immediate vicinity of the detonation source could be removed from the model as the overpressure wave propagates through the exchanger and once assumed strain limits are achieved. This simulation would be extremely computationally expensive and somewhat impractical. As stated, the overall predicted failure modes of the shell and tubesheet are in reasonable agreement with the actual damage observed in the field.

## 5. Conclusions

This study employs explicit finite element analysis to simulate a recent, real-life industry failure of a vertical heat exchanger subjected to an internal detonation. Realistic overpressure amplitudes are applied to 3D finite element models in accordance with the CONWEP blast loading model. Incident wave reflection is accounted for using this formulation without modeling the fluid medium and coupling it to the structural elements. The ensuing dynamic structural response and plastic deformations of both the vertical heat exchanger subjected to internal blast loading and the adjacent tower and support skirt subjected to external blast loading are quantified. This analysis permits the determination of an approximate amount of concentrated chemical product that caused the accidental explosion. Essentially, the approximate equivalent amount of explosives is back-calculated based on the predicted damage to the finite element model of the adjacent tower skirt. Computational iterations of varying charge amounts are performed and the predicted amount of plastic deformation is documented in each case.

Based on these iterations, the equivalent charge amount that most closely approximates the plastic damage of the nearby tower skirt is used to simulate internal blast loading on a 3D model of the vertical heat exchanger. An advanced element failure criterion is employed to simulate the damage of the pressure boundary observed in the field. According to this enforced formulation, elements lose stiffness and are removed from the model once a user-defined plastic strain threshold is reached. This dynamic simulation reasonably predicts the gross rupture and deflection of the pressure boundary observed and measured in the field. The explicit, three dimensional finite element based analyses discussed in this study permit the evaluation of the structural response and damage characteristics corresponding to this specific, recent industry failure. Explicit simulations, such as the ones discussed herein, can not only provide process safety engineers with a better understanding of the risks associated with accidental internal detonations, but can provide valuable insight into the mechanical design of fixed equipment that mitigates the risk of catastrophic rupture or failure due to internal or external blast loading.

## 6. References


1. Smith, P., and Hetherington, J., "Blast and Ballistic Loading of Structures, 1st ed," *Butterworth-Heinemann*, Oxford, 1994.
2. Kinney, G., and Graham, K., "Explosive Shocks in Air, 2nd ed," *Springer-Verlag*, Berlin, 1985.
3. U.S. Department of Defense, "Unified Facilities Criteria (UFC) 3-340-02: Structures to Resist the Effects of Accidental Explosions," *Department of Defense, United States of America*, 2008
4. Dusenberry, D., "Handbook for Blast-Resistant Design of Buildings, 1st ed," *John Wiley and Sons, Inc.*, Hoboken, New Jersey, 2010.
5. Bulson, P., "Explosive Loading of Engineering Structures, 1st ed," *Taylor and Francis Group*, London, 1997.
6. ASCE, "Design of Blast-Resistant Buildings in Petrochemical Facilities, 1st ed," Task Committee on Blast-Resistant Design, *American Society of Civil Engineers.*, Reston, Virginia, 1997.
7. Abaqus 6.10-1, DSS Simulia, "Analysis Users' Manual," 2010.
8. Hyde, D., "Users' Guide for Microcomputer Programs CONWEP and FUNPRO – Applications of TM 5-885-1," *U.S. Army Engineer Waterways Experimental Station*, Vicksburg, VA, 1988.
9. Duffey, T.A., Rodriguez, E.A., and Romero, C., "Detonation-Induced Dynamic Pressure Loading in Containment Vessels," Los Alamos National Laboratory: LA-UR-03-0741, 2003.
10. Dong, Q., Li, Q.M., and Zheng, J.Y., "Interactive Mechanisms Between the Internal Blast Loading and the Dynamic Elastic Response of Spherical Containment Vessels," *International Journal of Impact Engineering*, Volume 37, 4, 2010.
11. Prueter, P.E., Dewees, D.J., and Brown, R.G. "Examining Computational Techniques to Simulate the Effects of External Chemical Explosions Using Explicit Finite Element Analysis," *In ASME Pressure Vessels and Piping Conference*. PVP2012-78566, Toronto, Canada 2012.
12. Prueter, P.E. and Dewees, D.J., "Using Explicit Finite Element Analysis to Simulate the Effects of External Chemical Explosions on Single and Double-Walled Storage Tanks," *In Proceedings of the AIChE Spring Meeting and 9th Global Congress on Process Safety*, San Antonio, Texas, 2013.
13. Huntington-Thresher, W., and Cullis, I., "TNT Blast Scaling for Small Charges," *19th International Symposium of Ballistics*, 2001.
14. Baker, W., "Design Considerations for Toxic Chemical and Explosives Facilities – Chapter 1, Blast Pressure Effects: An Overview," *American Chemical Society*, 1987.
15. HSE, "Offshore Technology Report – OTO 1999 018: The Behaviour of Carbon Steels at High Strain Rates and Strain Limits," Health and Safety Executive, 1999.
16. Prueter, P.E., Macejko, B.R., Brown, R.G., and Dewees, D.J., "Assessing Flanged Joint Leakage and Tube-to-Tubesheet Joint Failure Due to Fluid Temperature Differentials in a Floating Head Heat Exchanger using Finite Element Analysis," *In ASME Pressure Vessels and Piping Conference,* PVP2013-97576, Paris, France 2013.

# Analysis of the Potential Energy Sources of Risk of Hand Tools in Presence of Hazardous Area in Segment of Oil and Gas

**Leandro Erthal**
**Petrobras S.A.**
**Av. Almirante Barroso, 81 suite 23 - Centro - Rio de Janeiro - Brazil**
**leandroerthal@petrobras.com.br**
**Universidade Federal do Rio de Janeiro (UFRJ) - Chemical Engineering Department (EQ)**
**Av. Um s/n - Cidade Universitária UFRJ - Rio de Janeiro - Brazil**

**Caetano Moraes**
**Universidade Federal do Rio de Janeiro (UFRJ) - Chemical Engineering Department (EQ)**
**Av. Um s/n - Cidade Universitária UFRJ - Rio de Janeiro - Brazil**
**caetano@eq.ufrj.br**

**Denize D. Carvalho**
**Universidade Federal do Rio de Janeiro (UFRJ) - Biochemical Engineering Department (EQ)**
**Av. Um s/n - Cidade Universitária UFRJ - Rio de Janeiro - Brazil**
**denize@eq.ufrj.br**

Prepared for Presentation at
American Institute of Chemical Engineers
2014 Spring Meeting
10th Global Congress on Process Safety
New Orleans, LA
March 30 – April 2, 2014
UNPUBLISHED

# Analysis of the Potential Energy Sources of Risk of Hand Tools in Presence of Hazardous Area in Segment of Oil and Gas

**Leandro Erthal**
**Petrobras S.A.**
**Av. Almirante Barroso, 81 suite 23 - Centro - Rio de Janeiro - Brazil**
**leandroerthal@petrobras.com.br**
**Universidade Federal do Rio de Janeiro (UFRJ) - Chemical Engineering Department (EQ)**
**Av. Um s/n - Cidade Universitária UFRJ - Rio de Janeiro - Brazil**

**Caetano Moraes**
**Universidade Federal do Rio de Janeiro (UFRJ) - Chemical Engineering Department (EQ)**
**Av. Um s/n - Cidade Universitária UFRJ - Rio de Janeiro - Brazil**
**caetano@eq.ufrj.br**

**Denize D. Carvalho**
**Universidade Federal do Rio de Janeiro (UFRJ) - Biochemical Engineering Department (EQ)**
**Av. Um s/n - Cidade Universitária UFRJ - Rio de Janeiro - Brazil**
**denize@eq.ufrj.br**

**Keywords:** Hazardous Area, Oil and Gas, Hand Tool

## Abstract

One of the methods adopted to protect against explosion in the oil industry refers to the use of hand tools for non-ferrous materials, usually bronze, also known as non-sparking hand tools. Studies conducted by the American Petroleum Institute (API/USA) and European standards disagree as to the risks involved in the energy of sparks produced by mechanical tools. The API advises that operations with mechanical equipment at high speeds and high contact pressures are extremely difficult to produce sparks capable of causing ignition of gases and vapors in the oil industry, and concludes that the risks involved are low for hydrocarbon vapors. The IEC (International Electrotechnical Commission), however, has begun to formulate a set of standards on this topic. These standards will be adopted and may affect the operations in the oil and gas industry.

Concerning the incident in the British Petroleum Refinery, Texas City, USA, the "Report of the BP US Refineries independent Safety Review Panel" or "Baker Panel Report" [1] - a report produced by an independent commission of inquiry - recommends that the accident prevention process require permanent vigilance. Moreover, it states that the lack of accident for a long period is not an appropriate indicator of high level process safety and can contribute to a dangerous and growing sense of complacency.

Using data from the literature and laboratory tests, this paper aims to clarify and demonstrate the actual risks to the energy produced by ferrous and nonferrous hand tools.

# 1. Introduction

In the late 20s, seeking to discover the causes of fires in the mining industry, was published the article entitled "Ignition of Gases and Vapors - Review Ignition of Flammable Gases and Vapors by Friction and Impact" by F. Powell [2]. He reports the work done in investigating more than 82 accidents, but only some of the references involving hand tools and the author avoided making any conclusions about it. A study presented to the Group Session on Fire Protection at the annual meeting of the American Petroleum Institute, in 1941, described a series of test conducted over 15 years. According to the test, it would be very unlikely that the ignition of vapors of oil could be caused by sparks produced by the contact between steel and steel abrasive wheel, or a given mechanized equipment. The nature of the sparks was discussed and it was demonstrated that any material harder than iron could produce sparks when struck against it. Thus, it was concluded that the insistence on the use of special hand tools called "non-sparking tools" creates a false sense of security at the expense of other more important measures of security protection against fire. Comprehensive rules covering the use of such tools were recommended, many of which are against the industry interests. The use of non-sparking hand tools was not universal throughout the industry until 1941, but then, after the presentation of the article of Group Session on Fire Protection, many companies started a program to phase out the use of special tools. However, some companies associated with API believed that additional research was needed to obtain a definitive opinion, and an independent institution should do it. In 1948, the necessary funds had already been obtained and contracted in 1950 to conduct a series of tests at Underwriters Laboratories Inc, in Chicago, sponsored by the American Petroleum Committee on Accident Prevention and Fire Protection. Little progress, however, was done [3].

The Institute of Petroleum (UK) began to consider the importance of sparking hand tools as an ignition source in the early 60's through the Committee on Industrial Fires and Explosion of the Fire Research Board. These studies [4] included a comprehensive review of published information, but they did not contain any new experimental evidence. An introductory note prepared by Engineering Committee of the Petroleum Institute indicated that additional experimental studies were not necessary; accepting the conclusion that non-sparking tool materials do not significantly reduce the risk of ignition of oil vapors by sparks due to friction force, compared to coming from ferrous tools.

The American Petroleum Institute (API) issued in July 2004, the fourth edition of the report API RP 2214 - Sparks Ignition Properties of Hand Tools [5], concludes that nothing essentially new has been learned since the publication of the first document in 1956. This document complements recent publications such as the NFPA's Fire Protection Handbook, with the same conclusion of 40 years ago. Sparking produced by violent impact between some materials, including metals commonly termed as non-sparking, can actually cause the ignition of gases and vapors if generated enough energy. However, such conditions are rare to occur in operations using hand tools. The records on the fire at companies, which have never used or ceased to use non-sparking tools, confirm the position of the Safety Committee of API's Board of Directors in 1956.

According to PROUST [6], it has been well known for more than a century that the combustible dust suspended in air are responsible for part of the blasts, but available knowledge and practices about the explosions of flammable gases and vapors still contain a significant part of empiricism [7].

Over the last decade, many investigation committees have been created with a map road including several patterns with a set of standards to be developed for non-electrical equipment and components.
The aim of this work is to investigate what the real risk is and what parameters should be taken into account when manual hand tools are used in hazardous locations within the oil and gas industry.

## 2. Definitions

### *2.1 Hazardous locations*

The manufacturing process, according to Garside [8], often involves the use, processing or storage of hazardous materials (flammable). Evidently, industrial plants associated with these operations need to be arranged and controlled so that these materials do not ignite or cause explosion, generating damage to people or the plant itself. Thus, the presence of gases, vapors, combustible dusts and fibers associated with the air, in optimal proportions, may occur in certain places and is referred as "explosive". From a chemical standpoint, oxidation, combustion and explosion are reactions of different speeds, being initiated by ignition or detonation.
In the "definitions" of API RP 500 [9], it is mentioned that any area where gases and vapors are present or may be present in the atmosphere in sufficient quantities to produce a flammable mixture concentrations is a "hazardous (classified) location."
In industrial plants, a technique of risk analysis called "classification of areas" is usually used.
An area classification or hazardous location is a study that analyzes the local industrial plant during the normal operation process that can contain flammables within its flammable range, with some frequency and therefore, requires that special care to be adopted to prevent that electrical or mechanical equipment be a source of ignition in such explosive atmospheres [10,11].

### *2.2 Risk of hazardous locations*

There are numerous potential sources of ignition of an explosive atmosphere, according to Garside [8], such as the uncontrolled use of open flames, hot surfaces of the process itself, sparking, static electricity, including those generated by clothing and electrical discharges. Many of these sources can be controlled or minimized by a good arrangement process, management and control. However, there is an additional issue that are the electrical and mechanical equipment necessary to operate an industrial plant, such as pumps, fans, lights, instrumentation and other equipment are themselves potential sources of ignition.
Directive 1999/92/EC of European Committee [11] states that, "The installation, equipment, protection systems and associated connecting devices must only be put into service if the documents about the protections explosion indicate that it may be safely used in presence of explosive atmospheres". Another important point that determines Directive 1999/92/EC [11] is that the same conditions above must be:

> [...] It also applies to work equipment and associated connecting devices which are not considered equipment or protective systems in the design of Directive 94/9/EC if their incorporation into an installation can in itself lead to a risk of explosion. The necessary measures should be taken to avoid any incompatibility between connecting devices. "

Directive 1999/92/EC [11] provides as a protection against explosions:

> "All necessary measures should be taken to ensure that the workplace, work equipment and connecting devices made available to workers have been designed, manufactured, assembled and installed, and are maintained and operated as to minimize risk of explosion".

### *2.3 Certification*

A certificate of conformity is to issue a statement based on a decision made after the critical analyses of the service to specified requirements has been demonstrated

### *2.4 Studies on the mechanical ignition risk*

For the development of standards methods of explosion protection for non electrical equipment, according to PROUST [6], there is a significant gap in knowledge of their mechanisms of explosion. However, the key parameters were not identified in a precise manner.
There is some information provided to workers about the risks of fire and explosion of tools called "sparking", made from "materials harder", made of steel with index Vickers hardness greater than 230 HV, according to ASTM E92 (Standard Test Method for Vickers Hardness of Metallic Materials) [12].
For SCHUTH [13]:

> "When we refer to surveys conducted in Europe and USA, it was found that the source of the spark that causes more explosions relates to the use of common tools (steel) in places where there is risk of explosion ... According to experts, a spark from a tool steel reaches temperatures of 1550 °C to 1850 °C, being the product of a conversion of energy, i.e. the kinetic energy resulting from force applied to the tool or consequence of free fall, becomes the impact on thermal energy in both materials in collision ... It is clear that the risk of explosion exists and should be considered rigorously, and it is therefore avoided the use of steel tools where such risk may exist.

A study carried out by MADAKSON [14] shows that the friction of a material also depends on the test system. Replicate samples of the same material were distributed in different laboratories to measure friction under certain conditions. Using different measuring systems, each laboratory reported a value different from the friction. Between January 2001 and May 2005, the European Community developed a project called "MECHEX" (Mechanical ignition hazards in potentially explosive gas and dust atmospheres), with the French laboratory INERIS, to review of the risk of mechanical ignition (Ignition Hazard Assessments) and be able to apply the European standards (EN). This project sought to develop knowledge of the risks associated with these devices through two research projects: characteristics of sparks and hot spots produced under impact, friction or grinding. Simulated ignitions with the same characteristics of those found in industry to measure the sensitivity of an explosive environment ignition atmospheres and characterization with the goal of standardizing the ignitability sources of explosives were used. The results of these two projects were combined in order to categorize the equipment to use in explosive atmospheres based on the type of mechanical ignition sources expected to appear in a specific project, the method of operation and the environment capable of suffering ignition. Thus, the process of degradation of the mechanical energy into heat for the friction and impact was studied.

# 3. Overview of Associations and Companies

## *3.1 Certification of hand tools*

The operating procedure called "Approval Standard for Sparks Resistant Tools" [15] was published in December 2007 by the certification body FACTORY MUTUAL APPROVAL (FM). The scope says that operating certification procedure is applied to any tool or device whose purpose is to function as a spark resistant. These devices can be manual or electric, such as hammer, screwdriver, pliers, sander or any other. It mentions that this procedure assesses the potential of these tools and devices, or material, in which it can be manufactured to generate a visible or cause ignition of flammable materials under certain conditions in laboratory test spark. It restricts the condition that it does not assess the practicality or suitability of the purpose of using the device, but it is limited to determining the potential to cause ignition of materials in normal operation.

According to WANG, in NFPA Memorandum [16], FM conducted tests described in "Report n° IN4A0.AF Job identification" in accordance with the requirements of the "Safety Tools Approval Test Procedure - 7910 class". It is used as a guide to assess tools designed for use in environments where the risk of ignition of flammable materials, dust or vapors by mechanical sparks generated by use of hand tools, iron and steel that may slip by or affecting a surface exists. This report has shown that:

- ✓ Sparks produced by steel plates caused explosion of the pad saturated with carbon disulfide in eight of eight tests, ethyl ether pads, five of eight tests and naphtha in four of eight tests;
- ✓ No sparks were observed and no ignition of flammable liquids mentioned above was certified security tools.

This document states that its requirements reflect tests and practices used to analyze the metallic characteristics to allow their use in the manufacture of spark resistant in order to obtain a certification tool.

The above-mentioned requirement to meet compliance of this operating procedure is that the tool should produce no visible spark when struck on a variety of surfaces. According to FM, this demonstrates a lack of ability to produce sparks, and thus reduces the risk of causing ignition of the material or atmosphere around them when the tool is used. In order to meet this requirement, no visible spark should be identified in the video recording of high speed. The spark must be generated by the impact of the alloy tested on a range of surfaces like those on which are expected in service. To meet the requirement of the test for flammable chemicals, particles generated by applying a sample of the material on the grinding wheel must not cause ignition of flammable vapors. The particles are released on a wad of cotton soaked with a flammable chemical. The cotton swab should not be ignited.

## *3.2 International Association for Hydrogen Safety (HySafe)*

According to HySafe, the use of non-sparking tools is often recommended; however, the energy to ignite a flammable mixture of hydrogen in air is so small that these tools are still capable of causing ignition. Consequently, all tools need to be used judiciously to prevent landslides, shocks or falls because all these conditions can cause sparks. According to HAWKSWORTH [17] one, regardless of their material tool, dropping the height of a man is capable of causing ignition of an explosive atmosphere of hydrogen gas.

### *3.3 International Eletrotechnical Commission - IEC*

#### 3.3.1 Assessment of single impact sparks as Potential Ignition Sources

According to the criteria of the standard PT ISO/IEC 80079-36 [19] "single impacts between metal parts need not to be considered as potential ignition sources in the ignition hazard assessment if the following conditions are met.
Either:
a) the impact velocity is less than 1 m/s and the maximum potential impact energy is less than 500 J, and

- aluminium, titanium and magnesium in combination with ferritic steel is not used, or
- aluminium in combination with stainless steel (≥ 16,5 % Cr) can be only used if the steel cannot corrode and no iron oxide and/or rusty particles can be deposited on the surface (appropriate reference to the properties of the stainless steel shall be given in the technical documentation and instructions for use), or
- hard steel in combination with hard steel is not used, or
- hard steel is not used where it can impact on granite, or
- aluminium in combination with aluminium is only used if no iron oxide and/or rusty particles can be deposited on the surface.

NOTE 1 Hard steel is understood as being either all kinds of hardened steel (surface hardened or heat treated in another way to improve surface hardness) or other steel types with Vickers Hardness greater than 230 HV (acc. to ISO 6507 with test load ≥ 98 N).

Or

b) where a combination of non-sparking metals is used and the impact velocity is less than or equal to 15 m/s and the maximum potential energy is less than 60 J for gas/vapour-atmospheres or less than 125 J for dust-atmospheres.

#### 3.3.2 Assessment of single impact sparks as effective ignition sources

Ignition sources generated by impact need not be considered as effective ignition sources if the impact velocity is less than 15 m/s and the maximum possible potential energy is less than the values given in Table 1

On the other hand, if the energies exceed the values in the Table 1, this does not necessarily mean that the ignition source will become effective. In this case, the ignition hazard assessment needs to assess all aspects and may show that the likelihood of the impact is low enough to be acceptable.

The impact energies need to be assessed if they are larger than those shown in the table. In this case, consideration shall be taken of when they occur and whether they are able to ignite the explosive.

**Table 1. Single impact energy limits**

| Explosion group | Single impact energy limits | |
|---|---|---|
| | Non-sparking metals: | Other materials |
| IIC | 60 J | 5 J (Hydrogen)<br>3 J (Hydrocarbons incl. acetylene) |
| IIB | 125 J | 10 J |
| IIA | 125 J | 20 J |
| NOTE These criteria do not apply for atmospheres with flammable gases such as carbon disulfide, carbon monoxide and ethylene oxide | | |

### *3.4 Health and Safety Executive (HSE-UK)*

In the publication HS (G) 140 [19], refers to the need not to rely too much on the use of non-sparking tools in areas of flammable, being necessary to remove all flammable materials before the service is performed, if possible, and if otherwise, the need to keep all the tools free of fouling.

### *3.5 National Fire Protection Association – NFPA*

NFPA 30 Flammable and Combustible Liquids Code: 2003 [20] as well as the NFPA 1 Fire Code: 2008 [21], there was no recommendation on whether to use non-sparking hand tools.
The NFPA Technical Committee on Fundamentals cites in its opinion on the technical review conducted in consultation with the NFPA 30 - Flammable and Combustible Liquids Code:2011[22] that the work of the American Petroleum Institute (API 2214) [23] refutes the need for using non-sparking hand tools, except in special circumstances. After the technical argument against the representative of a company non-sparking tools on the proposal no. 30-58, the Committee decided to reject the suggestion of inclusion of the proposed item: "6.5.6 Heating by friction or mechanical sparks, with the following text: iron and steel hand tools may produce sparks that can be a source of ignition of surrounding flammable substances. Where this risk exists, non-sparking tools should be used. "Thus, the decision was to include a new text A.6.5.1 in which to control ignition sources precautions should be taken to prevent ignition of flammable by sources such as heating by friction or sparks".
This same text was copied and expressed in NFPA 1 Fire Code: 2012 [24], and complemented the informational item related to this subject, A.66.6.5.1 item. It cites that for frictional heating or sparking, it is recognized that there is the need to control sources of ignition, including sparks mechanic hand tools; we have enough for the ignition of flammable vapors energy.

### *3.6 U.S. Occupational Safety & Health Administration - OSHA*

In their training program "Office of Training and Education":1996, entitled " Construction Safety and Health Outreach Program" [25], in the chapter " Recognition of Risks ", it is mentioned that around flammable substances, sparks produced by the hand tools of iron and steel can be a dangerous ignition source. Where such hazards exist, non-sparking tools made by brass, plastic, aluminum or wood supply security.

Recognizing that steel tools can be a potential source of ignition in explosive atmospheres, OSHA has prepared a guidance through the booklet 3080 - Hand Tools and Power , Review 2002" [26], which describes that the steel hand tools and iron can produce sparks and that these can be a source of ignition of surrounding flammable materials. Where this hazard exists, non-sparking hand tools need to be used.

The OSHA inspection reports #309178523, #122021017, #313996118, #302920400 and #309178523 as reported security flaws lack of use of non-sparking hand tools. The OSHA report #313523375, for example, cites a violation non-compliance OSHA 29 CFR 1910.106 (h) (7) (i) (a) - Flammable Liquids and Combustive [27] that "precautions were not taken to prevent the ignition of flammable vapors. The employer did not use non-sparking hand tools in hazardous location classified as Class I Division I".

OSHA Regulation 29 CFR 1910.146 - Permit-Required Confined Spaces, appendix D - Confined Space Pre -Entry Check List [28], non-sparking hand tools are on the checklist of equipment, in clause 9, to enter into confined spaces.

OSHA 29 CFR 1910.146:1996 "Standard Interpretations, Interpretative guidance and Tackle approved for hazardous locations relative to the Permit -Required Confined Space", states that "... electrical equipment as well as other sources of ignition (non-sparking hand tools) needs to be considered and adopted to go into hazardous areas".

OSHA Office of Training and Education has a PowerPoint presentation about Hand and Power Tool Safety [29] that reports that environments around the flammable sparks produced by iron in contact with steel hand tools can be a dangerous ignition source. Moreover, it cites that non-sparking hand tools, made of brass, plastic, aluminum or wood, should be provided for safety where this risk exists.

### *3.6 Petroleum Companies*

**Table 2. Company versus recommendation to use non-sparking hand tools**

| *Company* | *Where* | *Recomendation* |
|---|---|---|
| **ExxonMobil** | MSDS-USA "gas" in land spill | Use |
| **Chevron** | MSDS-USA "gas", section 6 | Use |
| **Valero** | MSDS-USA "gas", section 6, Accidental Release Measures | Use |
| **CITGO Petroleum Corp.** | MSDS conform OSHA 29 CFR 1910.1200. Section 6 - Accidental Release Measures [30] | Use |
| **Marathon/Petrocom** | MSDS-USA "gas" | Not informed |
| **Shell** | MSDS-USA "gas" | Not informed |
| **ConocoPhillips** | MSDS-USA "gas" - Section 7: Handling and Storage: Precautions for safe handling | Use |
| **British Petroleum (BP)** | MSDS-USA "gas" - Section 7: Handling and Storage: Handling: | Use |
| **Total** | MSDS-USA "gas" - Section 7: Handling and Storage: Precautions for safe handling | Not informed |
| **Statoil** | MSDS "gas" - Section 7: Handling and Storage: Technical measures | Use |
| **Petrobras** | MSDS "gas" - Section 7: Handling and Storage | Use |

# 4. Method and type of search

## *4.1 Test method*

Regarding the purposes of this project, the test method to be adopted is an adaptation of the ASTM E680-79 standard, revalidated in 2011[31]. This standard has original purpose as being a standardized test method to evaluate the sensitivity of hazardous materials in the solid phase to the impact of falling weights. Impact tests by falling weight specified in this standard will be performed using the method of Bruceton up and down.

Laboratories: At the first moment, the test will be conducted in a mechanical laboratory to verify "vibration" and energy. Next, the test will be in a laboratory to atmosphere explosive.

The research of the potential risk of power tools will be in relation to:

Hammer head (new and used) by stainless steel, steel carbon and others material called non-sparking.

Objective: To evaluate the ability of new tools and steel rust capable of causing ignition of gases from petroleum environments.

Phases:

1) Chemical analysis of the hammer's material;

2) Conducting impact tests for samples according to test procedure.

Materials:

1 ) Hammer head steel with Rockwell hardness according to ASTM E680-79 [31] (surface between 55 HRC and 59 HRC ), industrial line without oxidation;

2 ) Model wooden handle;

3 ) Mass hammer considering only the mass of the steel head: 1.0 kg, 2.0 kg, 3.0 kg, 4.0 kg and 5.0 kg;

4 ) Cash test with pressure relief through plastic film covering;

5 ) Gases and reference concentrations for tests . The criteria set forth by the IEC 60079-1 [32] will be adopted: $(4,6 \pm 0,3)\%$ de propane and $(8 \pm 0,5)$ % de ethylene;

6) Three samples of hammer;

7 ) Mechanic impact;

8 ) High-speed video record camera;

9 ) Equipment for measuring temperature;

10) Humidity between 50% up 90%;

11 ) Temperature between 30 °C up 40 °C;

12 ) Atmospheric pressure between 86 kPa up 106 kPa (860 mbar up 1060 mbar);

13 ) Setting of test laboratory accredited by Inmetro (Brazil's official accreditation) which scope is to evaluate electrical equipment for explosive atmospheres.

Impact test :

1) The hammer shall be tested by running 300 strong , lilting and consecutive blows , with a difference of 10 seconds between impacts, simulating the use by a man of average height with a mass about 80 kg. The impact energy shall be up to 8.0 Kgf.m;

2) Address the impacts against a new target hardened steel with Rockwell hardness in the range of ( 55-59 ) HRC . The roughness of the target to be in the range of 6.3 to 8.0 μ μ .

3) Run the impacts against the surface of different materials target, like steel stainless, steel carbon and aluminum;

4 ) Evaluate if ignition up to 300 impacts occurs

### *4.2 Criteria for assessing the likelihood of ignition of hand tools*

Four techniques will be used to evaluate the capacity of sparks:
a) Measurement of vibration;
b) Thermography;
c) Video recording in high speed;
d) Testing in explosive atmospheres.

4.2.1 Target for each technique

a) Measurement of vibration – by each type of hammer and height, a series of twelve tests (12) will be done to obtain maximum real value of energy caused by impact (taking into account the energy lost by friction resistances of the system) and adjust the scale of the equipment to test the values of energy. These values are measured by accelerometers installed on the hammer;
b) Thermography – to verify the maximum temperatures generated by impacts and to compare them with the ignition temperatures of flammable substances in the test;
c) Video recording in high speed – to check the capacity of impacts of a hammer to generate light energy on plates of different materials (stainless steel, carbon steel and aluminum);
d) Testing in explosive atmospheres – through a type of pass/no pass test to verify if the luminous and thermal energy released is sufficient to ignite an explosive atmosphere may be present in an environment of the oil industry.

4.2.2 Equipment test

The equipment is three-meter high, has a target that can change the position of impact, has an electrical system (drive keys, electrical protection and a frequency converter that drives an electric motor) and an instrumentation system (sensors associated with the frequency converter). When the system is turned on, the magnetic sensor at the end of the path identifies the presence of the hammer and fires the frequency converter. The motor starts slowly lifting the hammer. As the desired time is reached, another sensor to the converter indicates this position. The motor is turned off causing the free fall of the hammer. Upon reaching the end of the course, the first sensor detects the hammer and resumes operation. This condition can be programmed to repeat automatically the test for a fixed number of times and can be triggered manually or perform the test continuously.

**Table 3. Photos of the equipment test**

The path to the target

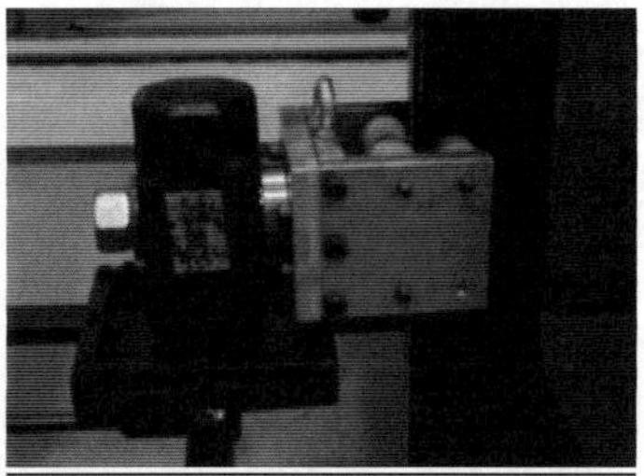

The hammer and the target
(with regulators of angles)

The hammer at falling (freedom)

## 5. Discussion and critical analysis

According to PROUST [6], explosions of flammable gases and vapors have mechanisms that are not well known, and practices still seem to contain a significant portion of empiricism.

There are many different opinions about this matter but no confidence form to prove that a steel hand tool can be a source of ignition in an oil industry. The unique report confidence is the European program MECHEX that was dedicated to study the process of ignition during friction and impact with gas, improved by Ineris Laboratory, to attend the European Norm. This report concluded that "for "impacts", the relevant parameter for ignition is not the kinetic energy of the projectile, but its velocity and the nature of the materials. The experimental data are not enough to validate the model and more experiences need to be lead."

Regarding the document API 2214:1956 [3], it can be observed that experiments carried out by consolidated companies and research conducted in the 50's demonstrate that these called non-sparking tools do not improve the security of the premises. Initially, this document advised not to use this type of tool. In the 2004 version [5], it stated that the non-sparking tools had less energy and at certain points showed a dubious position about the topic, without being conclusive. This version cites the position of the NFPA to support its point of view, but NFPA is motivated by a demand for manufacturers of non-sparking tools, without scientific support, mainly based on the U.S. law. It can be seen that this legislation is guided by the historical results of accidents in mining segment in which the chemicals involved are different, extrapolated to all activities. It is not possible to say whether the manufacturers of this type of tool influenced the U.S. legislation, but by acting with NFPA, presumably, they act in the interests of their industry.

Other investigations also contradict API's conclusion on non-sparking tools. A paper published in the Fire Protection Journal, "The Danger of Incendive Sparks" [16], stated that after API's conclusion on "non-sparking" tools, the Institute of Petroleum sought the advice of the Fire Research Board Committee on Industrial Fires and Explosion. This committee came to certain conclusions, one of which was that "Until...evidence is obtained to show that the impact of steel hand tools on steel is safe, it is advisable to use non-sparking tools in situations where flammable atmospheres are unavoidable".

Another point worth mentioning is that during the period of studies and evaluations to issuing the first paper of this issue, the oil industry began operations with hydrogen, and it is not expected that this chemical element was being considered at that time. In tests conducted by BOEMRE, the impact between rusted steel and steel with no rust were capable of causing ignition of a flammable mixture; however, OLSON [33] showed that the threshold pressure for this fact was 100 psig, which corresponds to projectile speed 20 m/s, incompatible with the operating conditions of hand tools in the oil and gas industry. This fact was confirmed in the tests and procedures performed by Foyn and Moe, in PROST [6], where no ignition was recorded for impact between rusted steel and steel.

Companies in the oil and gas segment diverge on the topic for several reasons:

1) It must meet local legislation. In the United States: OSHA 29 CFR 1910.1200 (g) (2) (viii) standard [30];

2) The API reference shows little conviction in its recommendation without theoretical support and recent laboratory tests, repeating observations only practices a very different period. Professionals responsible for this activity tend to be conservative when they are not supported by laboratory tests or certifications of third party;

3) The oil industry' 50s had few processes that used hydrogen, unlike the current processes;

4) Europe has a long tradition in mining, which influences European companies and their employees;
5) World companies such as Shell and Total, as well as other companies do not recommend non-sparking tools in their MSDS. Others have recommended its use. Considering Petrobras, there is an orientation not to recommend non-sparking tools, but it is regarded in its large-scale application.

## 6. Conclusion

The tests with mechanical systems have a great variability of results due to the complexity of the reactions involved. The initial idea was to capture the thermal energy released by the decomposition of the system energy: total energy - (energy transferred in the form of vibration + energy transferred by deforming + sound energy + light energy) = thermal energy, but as the amount of thermal energy is low (in the order of milijoules) only mechanical variations unfeasible this analysis by decomposition of factors.
Measurements of thermograph are not sufficient for obtaining reliable data accuracy. For tests, the measured values were much lower than the ignition temperature of the gases. It is not possible to identify visually the light energy being released in impacts. During the tests in the laboratory of explosive atmospheres, a high-speed camera will assist to identify possible release of light energy.
At this moment, there is no evidence that a steel hand tool can cause ignition of an explosive atmosphere in an oil industry. More tests are being carried out at this moment.

## 7. References

[1] THE REPORT OF THE BP U.S. REFINERIES INDEPENDENT SAFETY REVIEW PANEL - *2006 - Process Safety Culture Survey.*

[2] POWELL, F., “Ignition of Gases And Vapor – Review Of Ignition Of Flamamable Gases And Vapors By Friction And Impact.” Industrial and Enginnering Chemistry. (1969). vol. 61 (12) pp. 29-37.

[3] AMERICAN PETROLEUM INSTITUTE. API 2214 - Spark Ignition Properties of Hand Tools. USA: API, 1956.

[4] RIDDLESTONE, H.G., BARTELS, A., “The Relative Hazards In The Use Of Ferrous And Non-Sparking Tools In The Petroleum Industry. American Petroleum Institute. 51 (495), 106-10.1965.

[5] AMERICAN PETROLEUM INSTITUTE, API RP 2214 - Spark Ignition Properties of Hand Tools. 4. ed. USA: API, 2004.

[6] PROUST, Ch., HAWKSWORTH, S., ROGERS, R., BEYER, M., LAKIC, D., RAVEAU, D., HERVÉ, P., PINA, V., PETITFRÈRE, C., LEFEBVRE, X., “Development Of A Method For Predicting The Ignition Of Explosive Atmospheres By Mechanical Friction And Impacts (MECHEX)”. Journal of Loss Prevention in the Process Industries. (2006). Vol 19 p. 104-120. Elsevier Ltd.

[7] INTERNATIONAL CHAMBER OF SHIPPING (ICS) & THE OIL COMPANIES INTERNATIONAL MARINE FORUM (OCIMF) & INTERNATIONAL ASSOCIATION OF PORTS AND HARBORS (IAPH). International Safety Guide For Oil Tankers And Terminals (ISGOTT). London: ICS/OCIMF/IAPH. WITHERBY & CO. LTD. 5. ed., 2006.

[8] GARSIDE, R., "Electrical apparatus and hazardous areas". Aylesbury, Buckinghamshire, UK: Hexagon Technology Limited, 2.ed.rev., 1995a. 395 p.

[9] AMERICAN PETROLEUM INSTITUTE, API RP 500 - Recommended Practice for Classification of Locations for Electrical Installations at Petroleum Facilities Classified as Class I, Division 1 and Division 2. 2. ed. USA: API. 1997 and reaffirmed 2002. 131 p.

[10] ASSOCIAÇÃO BRASILEIRA DE NORMAS TÉCNICAS. ABNT NBR IEC 60050-426 - Vocabulário eletrotécnico internacional parte 426: Equipamentos para atmosferas explosivas. Rio de Janeiro. 2011. 53 p.

[11] PARLAMENTO EUROPEU. DIRECTIVA ATEX 1999/92/EC: EN Official Journal of the European Communities. 28.1.2000, L 23/57. Site: http://eur-lex.europa.eu/LexUriServ/LexUriServ/LexUriServ.do?uri=OJ:L:2000:023:0057:0064:EN:PDF. (accessed Jan. 15, 2012 at 13h40).

[12] AMERICAN SOCIETY FOR TESTING AND MATERIALS. ASTM E92-82 (2003) - Standard Test Method for Vickers Hardness of Metallic Materials. USA: 1982 reaffirmed at 2003. 9 p.

[13] SCHUTH, K. H., Palasio, C.M. Jr., "Ferramentas Anti Faiscantes". Net, Rio de Janeiro, Site: http://www.ecivilnet.com/artigos/ferramentas_antifaiscantes.htm. (accessed Jan. 02, 2012 at 13h53).

[14] MADAKSON, P.B., "The Frictional Behavior Of Materials". Wear. (1983). pp. 191-206.

[15] FACTORY MUTUAL APPROVAL. Procedure Class Number 7410 - Approval Standard for Sparks Resistant Tools. 2007. 18 p.

[16] NATIONAL FIRE PROTECTION ASSOCIATION: Memorandum: NFPA 30 ROC TC Letter Ballot (A2011). USA, Massachusetts, Quincy, 2010.

[17] HAWKSWORTH S., ROGERS R., PROUST, C., BEYER, M, SCHENK, S., GUMMER, J., RAVEAU, D., "Mechanical Ignition Hazards in Potentially Explosive Atmospheres - EC Project MECHEX". In: Int. ESMG Symposion on Process Safety and Industrial Explosion Protection (ExplosionRisk), Nuremberg, 2004.

[18] INTERNATIONAL ELECTROTECHNICAL COMMISSION: PT ISO/IEC 80079-36: Explosive atmospheres — Non-electrical equipment for explosive atmospheres – Basic methods and requirements, draft

[19] HEALTH AND SAFETY EXECUTIVE . Health and Safety Guidance: HS(G)140 - Safe use and handling of flamable liquids. United Kington. HSE, 1996. Site: http://www.hse.gov.uk/pubns/books/ hsg140.htm. (accessed Jan. 13, 2012 at 16h).

[20] NATIONAL FIRE PROTECTION ASSOCIATION. NFPA 30 - Flammable and Combustible Liquids Code, USA, Massachusetts, Quincy, Ed. 2003.

[21] NATIONAL FIRE PROTECTION ASSOCIATION. NFPA 1 – NFPA Fire Protection Handbook, USA, Massachusetts, Quincy, 15. ed., (2008).

[22] NATIONAL FIRE PROTECTION ASSOCIATION. NFPA 30 - Flammable and Combustible Liquids Code, USA, Massachusetts, Quincy, Ed. 2011.

[23] AMERICAN PETROLEUM INSTITUTE. API RP 2214 - Spark Ignition Properties of Hand Tools. 3ed. USA: API, 1989

[24] NATIONAL FIRE PROTECTION ASSOCIATION. NFPA 1 – Fire Code, USA, Massachusetts, Quincy, Ed. 2012.

[25] U.S. OCCUPATIONAL SAFETY & HEALTH ADMINISTRATION (U.S.OSHA). OSHA Office of Training and Education: Construction Safety And Health Outreach Program. (1996). Site: http://www.osha.gov/pls/imis/establishment. inspection_detail?id=304994304, (accessed Jan. 11, 2012 at 10h22).

[26] U.S. OCCUPATIONAL SAFETY & HEALTH ADMINISTRATION (U.S.OSHA). OSHA Office of Training and Education: OSHA book 3080 - Hand And Power Tools. (2002). Site: http://www.osha.gov/Publications/osha3080.html, (accessed Jan. 11, 2012 at 10h30).

[27] U.S. OCCUPATIONAL SAFETY & HEALTH ADMINISTRATION (U.S.OSHA). OSHA 29 CFR 1910.106: Flammable and combustible liquids. Site: http://www.osha.gov/pls/oshaweb/owadisp.show_document?p_id=9752&p_table=STANDARDS; (accessed Jan. 11, 2012 at 10h22).

[28] U.S. OCCUPATIONAL SAFETY & HEALTH ADMINISTRATION (U.S.OSHA). OSHA 29 CFR 1910.146: Permit-Required Confined Spaces, Appendix D - Confined Space Pré-Entry - Check List (1996). Site: http://www.osha.gov/pls/oshaweb/owadisp.show_document?p_table=INTERPRETATIONS&p_id=22234. (accessed Jan. 11, 2012 at 14h40).

[29] U.S. OCCUPATIONAL SAFETY & HEALTH ADMINISTRATION (U.S.OSHA). OSHA Office of Training And Education: Apresentation: *Hand and Power Tool Safety-Spark Resistant Tools*. (2008). Site: www.osha.gov/dte/.../module_5.ppt, (accessed Jan. 11, 2012 at 14h50)

[30] U.S. OCCUPATIONAL SAFETY & HEALTH ADMINISTRATION (U.S.OSHA). OSHA 29 CFR 1910.120: Hazardous waste operations and emergency Response. Site: http://www.osha.gov/pls/oshaweb/owadisp.show_document?p_table=STANDARDS&p_id=9765. (accessed Jan. 11, 2012 at 9h27)

[31] AMERICAN SOCIETY FOR TESTING AND MATERIALS. ASTM E680 - 79 - Standard Test Method for Drop Weight Impact Sensitivity of Solid-Phase Hazardous Materials. USA: 2007. (Reapproved 2011) 7p. Site: http://www.astm.org/Standards/E680.htm. (accessed Fev. 18, 2012 at 10h30).

[32] INTERNATIONAL ELECTROTECHNICAL COMMISSION: IEC 60079-1: Explosive atmospheres: Part 1: Equipment protection by flameproof enclosures “d”. Geneve, 2007. 73 p.

[33] OLSON, D.L., MAROEF, I., and PARK, Y.D., “Feasibility Of Thermite Sparking With Impact Of Rusted Steel Onto Aluminum Coated Steel”, Colorado School of Mines (2002). p66.

# Increase Hazard Discovery and Minimize Errors in your Process Hazard Analyses

## A Graph Theoretical Approach

**R. Maqbool Qadir, PE**
**Enpro Solutions, Inc.**
**6500 Dublin Blvd., Suite 215, Dublin, CA 94568**
**Qadir@enprosolutions.com**



Prepared for Presentation at
American Institute of Chemical Engineers
2014 Spring Meeting
10th Global Congress on Process Safety
New Orleans, LA
March 30 – April 2, 2014

UNPUBLISHED

# Increase Hazard Discovery and Minimize Errors in your Process Hazard Analyses

## A Graph Theoretical Approach

**R. Maqbool Qadir, PE**
**Enpro Solutions, Inc.**
**6500 Dublin Blvd., Suite 215, Dublin, CA 94568**
**Qadir@enprosolutions.com**

**Keywords:** PHA, HAZOP, FMEA, FTA, graph theory, hazard

## Abstract

Process Hazard Analysis (PHA) methods and techniques are constantly being developed. Yet, PHAs often fail to recognize discoverable, credible hazard scenarios. Unrecognized hazards were implicated in the BP Texas City disaster, Formosa Plastics explosion, Chevron Richmond Refinery fire and other high profile incidents. Unrecognized scenarios represent a gap in understanding of process hazards and keep us from preventing accidents and disasters. This paper focuses on improving the completeness of the hazard identification. The PHA process is visualized by mapping it in the form of graphs. Ammonia tank hazard and operability study (HAZOP) and Failure Modes and Effects Analysis (FMEA) examples are mapped as graphs to reveal tree structures. These tree graphs or inference maps directly allow us to visualize causal connections and inferences. Chains of cause and effect are mapped as unique pathways that can be represented as both graphs and matrices. For example, one can graphically trace the pathways to hazardous events such as loss of containment.

This paper investigates the role of errors, omissions and constraints in reducing the discovery of process hazards. What are the sources of errors and omissions (E/O)? What are the effects of constraints like budget and time? How can PHAs be audited efficiently to identify and correct E/O? A fault tree analysis (FTA) of the generic HAZOP process is performed to trace the sources of the E/O and how they may be minimized. The FTA identifies seventeen distinct sources of E/O. The sources of E/O are mapped to the issues they affect and a quality assurance checklist is derived from the FTA of the HAZOP process. CSB case studies are cited showing the impact of the E/O on hazard discovery in real incidents.

# 1. Introduction

Process Hazard Analysis (PHA) is an analytical discipline whose purpose is to discover process hazards and initiating causes that if uncorrected could lead to process upsets, accidents and disasters. CSB and OSHA have found that a large number of incidents could have been prevented if the hazards implicated in the incidents had been recognized and mitigated. PHA is a fundamental part of any Process Safety Management (PSM) program and is one of the keys to preventing accidents. Furthermore, PHAs are required for facilities with hazardous materials subject to the OSHA PSM Standard [1] and/or the EPA Risk Management Program or RMP [2] as well as state and local regulations.

In the past decades improvements have been made to existing PHA methods and new techniques are constantly being developed. Yet, PHAs conducted often fail to recognize discoverable, credible hazard scenarios. Unrecognized hazards and other PHA deficiencies were implicated in the BP Texas City disaster [3], Formosa Plastics explosion [4], Chevron Richmond Refinery fire [5] and other high profile incidents. Unrecognized scenarios represent a gap in understanding of process hazards and limit our ability to prevent accidents and disasters. There is a need therefore to improve the effectiveness of PHA to maximize the discovery of credible hazard scenarios. This would improve our ability to identify measures that can reduce risk and increase safety performance. It is also essential to understand the limitations of analysis - that all hazards cannot necessarily be identified even with the best teams and methods. Constant vigilance and periodic re-evaluation of existing PHAs are required to operate safely.

This paper investigates the role of errors, omissions and constraints in reducing or limiting the discovery of process hazards by PHA. Specifically I explore the following questions – Why do PHAs fail to identify credible hazard scenarios? What are the sources of errors and omissions (E/O)? What is the effect of constraints like budget and time? How can PHAs be audited efficiently to identify and correct errors and omissions? I perform a fault tree analysis (FTA) of the PHA process to trace the sources of the E/O, show how they arise, and how they may be minimized. I cite CSB reports where the E/O identified in the FTA occurred in real incidents. The paper will also discuss regulatory requirements, standards/ guidelines and best practices that apply to PHA. This article will focus mainly on the hazard and operability (HAZOP) method but the arguments presented are applicable to other techniques as well.

# 2. PHA Methods and discoverability of hazards

There are many structured brainstorming methods used by multidisciplinary teams of analysts to conduct hazard analyses on complex systems such as process plants, aviation systems and manufactured components. The HAZOP method is arguably the most popular in the chemical process industries. Another popular method is the failure modes and effects analysis (FMEA). Both methods are recognized in OSHA and EPA regulations [1, 2]. Each method was developed for a specialized purpose and its evolutionary history makes one method more suited to a particular situation than another. For instance the HAZOP method is particularly well suited to the analysis of process plants while the FMEA was evolved to study the failure modes of aircraft components. Because of the variations in approach and methodology among the various PHA methods, discovery of hazard scenarios depends to an extent on the method selected. A HAZOP and FMEA conducted on the same system will tend to identify a similar but not identical set of hazard scenarios. Furthermore, the combinatorial possibilities in complex systems are far too many for a team applying a single PHA method to identify, much less evaluate *all* potential hazard scenarios in any reasonable timeframe. This is one reason why PHAs can fail to detect certain hazards. This dependence of hazard scenario discovery on PHA method is illustrated in Figure 1.

**Figure 1:Hazards discovery using various PHA methods**

## 3. Visualizing PHA - A Simple HAZOP and FMEA example

HAZOP teams study complete processes or plant sections within a process to predict how hazards may arise and how they might be mitigated. For this purpose they examine drawings and schematics, especially piping and instrumentation diagrams (P&IDs) to understand how individual plant equipment and control systems work together as complex systems. The use of HAZOP and FMEA methods is illustrated below to analyze hazards associated with an anhydrous ammonia storage tank depicted in the simplified P&ID in Figure 2.

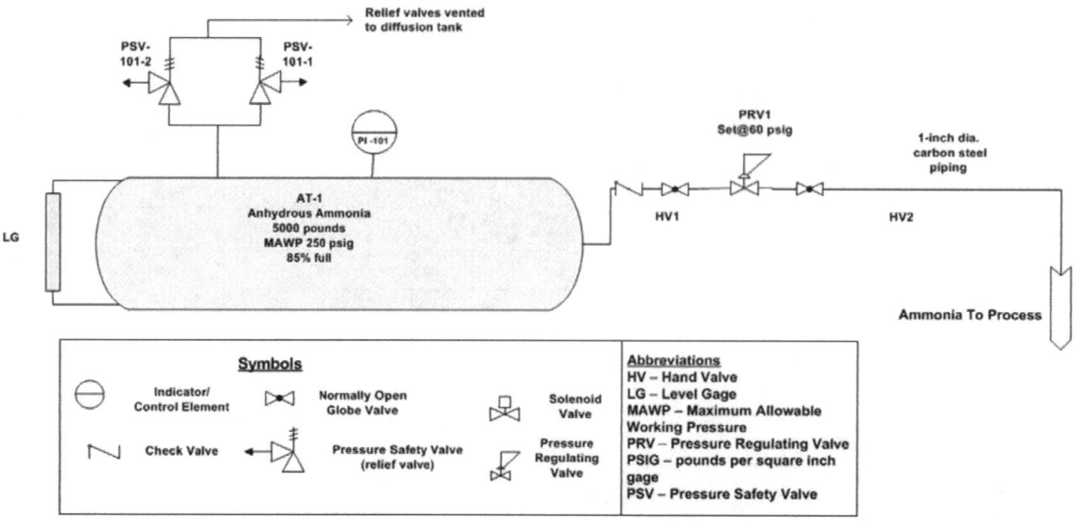


**Figure 2: Simplied P&ID for an Anhydrous Ammonia Storage Tank**

Any PHA examines a process step by step and comes up with a series of inferences that lead to identifying and evaluating hazard scenarios, initiating causes and safeguards. PHA team findings are typically documented in spreadsheet or tabular format. However, this mode of presentation obscures an underlying structure to the PHA. Therefore, this section introduces the use of graphs for visualizing the sequential steps of a PHA. The graphs shown in Figures 3 and 4 illustrate the HAZOP and FMEA analyses as applied to the anhydrous ammonia tank system of Figure 2.

The illustrations are not meant to be complete analyses or full explanations of the use of HAZOP/FMEA methods. Rather the main point of these graphs is to reveal a tree structure that will allow us to gain some useful insights into the analytical process. In the language of graph theory the HAZOP analysis can be modeled as connected, acyclic graphs, with the "root vertices" at the nodes and "leaves" at the terminal ends ("recommendations" in Figure 3). These tree graphs or inference maps directly allow us to visualize causal connections and inferences. For example, one can graphically trace the pathways to hazardous events such as loss of containment (LOC) as discussed below.

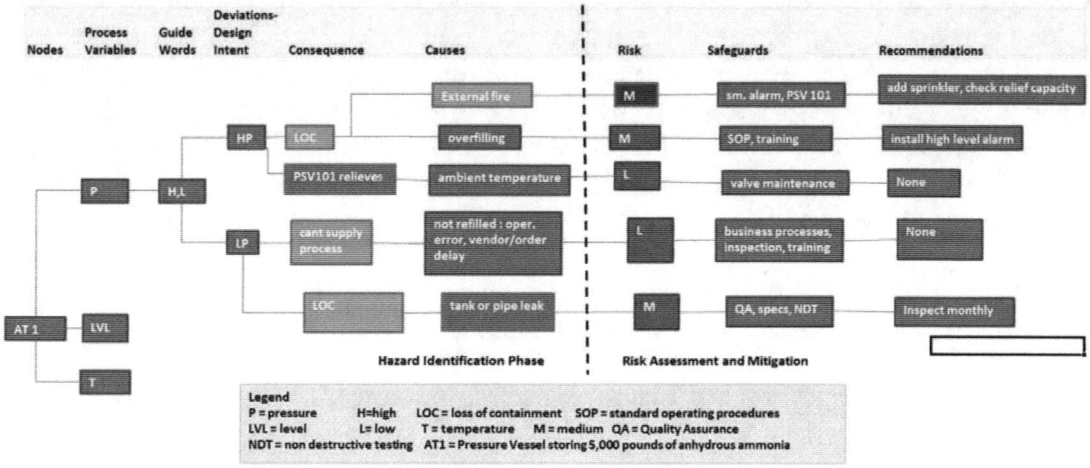


**Figure 3: HAZOP GRAPH**

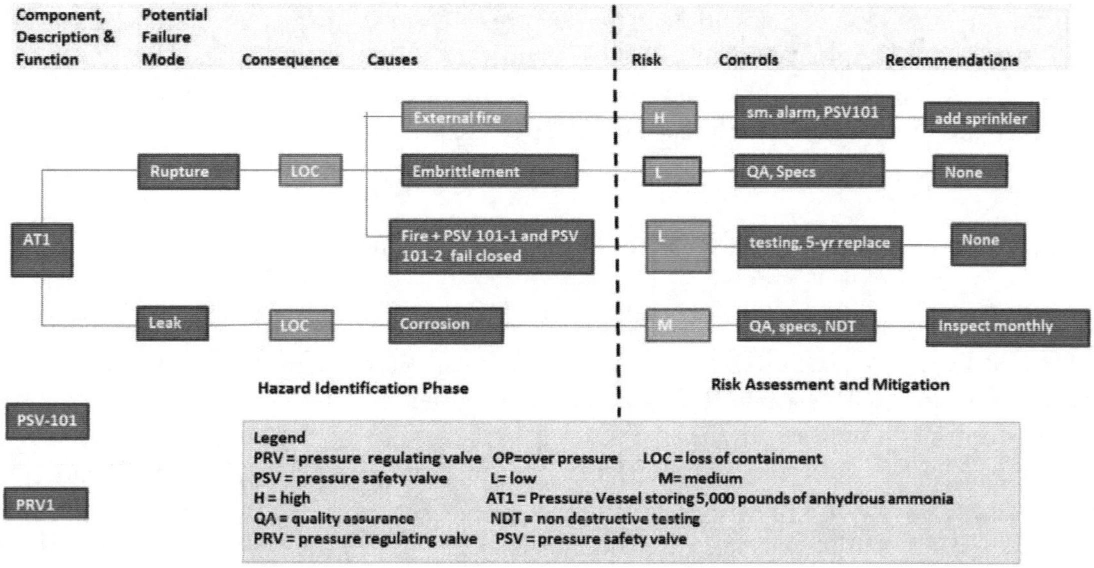


**Figure 4: FMEA GRAPH**

# 4. Hazard Identification and Consequence Assessment

The steps "nodes" through "causes" shown in the area to the left of the dotted line in Figure 3 and "components" through "causes" in Figure 4 might be termed the "hazard identification" or "discovery" phase of the HAZOP or FMEA analysis. The HAZOP method [6] combines key process variables such as pressure, flow and level with guide words such as less, more, high and low to explore deviations from design intent such as high pressure, high level, no flow etc. For example in Figure 3 the deviations high pressure (HP) and low pressure (LP) were identified for the ammonia tank AT1 of Figure 2.

The team then brainstorms what could cause the deviations and how the system would behave when perturbed from steady state design conditions. Hazards may arise from the effects that would occur as a result of these initiating causes. The team documents the plant section or equipment studied (nodes); process deviations, potential causes and their effects or consequences - credible hazard scenarios are identified as cause-consequence pairs. For example the hazard identified in the topmost branch of figure 3 is LOC from tank-AT1 caused by fire. The complete pathway is - fire causes high pressure in Tank AT1 resulting in LOC. This pathway is easily seen in the color coded graph. A detailed consequence assessment could be required for the LOC such as estimating exposure of employees to a toxic release and offsite consequences to nearby receptors.

FMEA is similar to HAZOP in some ways but focuses on failure modes of major components [6] or equipment and their causes rather than deviations in process variables and the causes of these deviations. For example, the FMEA in Figure 4 identifies rupture and leakage as modes by which AT1 and the associated pipeline might fail. Like the HAZOP the FMEA also identifies LOC pathways for tank AT1. However, the causes identified are corrosion, embrittlement, and stress from high pressure sustained due to fire. Corrosion and embrittlement were not identified as potential causes of a LOC in this example HAZOP (although the tank or pipe leak mentioned as a cause should lead one to consider corrosion). This illustrates the effect of differences in PHA methods on hazard scenario discovery.

## 5. Hazard Scenarios - Understanding Cause and Effect

Investigating deviations and understanding their causes and effects is at the heart of the hazard discovery process. This is where much of the team brainstorming effort is directed. To further investigate this portion of the analysis one can abstract the chain of cause and effect shown as graphs in Figure 5 below with the following notations: D for deviations, C for causes and E for effects or consequences. Some common cause and effect patterns signifying hazard discovery pathways are depicted as a series of graphs in Figure 5. These pathways can also be represented as matrices - which we will call Deviation Matrices - as shown in Figure 5. Four common patterns are demonstrated with examples that are not necessarily related to the system shown in Figure 2.

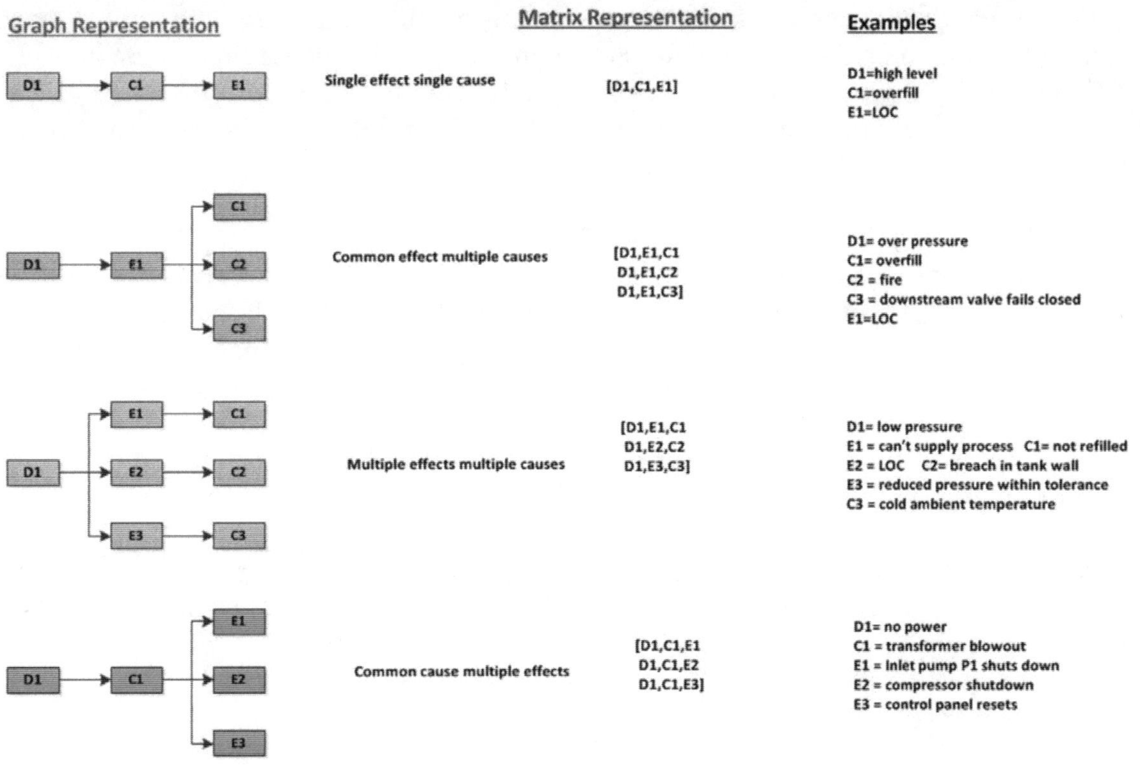


**Figure 5: Cause and Effect Patterns**

The first case "deviation, single effect single cause" depicts the general case of cause and effect and may be regarded as the fundamental unit of hazard scenarios. Logically, the cause always precedes the effect. The deviations can be regarded as intermediate effects, thus the logical sequence would be C-D-E e.g. an initiating cause such as overfilling leads to a deviation such as high level which if continued leads to LOC as the ultimate consequence or effect . However, the HAZOP process uses deviations from design intent to examine the behavior of a system perturbed from its design or steady state. The team reasons backward to find causes and forward to estimate effects. Therefore the brainstorming sequence is usually D-C-E or D-E-C. The two sequences are logically equivalent but there are advantages to each. It is easier to think about the cause leading to the deviation leading to the effect. However, for reasons of economy it can be advantageous to list effects first so that deviations leading to moderate or benign effects can be eliminated from further consideration. Thus the team's brainstorming efforts in search of causes can be reserved for the more serious or harmful effects.

The HAZOP process may first identify an immediate or proximate cause but the complete pathway may require going to a more primitive or root cause. For example in Figure 3, a LOC caused by a tank or pipe leak begs the question – what caused the leak? So a complete pathway would be for example, a LOC caused by a leak which is caused by corrosion. This may seem obvious but in the recent Chevron Richmond Refinery catastrophic pipe failure and fire of 2012 the CSB found [5] that a HAZOP team had looked at the deviation "leak/rupture" associated with piping from a crude distillation tower. However, they did not identify corrosion as a potential cause – and ultimately the tubes failed due to corrosion releasing hot hydrocarbons that caused the explosion! The CSB further determined that this apparent oversight occurred because the team had not identified a damage mechanism for the type of corrosion that occurred, perhaps because the team had no corrosion experts on it.

Similarly, in Figure 3 the LOC is identified as a consequence of a leak. But a complete consequence assessment would flesh out the scenario to look at the impacts of the LOC such as worker exposure or offsite consequences. In the case of BP Texas City explosion of 2005, the CSB found [3] that consequences for the deviations "high level" and "high pressure" in the raffinate splitter tower where the explosion took place were not fully identified. Specifically that safety relief valves would be over-pressurized and liquid would overflow from the tower into a blow down drum and then overflow from the blow down drum were not identified.

Looking at the graphs in Figure 5, there are three ways of failing to identify hazard pathways for a given process variable – miss a deviation, miss one or more causes, and miss one or more effects.

## 6. Risk Assessment and Risk Reduction

The steps to the right of the dotted line in Figures 3 and 4 are devoted to risk assessment and mitigation of the hazards identified in the first phase. The risk of an event is estimated from the combination of its estimated frequency of occurrence and severity of consequence. Existing safeguards are identified such as process controls, alarms and trips that would prevent a process deviation from turning into a hazardous event. If adequate safeguards are lacking the team comes up with recommendations for additional engineering and/or administrative controls to reduce the frequency or consequence of a potential event. This paper does not address risk assessment - risks can be estimated by various methods: qualitative risk assessment such as a Hazard Matrix, semi-quantitative methods such as Layer of Protection Analysis (LOPA) and quantitative risk assessment (QRA) methods. Risk reduction includes the study of inherently safe design, process controls and alarm systems, passive and administrative safeguards and other preventive measures and is also not the focus of this paper.

## 7. Mapping the Hazard Identification Phase

The more complete the hazard identification the more successfully the associated risks, safeguards and mitigation measures can be evaluated. This paper focuses on improving the completeness of the hazard identification phase. In order to do this we map and analyze its structure as discussed below. Figure 6 shows an abstract representation of a hypothetical PHA where the process contains three nodes (N), three process variables of interest (V) in each node, three potential guide words (G) for each variable and three causes (C) for each deviation (D) identified by combining variables and guide words. The example is constructed to show that consideration of a complex system can easily lead to a very large number of deviation-cause-consequence pathways to be evaluated. Referring to Figure 6, an example of a complete pathway is N1-V1-G1/D1-C1-E1, where the hazard scenario would consist of the cause-consequence pair C1-E1.

This formulation of the problem would lead to N*V*G*C potential pathways i.e. $3^4$ or 81 hazard scenarios. Of course in a real case many pathways would not be of interest and would not lead to "credible" scenarios. But the question is - what is the impact of not identifying a relevant guideword? In this example nine cause-consequence pairs or hazard scenarios are obtained for a single process variable from one node. Failure to recognize a cause- consequence pair at the terminal branches results in missing one scenario. Not recognizing a single valid guide word results in the failure to recognize three potential hazard scenarios. Not recognizing an important process variable would lead to a loss of information about nine potential hazard scenarios. Thus an omission at the front end or "upstream" in the analysis is more consequential than at the terminal end.

The above analysis has implications for auditing a PHA and performing completeness checks on a PHA. Suppose that a review of process variables indicated that loss of containment had not been analyzed. Then referring to Figure 6, one would add a variable "containment" as V4 for example. The guide word could be "no" and the deviation would be loss of containment and one would proceed to investigate causes and consequences. Our analysis shows that it is easier to review process variables and guide words to successfully detect an omission (Figure 6) rather than trying to find a missing cause or consequence among the many that might have been generated. Therefore, a gap analysis should first check process variables and guide words as control points when completing or revalidating a PHA.

The PHA tree or inference map allows us to perceive the structure of the hazard discovery process and perform the following:

- See paths leading to hazard scenarios for each node and mode of operation.
- Recognize and remove unimportant branches – saving effort for analysis.
- Add a branch back in if re-considered.
- Do Quality Assurance (QA) on PHA to do "completeness checks".
- More easily audit for errors and omissions in analysis.

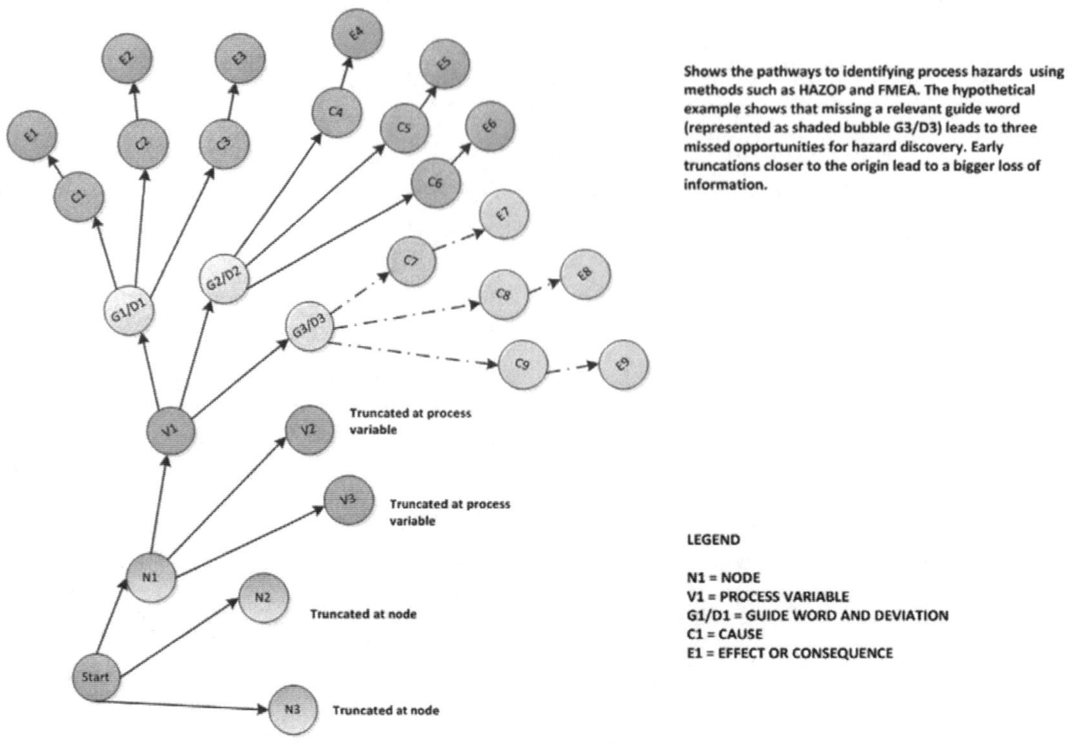


**Figure 6: PHA Tree**

# 8. Sources of Errors and Omissions in PHA - A Fault Tree Analysis

We mapped the PHA process using HAZOP as a template and illustrated how deviations from design intent are generated from a consideration of nodes, key process variables and guide words. At the next stage a consideration of causes and consequences of deviations results in identifying hazards as cause-consequence pairs. We demonstrated the loss of information that can occur at each stage of the analysis such as when a PHA team fails to identify a process variable or guide word. We also discussed the limitations of individual methods. Next we bring together all these factors along with a consideration of PHA input data to identify additional sources of errors and omissions that lead to non-recognition of hazard scenarios. For this purpose a FTA of the PHA process itself was performed as shown in Figure 7 to discover the root causes of errors and omissions.

For purposes of this analysis an error is a mistake in applying an analytical method, a calculation error, obsolete data, incorrect data entry etc. An omission is a failure to include any valid consideration or data in the analysis. A constraint is a limitation imposed by time, budget, scope, technical expertise, or industry state of knowledge. Through sheer reasoning alone the FTA identifies 17 distinct sources of errors, omissions and constraints that could result in a failure to identify a credible hazard. These findings are summarized in as a checklist in Table 1 below. To demonstrate the real world relevance of these findings, cases are cited where CSB found the E/O to have occurred. A further insight provided by the FTA is that the sources of E/O are mapped to the issue they affect e.g. missing process safety information (PSI) are mapped to missing deviations from design intent and ignoring incident history is mapped to failure to recognize hazards and incorrect evaluation of consequences.

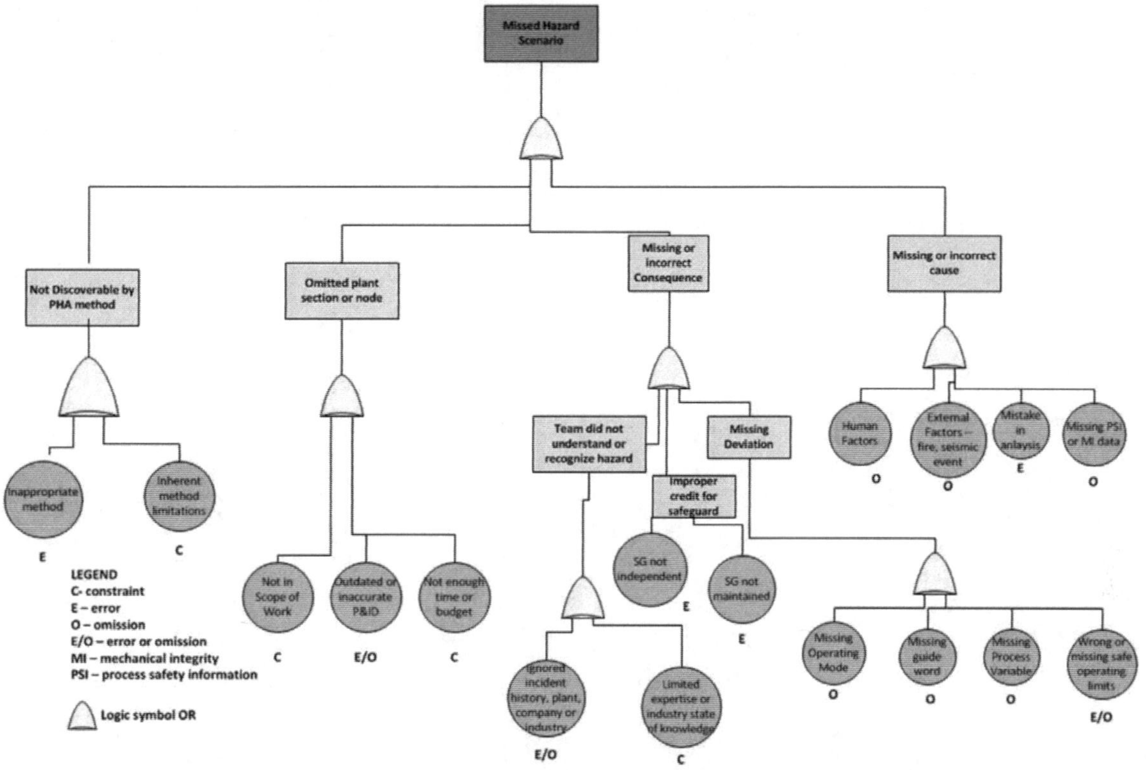


**Figure 7: Fault Tree, Errors and Omissions in Process Hazard Analysis**

**Table 1 – A Checklist for Reducing Errors and Omissions in PHA (Sources of E/O Identified by Fault Tree Analysis)**

| Item No. | FTA Finding | Constraint, Error, or Omission | CSB case study documented examples of occurrence |
|---|---|---|---|
| 1. | Inherent PHA method limitation | Constraint | None provided |
| 2. | Not in scope of work | Constraint | None provided |
| 3. | Not enough time or budget | Constraint | None provided |
| 4. | Limited Expertise or industry state of knowledge | Constraint | T2 labs [7], BP Amoco [9], process chemistry - reaction hazards not understood or poorly understood. |
| 5. | Inappropriate PHA method | Error | None provided |
| 6. | Outdated or inaccurate P&ID | Error | Oleum release [8], emergency power supply not included on P&ID |
| 7. | Ignored incident history – plant, company or industry | Error | BP Texas City [3], BP Amoco [9], Formosa Plastics [4], Chevron [5] – |

| | | | |
|---|---|---|---|
| | | | PHAs and revalidations did not take into account documented plant incidents or company incidents, and/or ignored recommendations from incidents. |
| 8. | Took credit for safeguard that is not independent | Error | Oleum [8], All safeguards required operator attention and action, no engineering controls independent of operator. Chevron [5] PHA only listed qualitative safeguards such as inspection and corrosion allowance. |
| 9. | Took credit for safeguard that is not maintained | Error | BP Texas City [3] – raffinate splitter alarm and blow down drum high level alarm did not work. Chevron [5] PHA only listed qualitative safeguards such as inspection and corrosion allowance. |
| 10. | Mistake in analysis | Error | Chevron Richmond [5], Did not identify failure mechanism for sulfidation corrosion |
| 11. | Incorrect safe operating Limits | Error | BP Amoco [9] Safe operating limits not identified. |
| 12. | Did not consider human factors | Omission | BP Texas City (3), Oleum [8], Formosa Plastics [4] – Human error not taken into account in evaluating safeguards. |
| 13. | Did not consider external factors like fire or earthquake | Omission | None provided |
| 14. | Did not have process safety information or mechanical integrity (MI) data | Omission | Oleum – PHA team lacked information on hazards of transferring oleum using emergency power supply. |
| 15. | Did not consider non-routine operating modes such as startup and shutdown | Omission | BP Texas City and BP Amoco did not consider startup and shutdown. |
| 16. | Did not consider valid guide word | Omission | Formosa cited by OSHA [10] for leaving out key guide words for batch processes. |
| 17. | Did not consider key process variable | Omission | CSB Sterigenics [11] No explosion scenarios evaluated for explosive concentrations. |

# 9. Deficiencies in PHA Data Sources

Incorrect or missing data are implicated in a number of sources of errors and omissions identified in Table 1 above. Table 2 maps data required for conducting a HAZOP to the output that is affected by deficiencies in those data. For example deficiencies in process chemistry data might cause reactive hazards to go unidentified. Similarly, a team working with old and outdated operating procedures might fail to identify hazards associated with starting up a process unit.

**Table 2 - Impact of Data Deficiencies on PHA Findings**

| INPUT | | OUTPUT |
|---|---|---|
| P&ID | | Nodes, upstream, downstream, loops, system effects |
| Equipment Specs, materials and safe operating limits | HAZOP | Deviations and consequences of deviations |
| Instrumentation, alarms and interlocks | | Safeguards, BPCS, ERP |
| Process Chemistry | | Reactive Hazards |
| Utilities | | Deviations, common cause failures, consequences |
| Labeling, siting, HMI | | Human Factors |
| Standard Operating Procedures | | Task design, batch operations, startup, shutdown, maintenance |

# 10. Regulations, Standards and Best Practices

Applicable regulations such as OSHA PSM [1] and EPA RMP [2] contain requirements and guidelines for conducting PHAs. OSHA has detailed requirements and guidelines including team composition, team expertise, recognized PHA methods, human factors, siting, evaluation of hazards and safeguards. There are also OSHA letters that address the interpretation of specific issues. CCPS has many guidelines for hazard evaluation [6] that include best practices to minimize deficiencies and produce robust PHAs.

# 10. Conclusion

What does the analysis presented in this paper show us about controlling or reducing errors and omissions in PHA to maximize the discovery of hazards based on?

***10.1 Team***: All of the errors and omissions identified in conducting a PHA are human errors. Therefore, picking the right team with the appropriate training, expertise and leadership is essential. PHA is a team exercise requiring a properly balanced and knowledgeable team with the appropriate composition. This is an OSHA PSM requirement [1]. A good team leader can help the team maintain cohesion and apply the proper discipline, skill, and awareness to the analysis. The leader and the team must be confident, curious, be willingness to question assumptions and have the ability to express disagreement without being disruptive.

***10.2 Errors***: Upfront preparation and a thorough review of the data sources should be performed prior to conducting the PHA to prevent errors from incorrect or outdated data. A review of the completed PHA documentation with appropriate checklists can aid in detecting errors. Tables 1 and 2 serve as checklists in addition to the many excellent literature references available to aid in successfully conducting PHAs.

***10.3 Omissions***: Omissions can be minimized with appropriate quality assurance procedures such as the checklists in Tables 1 and 2. A review of data sources prior to conducting the analysis should be conducted to ensure that PSI and MI data are current and available to the team. An incident history of the plant, company and industry should be compiled and reviewed to ensure that known problems are not ignored. Process incident databases such as the CCPS database can help to identify industry incidents and lessons learned. LOC should always be considered as a deviation. Our analysis shows that it is easier to review process variables and guide words to successfully detect an omission (Figure 6) rather than trying to find a missing cause or consequence among the many that might have been generated. Therefore, a gap analysis should first check variables and guide words as control points when completing or revalidating a PHA.

***10.4 Constraints***: Proper scoping is essential to correctly prioritize unit selection so that the most hazardous units and the most hazardous equipment are selected for analysis. For proper scoping the PHA team and the consumers of the PHA must take into account the limits imposed by time, budget, PHA method, and company or regulatory priorities.

***10.5 Insights Gained from graph theoretical analysis and FTA***

1. The logical steps of a PHA can be modeled as a connected, acyclic graph or tree. The tree structure provides useful insights into the chain of inference.
2. The tree structure shows a logical precedence of positions in variables for generating hazard scenarios or pathways - from nodes outward to cause-consequence pairs (Figure 6).
3. A PHA completeness check or audit should follow this precedence from node outward to branches. This means that omissions or errors are more likely to be found by first looking at the process variables and guide words and then proceeding to examine causes and consequences.

4. An FTA of the generic HAZOP process (abstracted for logical form) identifies seventeen distinct sources of E/O – examples of real world incidents are cited. The findings serve as a QA checklist of the PHA process and an aid to maximize hazard discovery.
5. Use checklists, best practices and incident databases to minimize E/O.
6. Recognize constraints that impose limits on analysis. Proper project scoping will ensure that safety is enhanced by prioritizing plant areas of concern for PHA analysis.

## 11. References


[1] OSHA. "Standard for Hazardous Materials - Process Safety Management of Highly Hazardous Chemicals", 29 CFR, 1910.119, U.S. Occupational Safety and Health Administration

[2] USEPA. "Chemical Accident Prevention Provisions", 40 CFR, 68, Subparts C and D. U.S. Environmental Protection Agency

[3] CSB. "Refinery Explosion and Fire, (15 Killed, 180 Injured), BP Texas City, Texas, March 23, 2005", Investigation Report No. 2005-04-I-TX, U.S. Chemical Safety and Hazard Investigation Board, Washington DC (March 2007).

[4] CSB. "Vinyl Chloride Monomer Explosion, (5 Dead, 3 Injured, and Community Evacuated), Formosa Plastics Corporation, Illiopolis, Illinois, April 23, 2004", Investigation Report No. 2004-10-I-IL, U.S. Chemical Safety and Hazard Investigation Board, Washington DC (March 2007)

[5] CSB. "Chevron Richmond Refinery Fire, Chevron Richmond Refinery, Richmond, California, August 6, 2012", Interim Investigation Report , U.S. Chemical Safety and Hazard Investigation Board, Washington DC (April 2013)

[6] CCPS. *Guidelines for Hazard Evaluation Procedures.* 2nd ed., Center for Chemical Process Safety, American Institute of Chemical Engineers, New York, NY, 1992.

[7] CSB. "T2 Labs, Inc., Runaway Reaction, Jacksonville, Florida, December 19, 2007", Investigation Report No. 2008-3-I-FL, U.S. Chemical Safety and Hazard Investigation Board, Washington DC (September 2009)

[8] CSB. "Uncontrolled Oleum Release, Petrolia, Pennsylvania, (Three towns evacuated)", Case Study No. 2009-01-I-PA, U.S. Chemical Safety and Hazard Investigation Board, Washington DC (September 2009)

[9] CSB. "Thermal Decomposition Incident, (5 Killed), BP Amoco Polymers, Inc., Augusta, Georgia, March 13, 2001", Investigation Report No. 2001-03-I-GA, U.S. Chemical Safety and Hazard Investigation Board, Washington DC (June 2002)

[10] OSHA. "Citation and Notification of Penalty – Formosa Plastics Corporation, 19800 Old Route- 36, lllliopolis, IL 62539." U.S. Occupational Safety and Health Administration, Inspection No. 305893679, Issuance Date 10/22/2004

[11] CSB. "Sterigenics, (4 employees hurt), Sterigenics, Ontario, California, August 19, 2004", Investigation Report No. 2004-11-I-CA, U.S. Chemical Safety and Hazard Investigation Board, Washington DC (March 2006)

# 349055 Deflagration Incident Case Review

**Abstract:**

A deflagration incident occurred during the cleaning procedure of a drum that was used in a coating process. The pressure generated from the incident damaged the drum as well as several pieces of equipment that were connected through ductwork for the dust collection system. Thankfully, no personnel were injured during the incident. This case study will describe the coating process and the incident, provide an incident analysis, review the process safety information and evaluate results from additional testing that was performed. Lessons learned will be summarized including hazards related to material compatibility, thermal stability and combustible dust.

# 349382 Enhanced Lessons Learned Approach from the Bscat Investigation Approach

**Abstract:**

A key aspect of process safety management is the effective transmission of lessons learned both around the organization and specifically to those who most need to know. This often occurs poorly.

Most investigation methods have been optimized for occupational accident investigation. The BSCAT approach – Barrier-based SCAT – builds on an earlier root cause method, but integrates bow tie barrier structure into the investigation and provides a better focus for major accidents and their pre-cursor near-miss events. All incidents can be converted into a bow tie sequence of barriers that were challenged and which either failed (and gave the incident) or some succeeded (terminating or mitigating the incident).

The BSCAT approach provides many benefits for lessons learned. Its structure matches the API754 Tier 3 Process Safety Indicator as it records clearly challenges to safety barriers – and whether they worked or failed. A feature of bow ties is that barriers are frequently repeated between different Threats and Outcomes and different bow tie Top Events. Thus a barrier that was identified as failed or degraded from one investigation will warn that the same barrier used for other threats or appearing in different bow ties affecting other plants are also likely to be degraded. Software that records all incidents into a database, can show barrier failures on all bow ties in which that barrier appears and thus automatically warn the barrier or bow tie owners throughout the orgnaization. This does not require proactive communication from the plant where the incident occurred or from the Safety Department – as this often is incomplete.

This paper also compares the application of BSCAT to other traditional root cause approaches and demonstrates the clarity of communication possible with the bow tie structure. It has the added advantage that the focus on barrier failures tends to screen out "motherhood" recommendations that may be good practice but did not contribute to the incident.

# 349552 Death of "Landlord" or Collapse of "Tomb", Which Matters More? - Some Perspectives of Engineering Ethics and Engineering Philosophy on Enterprise Global Risk Management

**Abstract:**

Within the paper presented at 9th GCPS, the author named CCPS RBPS Temple the house for a "landlord" i.e. core value to live within and also stated that the 'landlord" could even buy a new house with new style of design to live within. In other words, a "landlord" could own a couple of, even many a few, houses at same time to enjoy life. Should "landlord" die withnin, however, the house may seem still a house outwardly, but substantively with a purpose of tomb.

And the author argues further that inhabited house could be timely repaired and maintained to good condition, otherwise uninhabited house whose "landlord" died within is more likely to degrade neglectedly, and eventually collapse. In context of enterprise global risk management, it seems that the structural soundness of the houses has been attracting more attention than should deserve, whereas the death of "landlord" neglected to some extent within process industry and beyond. And for the fact that the lack of correct core value and healthy process safety cultural communication was arguemented by the author to be the ultimate root cause for all the catastrophic process safety accidents, some attempts are made to explore and prospect if the absence of alive correct core value should be inherently of organizational and episodic conflicts of interest i.e. 'natural' multi-chain of command in light of Engineering Ethics.

Then the author goes one step further to characterise RBPS Core Value and Process Safety Cultural Communication with new implications in light of the emerging subject Engineering Philosophy, so that it could be prospected that how tradeoffs between different functions, stakeholders, regions and businesses could be negotiated, potential failure diagnosed and prevented, design style of house modelled, and ultimately how more critical "landlord" enjoying life could be than tomb's structural soundness, which is in fact of new implications for addressing some of the biggest challenges from enterprise global risk management, especially of merger and acquisition.

Finally, upon all of above, a few of important recommendations would be given to process industry and beyond so that a brighter future of enterprise global risk management could be prospected.

# 349837 New Tools to Aggregate Operational Risk Across an Enterprise of Assets and to Help Govern This Risk through Policy Which Can be Directly Linked to Front Line Decision Making

**Abstract:**

**Major catastrophic incidents typically result in a whole spectrum of consequences, the highest being the loss of life and serious injury. Beyond the harm to people, the financial cost can be extreme to the extent that it can jeapordize the future of the corporation responsible. Following recent headline grabbing incidents, board members who are responsible for the secure and prosperous stewardship of corporations are asking the question, how do we properly govern such risks in our organisations.**

**This is the challenge most organisations now face, how to make sure that the major Operational Risks with our plant, equipment and people are being properly managed and how within a large global organisation a picture of this aggregated risk can be presented. This paper will explore the nature of operational risk and its contribution not just from Process Safety Hazards but also from ongoing work hazards. It will propose that the combination of compromised protection layers and the risk from new hazards introduced as a result of ongoing work represent the cumulative Operational risk which needs to be managed. The day to day status of these protection layers and the ongoing work schedule represent the dynamic involved in Operational risk Management. Currently we lack proper tools to effectively measure this risk in pace with the dynamic of operations where every day is different. Similarly we lack the ability to aggregate this risk across many units in a large plant or across many plants in different regions to provide an overall picture of operational risk exposure to the organisation. There are, however, new technology solutions emerging which will allow this to be achieved. This new category of application platform helps manage ongoing work from a risk perspective taking into account real time data on the status of protection layers in the plant. Working in concert with the maintenance management system, a fuller picture of protection layer status can be presented and for the first time a picture showing maintenance and repair work efficiency along with risk status can be achieved. These tools allow better decisions to be made at the frontline. Furthermore, they are designed to be deployed across large organisations and with the ability to aggregate risk data across the corporation effectively in real time. They also provide tools to allow organisations to better manage the common operational risks helping directly drive policy in front line decision making.**

**The paper will present some realistic scenarios and explain how future operations can be not only be manged more safely but that governance of the enterprise organisation in terms of operational risk management is now more achievable.**

# Inherently Safer Design of Stirred Reactors and VisiMix® Modeling Software

**Y. Nekhamkin**
**VisiMix Ltd.**
**8 Hamarpe Street**
**7th Floor, Bynet Bldg.**
**Jerusalem 97774 Israel**
**email: info@visimix.com**

**L. Braginsky, Y. Kokotov**
**VisiMix Ltd.**
**8 Hamarpe Street**
**7th Floor, Bynet Bldg.**
**Jerusalem 97774 Israel**
**email: info@visimix.com**

Prepared for Presentation at
American Institute of Chemical Engineers
2014 Spring Meeting
10th Global Congress on Process Safety
New Orleans, LA
March 30 – April 2, 2014

UNPUBLISHED

# Inherently Safer Design of Stirred Reactors and VisiMix® Modeling Software

**Y. Nekhamkin**
**VisiMix Ltd.**
**8 Hamarpe Street**
**7th Floor, Bynet Bldg.**
**Jerusalem 97774 Israel**
**email: info@visimix.com**

**L. Braginsky, Y. Kokotov**
**VisiMix Ltd.**
**8 Hamarpe Street**
**7th Floor, Bynet Bldg.**
**Jerusalem 97774 Israel**
**email: info@visimix.com**

**Keywords:** stirred tanks/reactors, inherently safer design, process safety technology, process simulation, VisiMix

## Abstract

Stirred reactors belong to the key type of equipment used in all branches of the chemical process industry (CPI). Risk of accidents during various chemical processes is caused by deviation from normal course of the technological process. It means that any calculation technique (software) permitting to design an apparatus (chemical reactor) for a required chemical process and to define a range of normal operations of this process can be considered as an ultimate element of the "inherently safer design" (ISD) tool of the corresponding complex "chemical reactor – chemical process". This paper demonstrates an approach based on application of VisiMix® software intended for analysis and calculations of various chemical processes in stirred tanks/reactors and to provide their enhanced safety.

## 1. Introduction

Stirred reactors belong to the key types of the process equipment used in various technological processes in practically all branches of the chemical process industry (CPI) – chemical, pharmaceutical, biotechnological, manufacturing cosmetics, food products, etc.

The current status of CPI features considerable number of various accidents occurring during process operation [1…7] caused by different reasons. As regards to the stirred reactors, accidents in such equipment are associated with inadequate process analysis of heat transfer (23%), reaction problems (23%) and process contamination (16%) [2].

Providing process safety technology becomes now one of the basic requirements of the day and it resulted in the Inherently Safer Design (ISD) concept [8]. This concept asserts that the process can be "described as inherently safer if it reduces one or more hazards… associated with the materials and operations used in the process when compared to some alternative process, and this reduction or elimination is accomplished by characteristics which are permanent and inseparable parts of the process" [9]. Since risk of incident/accident is caused by any deviation or consequence of deviations from normal course of the technological process, adequate process analysis is a fundamental issue of the reactor safety design.

One of the main requirements to calculation tools used for the ISD is its potential to perform calculations linking equipment characteristics and process parameters. However the current state of the art in the calculation tools used for the stirred reactors design is mainly focused upon the reactive nature of the chemical processes [10, 11] aiming to predict thermal hazards (runaway reactions) and to diminish their impact ([12], [13]). This approach is based on the Well Stirred Reactor concept that eliminates mixing related processes from consideration.

## 2. Stirred Reactors

Stirred reactors, extensively used in the CPI, allow implementing various technological processes, including, in particular:

- homogeneous blending;
- liquid-solid mixing (among them suspension, dissolution of solid, liquid-solid extraction, etc.);
- liquid-liquid mixing (among them emulsification, liquid extraction, etc.);
- liquid-gas mixing;
- homogenization of multi-component mixtures;
- single phase chemical reactions in batch, semibatch and continuous flow reactors;
- heterogeneous reactions in liquid-liquid and liquid-solid mixing media;
- temperature-dependent chemical reactions in batch, semibatch and continuous flow reactors)

Stirred reactors' productivity and their quality should be based on thorough calculations, capable of covering all the unit operations implemented in the equipment.

Despite the complexity of processes in stirred tanks/reactors, the VisiMix® software [14] is capable of simulating these processes. This software has gained recognition because it provides adequate and complete description of process and equipment configurations based on reliable models verified in practice.

This paper describes the use of VisiMix® for ISD of various chemical processes in stirred reactors. Main features of the VisiMix® software are briefly described below.

# 3. VisiMix® Software. Main Features.

VisiMix® software was developed for process engineers as a universal tool for solving a wide range of technological problems. VisiMix® includes VisiMix Turbulent® for turbulent flow regime, VisiMix Laminar® for laminar flow regime, VisiMix Different Impellers® for simulating mixing devices with different impellers (to be used with VisiMix Turbulent®), VisiMix RSD® for simulation of rotor stator dispersers, VisiMix Excel® that integrates VisiMix reports in a standard Excel worksheet, and VisiMix Pipe-Line® for hydraulic calculations for low and high viscous and non-Newtonian liquids in plant pipe lines. These components perform simulation of different technological processes with respect to real equipment design and process regime parameters for both turbulent and laminar flow regimes. VisiMix® GUI allows for selecting the equipment types and to enter the equipment parameters and media properties (Fig.1).

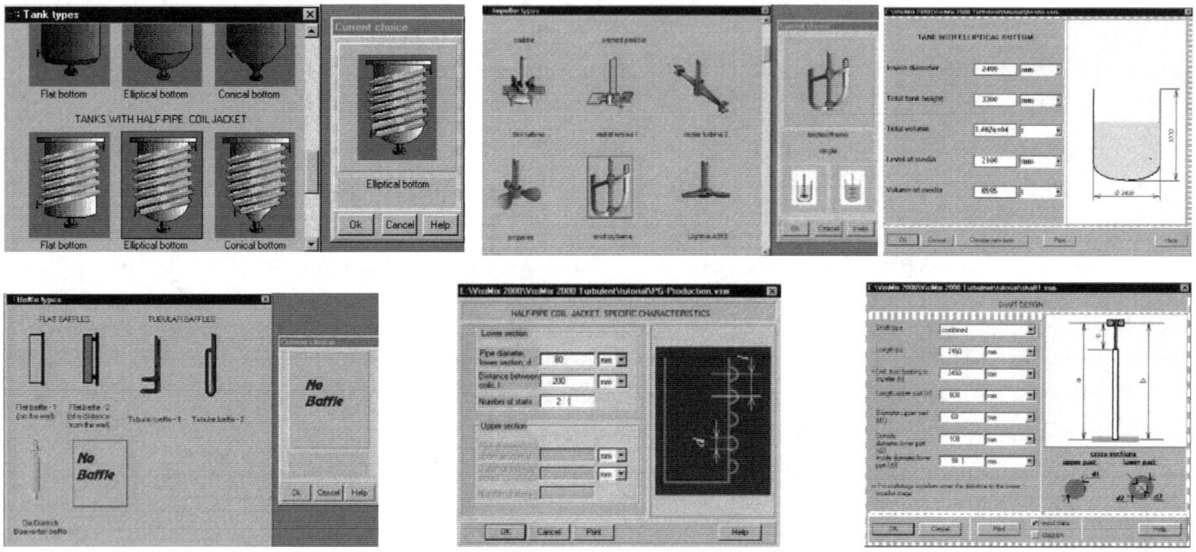


**Figure 1. Examples of the VisiMix® GUI Elements**

In addition to powerful calculation capabilities, VisiMix® has functions of an expert system: firstly, it allows for entering properties of materials from built-in databases, and secondly (and most importantly), it analyses the initial data and calculation results, and issues warning messages whenever the input results in unacceptable process course. These messages serve as a means for defining the safety range of the basic process parameters.

# 4. VisiMix® Application in ISD of Chemical Processes

This example demonstrates the use of VisiMix® for checking the safety of a process in a 5m$^3$ stirred jacketed reactors. The device has variable impeller rotational speed in the range 60-90 RPM, the motor rated power is 7500 W. The process is based on exothermal catalytic reaction with a solid catalyst.

The main risk for such a process is a runaway reaction that may start in the reactor because of a number of factors that we need to evaluate and prevent. First, we should check that there is no sedimentation of solid particles on the tank bottom, and that catalyst distribution inside the tank is uniform. Second, we need to check if the reactor's cooling system is capable of preventing the $2^{nd}$ order irreversible reaction from running away.

As input data for all calculations, Visimix® uses basic dimensions of the tank and baffles, type, dimensions, positions and rotation speed of the impeller, characteristics of the shaft, motor power and relevant properties of the media (both liquid and solid phases), as well as the jacket parameters and characteristics of cooling and heating agents.

### *4.1 Evaluation of Just Suspension Speed and Uniformity of the Catalyst Distribution*

In order to ensure absence of sedimentation, we will calculate Just Suspension Speed (JSS), which is the minimum rotational velocity of the impeller at which there are no stagnant zones at the bottom.

Although VisiMix® does not calculate JSS directly, it enables the user to easily determine this parameter.

When calculating any of the parameters of the Liquid-Solid Mixing, VisiMix® issues warning messages whenever settling of solid particles may occur. Therefore, to find JSS, it is enough to enter any reasonable value of rotational speed, and then to gradually increase it until VisiMix stops issuing the warnings. The first RPM value, which does not result in the warning message, will be the desired JSS value.

1. Before starting the calculations, enter the equipment parameters, including tank dimensions and impeller power and rotational speed, to display the diagram of your mixing system (Fig.2), and then enter the properties of solid and liquid phases (Fig. 3). Start with **60 RPM** as impeller rotational speed.

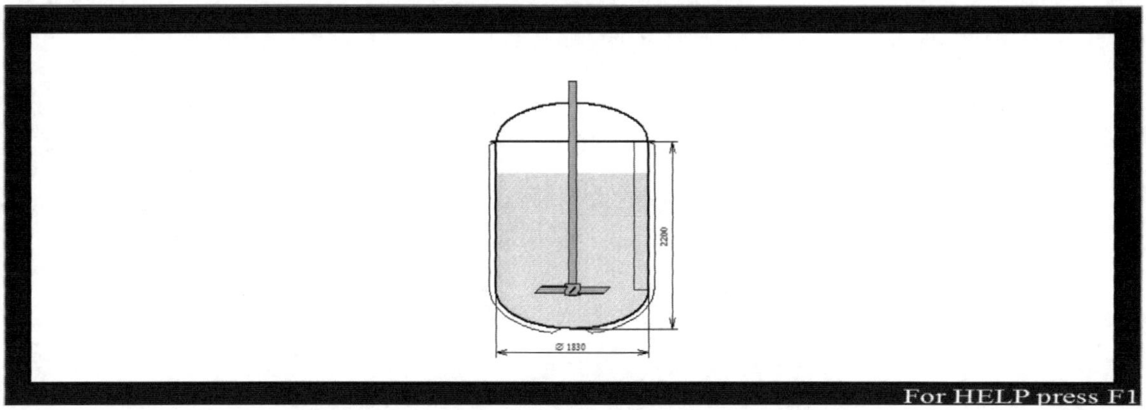


**Figure 2. The diagram of the mixing system**

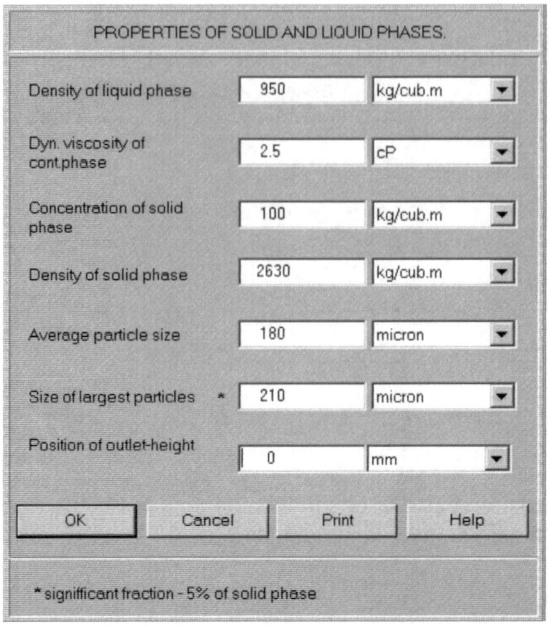


**Figure 3. Properties of solid and liquid phases**

2. To proceed to calculations, select **Calculate - Liquid-Solid Mixing**, then select any parameter, for instance, **Axial distribution of solid phase**.
3. For the data you have entered, VisiMix® issues a warning message (Fig.4).

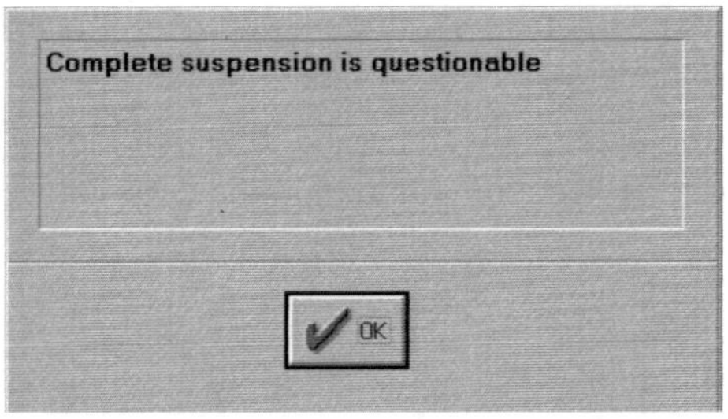


**Figure 4. The message warning of possible settling of solid phase**

This means that the entered value of rotational speed (60 RPM) is lower than JSS and does not ensure absence of sedimentation.

4. VisiMix® also checks uniformity of the catalyst distribution inside the tank. The resulting graph of axial distribution of solid phase is shown in Figure 5.

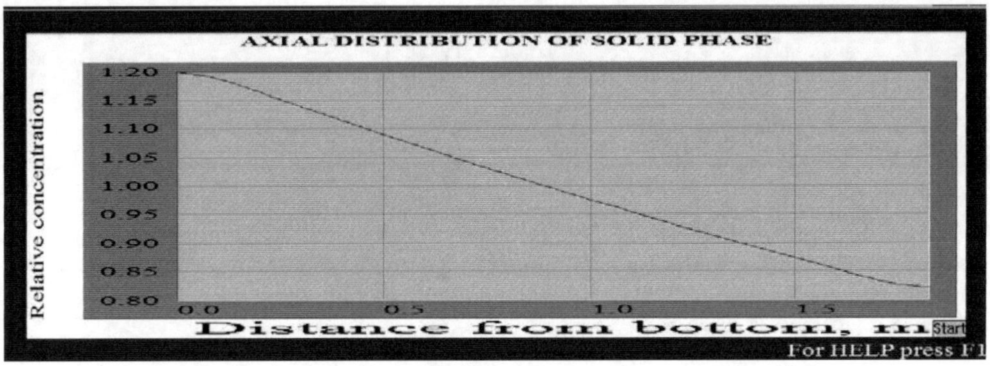


**Figure 5. Axial distribution of solid phase (60 RPM).**

This graph shows that at 60 RPM, the non-uniformity is about 20%. Catalyst non-uniformity may cause local overheating, which may also result in a runaway reaction.

5. Gradually increasing the impeller rotational speed, we see that at 80 RPM the message warning of possible settling of solid phase message appears no longer, so we can assume this RPM value as JSS. However, because we also want to improve our process configuration with regard to non-uniformity, we continue increasing the rotational speed, until, at 85 RPM, the axial distribution of the solid phase becomes more uniform, with maximum degree of non-uniformity of about 15% (Fig.6).

MAXIMUM DEGREE OF NON-UNIFORMITY - AXIAL, %

| Parameter name | Units | Value |
|---|---|---|
| Maximum degree of non-uniformity - axial, % | | 15.0 |

For HELP press F1

**Figure 6. Maximum degree of non-uniformity of catalyst axial distribution (85 RPM)**

Based on the above calculations, we will select **85 RPM** as safe rotational speed for our device and will now check the reaction thermal parameters.

### *4.2 Simulation 2nd Order Exothermal Reaction Carried Out in a Stirred Batch Reactor*

VisiMix® performs simulation of exothermal reaction based on the analysis of the equipment and process parameters to prevent runaway reaction that may take place when the energy generated by the reaction is greater than the energy removed from the reactor.

The reaction is run according to the stoichiometry

$$\boldsymbol{A} + \boldsymbol{B} \rightarrow \boldsymbol{C} \quad \text{[Eq. 1]}$$

The equation for the reaction rate is:

$$r = kC_A\ C_B \quad \text{[Eq. 2]}$$

where r is reaction rate, moles of A/(l·sec);
k is reaction rate constant, l/(mole·sec);
$C_A$ is concentration of reactant A, mole/l;
$C_B$ is concentration of reactant B, mole/l.

The reaction rate constant is a function of the system temperature and is given by

$$k = k_0 \cdot e^{-E/RT} \quad \text{[Eq. 3]}$$

where $k_0$ is Arrhenius constant, l/(mole·sec);
E is energy of activation, kJ/mole;
R is gas constant, J/(mole·K);
T is absolute temperature, K.

Kinetic reaction parameters were partially borrowed from [15].

For thermal calculations, additional input data are necessary.

The initial temperature inside the reactor is 20°C. Initial concentration of reactants A and B is the same and is equal to 5 g-mole/l. The actual process is performed in two stages: 1) initial heating which starts the reaction and 2) subsequent cooling required for removing excessive heat. In our case, the heating is with steam at atmospheric pressure supplied into the jacket, and the cooling is with ordinary water circulated through the jacket. The cooling water temperature is 20°C, and its volume flow rate is 40$m^3$/h.

The properties of the reacting mixture are identical to those of ordinary service water.

The maximum allowable temperature of the reactor is equal to the media vapor saturation temperature (143°C). The corresponding maximum working pressure of the reactor is 4 bar. Heat effect of reaction is 90 kJ / mole, energy of activation is 125.5 kJ/mole, and Arrhenius constant is $k_0 = 10^{17}$ l/mole·sec.

To evaluate the suitability of the reactor equipment to the exothermal process described above, we need to first check a number of important general parameters, such as **Mixing power, Macromixing time**, etc. Then we need to simulate the reaction to determine the process time before runaway regime starts, and make sure that efficient mixing is achieved. Thereafter, the cooling stage should be analyzed, and only then it will be possible to conclude whether the process safety is ensured.

1. In the VisiMix® **Hydrodynamics** menu, calculate **Mixing Power** (Fig. 7) to make sure that the consumed power does not exceed the motor rated power (7500 W). The calculated result shows that the power consumption is acceptable.

MIXING POWER

| Parameter name | Units | Value |
|---|---|---|
| Mixing power | W | 568 |

For HELP press F1

**Figure 7. Mixing power**

The second parameter, **Macromixing time** needed for the general check-up can be calculated using **Single-phase liquid mixing** submenu, the result is shown in Figure.8.

MACROMIXING TIME

| Parameter name | Units | Value |
|---|---|---|
| Macromixing time | s | 22.8 |

For HELP press F1

**Figure 8. Macromixing time**

We see that **Macromixing time** in our reactor is about 23 sec. This result is of great practical importance. Since acceptable mixing can only be achieved if the characteristic process time is at least 10 times greater than **Macromixing time**, the process time in the reactor should be at least 4 minutes.

2. We will now determine the process time before runaway regime starts. Select **Media temperature** in **Heat Transfer. Batch (BH) - Vaporous agent (VA).** At first we receive a warning message about a runaway reaction starting in about 7 minutes (Fig. 9). We can also see it in the graph shown in Figure10.

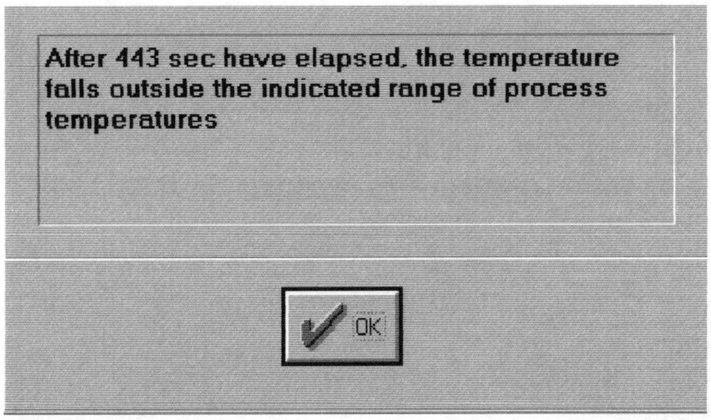


**Figure 9. The message warning of the temperature exceeding the prescribed limit**

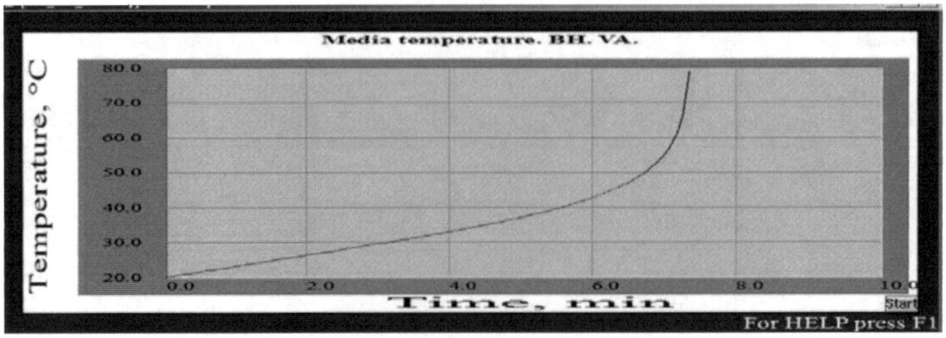


**Figure 10. Media temperature. BH. VA.**

To see the course of the reaction, select **Concentration of reactant A. BH. VA** (Figure 11).

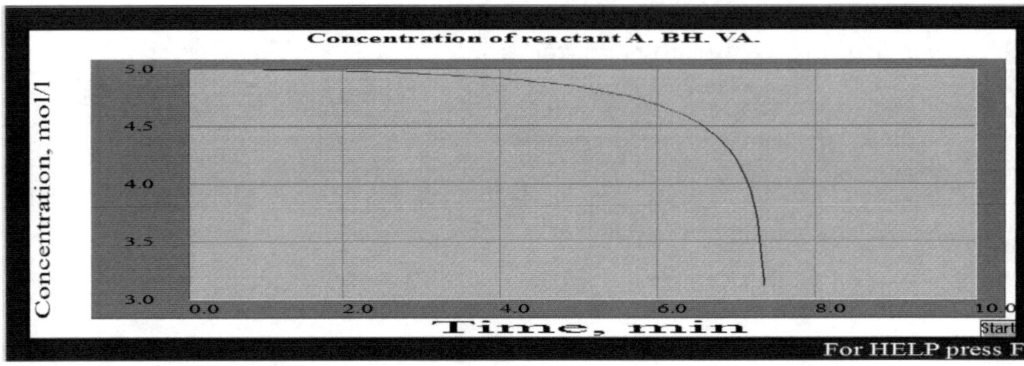


**Figure 11. Concentration of reactant A. BH. VA.**

Figures 10 and 11 show that runaway regime begins almost at the end of the reaction, when the reactant concentration is dramatically falling down, in about 7 minutes from the start of the process. This time is more than twenty times higher than the **Macromixing time** (23 sec in Figure 8). This corresponds to the basic requirement for the efficient mixing, and the reactor, therefore, may be considered as well stirred.

This information indicates that the second process stage – cooling should begin before the process approaches the runaway regime, i.e. about 5.5 minutes from the start of the process. At this moment the steam supply stops, the cooling water fills the jacket, and the second (cooling) stage starts. The results of the heating stage (temperature and reactant concentration values in Figures 10 and 11 corresponding to 5.5 minutes from the start of the process) will thus serve as the initial data for the cooling stage of the simulation, and the desired values are 40°C and 4.75 mole/liter (for both reactants).

3. Calculations for the cooling stage will be performed by means of the **Heat Transfer. Batch, Liquid agent (LA)** submenu. The results are presented in Figure 12 (**Media temperature. BH. LA**) and Figure 13 (**Concentration of reactant A. BH. LA.**).

   The results of the calculations show that we have started the cooling early enough to prevent the process from reaching the upper temperature limit (143°C) and, consequently, the reactor maximum working pressure (4 bar). The figures indicate that heat release stopped for the only reason that both reactants in the reactor had ended (about 2.8 minutes from the start of the cooling). The total duration of the process is about 8.3 minutes.

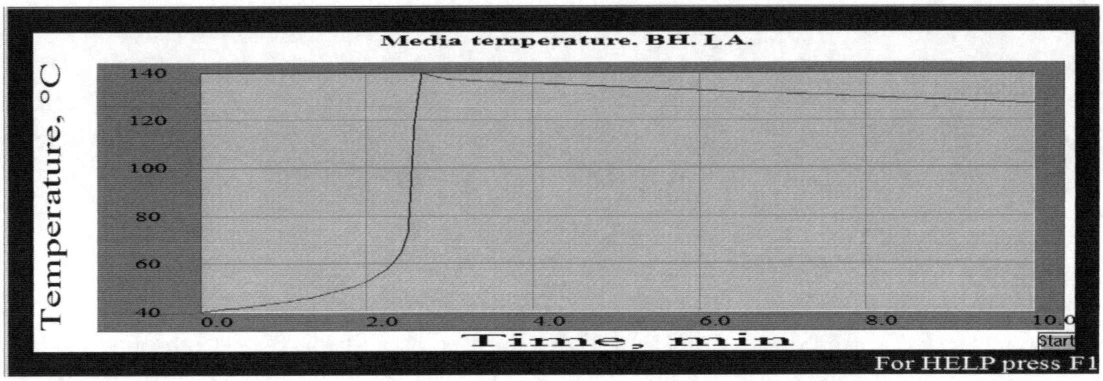


**Figure 12. Media temperature. BH. LA. The heating stage is 5.5 minutes.**

The results presented above show that if cooling starts 5.5 minutes from the start of the reaction, runaway regime will be prevented, and the reaction completed.

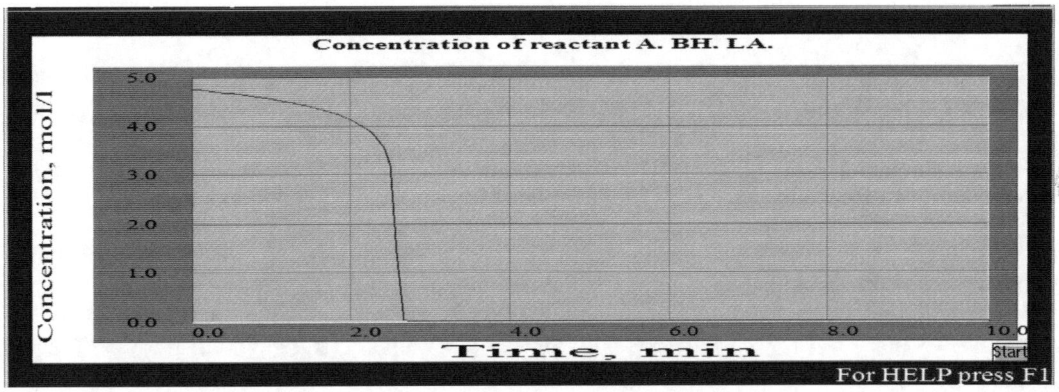


**Figure 13. Concentration of reactant A. BH. LA. The heating stage is 5.5 minutes.**

## 5. Conclusion

1. Process analysis based on valid technical calculations and simulation is a mandatory element of the ISD concept. The ISD software should meet the following requirements:
   - Provide adequate and complete description of the process and equipment configuration based on reliable models verified in practice.
   - Adequately detect any deviations from the process course which are generally regarded as main factors causing accidents in the process industries and provide quantitative analysis of the influence of these factors.
   - Be user friendly, accessible to plant engineers and allow quick results.
   - Combine a calculation tool with an expert system that analyses initial data and calculation results, and warns the user whenever the system approaches dangerous operation regimes.
2. Nuclear engineering experience can be used for advancing inherent safety in process industries. In the USA, for example, the Nuclear Regulatory Commission (NRC) grants a license to a plant if it submits, among other materials, the computer simulation results of the plant's expected behavior according to standard scenarios [17]. Computer codes for such simulation have been developed either by NRC or by its contractors. In the process

industries application of approved software tools for process analysis will improve safety status.

3. VisiMix® software which has proven capabilities in simulation of various technological processes in stirred tanks/reactors and estimation of their safety ranges can serve as a prototype for software tools applied for ISD in the process industries.

## 6. References


[1] Ta-Cheng Ho, Yih-Shing Duh, and J.R.Chen, Case Studies of Incidents in Runaway Reactions and Emergency Relief. Process Safety Progress, Vol.17, No.4, pp.252-262, (1998)

[2] Kamarizan Kidam and Markku Hurme, Analysis of equipment failures as contributors to chemical process accidents. Process Safety and Environmental Protection, 91 (2013), pp.61-78.

[3] C. Kirchsteiger, Trend in accidents, disaster and risk sources in Europe. Journal of Loss Prevention in the Process Industries, Vol.12, (1999), pp. 7…17.

[4] H. - J. Uth, Trend in major industrial accidents in Germany. Journal of Loss Prevention in the Process Industries, Vol.12, (1999), pp. 69…73.

[5] S. J. Kang, Trend in major industrial accidents in Korea. Journal of Loss Prevention in the Process Industries, Vol.12, (1999), pp. 75…77.

[6] S. Wakakura, Trend in chemical hazards in Japan. Journal of Loss Prevention in the Process Industries, Vol.12, (1999), pp. 75…77.

[7] Kamarizan Kidam and Markku Hurme, Origin of equipment design and operation errors. Journal of Loss Prevention in the Process Industries, Vol.25, (2012), pp. 937…949.

[8] T.A. Kletz and P. Amyotte, Process plants. A handbook for inherently safer design. 2nd Ed., Taylor CRC Press, Boca Raton. Fl. (2010).

[9] D.C. Hendershot, Inherently Safer chemical process design. Journal of Loss Prevention in the Process Industries, Vol.10, (1997), pp. 151…157.

[10] David J. Leggett, Safe process development from reaction hazards testing. Thermochimica Acta, 367-368, (2001), pp. 351…365.

[11] Theodor Grewer, David J. Frurip, and B. Keith Harrison, Prediction of thermal hazards of chemical reactions. Journal of Loss Prevention in the Process Industries, Vol.12, (1999), pp. 391…398.

[12] Valeria Gasson, David George Lister, Maria Francesca Milazzo, and Guiseppe Mascio, Comparison of criteria for prediction of runaway reactions in the sulphuric acid catalyzed esterification of acetic anhydride and methanol. Journal of Loss Prevention in the Process Industries, Vol.25, (2012), pp. 209…217.

[13] R. D. McIntosh and P. F. Nolan, Review and experimental evaluation of runaway chemical reactor disposal design methods. Journal of Loss Prevention in the Process Industries, Vol.14, (2012), pp. 17…36.

[14] www.visimix.com

[15] Safety, Health, and Loss Prevention in Chemical Processes. Problems for Undergraduate Engineering Curricula, (The Center for Chemical Process Safety of the American Institute of Chemical Engineers) pp. 162 - 164, Problem No. 81.

[16] A. Mak, S. Yang and N. G. Ozcan-Taskin, The Effect of Scale on the Suspension and Distribution of Solids in Stirred Vessels. In the book: Recents Progres en Genie des Procedes, (MIXING - 97. Multiphase Systems), Vol.11, number 52 (1997), pp. 97 - 104.

[17] L. M. Shotkin, "Development and assessment of U.S. nuclear regulatory commission thermal-hydraulic system computer codes," Nuclear Technology, vol. 116, no. 2, pp. 231–244, 1996

# An Easy Design of Inlet Piping Systems for Relief Valves in Gas/Vapor Service

**Guibing Zhao, Ph.D.**
**Chilworth Technology Inc.**
**113 Campus drive, Princeton, NJ 08540**
**Guibing.Zhao@dekra.com**
**www.chilworth.com**

Prepared for Presentation at the
American Institute of Chemical Engineers
2014 Spring Meeting
10th Global Congress on Process Safety
New Orleans, LA
March 30 – April 2, 2014

UNPUBLISHED

# An Easy Design of Inlet Piping Systems for Relief Valves in Gas/Vapor Service

**Guibing Zhao, Ph.D.**
**Chilworth Technology Inc.**
**113 Campus drive, Princeton, NJ 08540**
**Guibing.Zhao@dekra.com**

**Keywords:** Safety relief valve, inlet piping system, 3% pressure loss, flow capacity, allowable maximum velocity head loss

## Abstract

Safety relief valves that are used to protect against overpressure require well-designed inlet piping for proper operation. Good engineering practice requires that the un-recoverable pressure loss of the inlet piping of a safety relief valve be less than 3% of its set pressure to avoid chattering, while attaining its desired capacity. Therefore, the design of a proper inlet piping system may require several trial-and-error fluid-mechanical calculations to reach the goals of both the 3% rule and the capacity requirement. This article presents a simplified method for the design of inlet piping systems without trial-and-error and iteration calculations. To accommodate the 3% rule and the required flow capacity, the maximum allowable velocity head loss of an inlet piping system can be easily calculated from classical one-dimensional adiabatic compressible fluid-flow principles. Then the inlet piping system (including pipe length, diameter, elbows, tees, etc.), and the corresponding velocity head loss, can be easily adjusted to satisfy both the 3% rule and the capacity requirement. A case study involving the selection of a safety relief valve for a gas validates the proposed method.

## 1. Introduction

Safety relief valves are widely and effectively used to protect equipment from excessive overpressure. The proper design of a relief system requires not only determining the correct size for the valve, but also determining the proper size and configuration of the upstream and downstream piping, and providing a proper effluent-handling system. The inlet-line pressure loss in a safety relief valve system can be significant even when the piping is quite short, between the protected pressure vessel and the safety relief valve. These losses are due mostly to the length of the inlet piping, but also to friction losses in valves and other fittings, and the effects of changes in the direction of flow, as in elbows and tees.

A poorly-designed inlet piping system – having a large inlet-piping pressure loss – will cause (1) a reduction in the flow capacity of the valve, resulting from a reduced stagnation pressure at the valve nozzle inlet; thus, the pressure in the vessel might exceed the Maximum Allowable Working Pressure (MAWP), and/or (2) unstable operation of the valve due to a rapid cyclic opening and closing the valve, commonly known as "chattering", when the pressure loss in the inlet piping periodically exceeds the blowdown of the safety relief valve. The blowdown of a safety valve is the difference between the set pressure and the closing pressure of a safety valve, expressed as a percentage of the set pressure. When the inlet loss is low – as during the initial flow toward the valve – the valve will open at its set pressure. However, when the inlet loss is high – as during flow through the valve – the pressure at the inlet could decrease below the valve-closing [re-seating] pressure. Repetition of this cycle can lead to not only a reduction of the flow capacity, but also failure of the safety relief valve itself or the piping system, with possible rupture of the pressure vessel. Therefore, good engineering practices (API 520 [1] and ASME B&PV Code Section VIII [2]) requires that the "un-recoverable" pressure loss of the inlet piping of a safety relief valve be less than 3% of the set pressure, to avoid chattering and reduction in the flow capacity. Although the 3% rule has long been controversial [3] and been thought to be overly conservative [4], a discussion the 3% rule is beyond the scope of this article.

The methods for designing the inlet piping for a safety relief valve – to satisfy both the 3% rule and the required capacity- have been well established and are used in industrial practice [5-9]. However, as discussed below, the conventional method is quite lengthy and complicated, and requires several fluid-mechanic and trial-and-error calculations to reach the goals of the 3% rule and the required capacity. Thus, this article presents an easy and objective-oriented method for the chemical and petrochemical industry.

The method presented in this article directly incorporates the 3% rule and capacity requirement in the design calculations, with an assumption of choked [sonic] flow at the exit nozzle of a relief valve. The maximum allowable velocity head loss of the inlet piping system can be easily calculated from classical one-dimensional adiabatic compressible-fluid-flow principles for gas or vapor relief. Therefore, any inlet piping system (including pipe length, diameter, elbows, tees, etc.) with a velocity head loss that is less than the calculated maximum allowable velocity head loss will automatically satisfy both the 3% rule and capacity requirement.

## 2. A basis for the design of inlet piping for safety relief valves

A schematic for a typical inlet piping system for a safety relief valve is shown in Figure 1. It is assumed that the characteristics of the protected tank or pressure vessel and its contents – such as the MAWP, and the temperature and pressure of the gas or vapor – in an emergency-venting situation are known. The calculations involve the pressure, temperature, density, and the velocity (Mach number) at the entrance to the piping between the protected vessel and the relief valve, at the entrance to the relief valve, and the nozzle exit of the relief, with the assumption that flow at nozzle exit of the valve is choked.

It is also assumed that the flow throughout the system is turbulent and one-dimensional, and that the flow through the valve-inlet piping is adiabatic, with no heat losses from the gas/vapor stream. Adiabatic flow can be either frictional or frictionless, and reversible adiabatic flow (without friction loss) is isentropic flow. Also, the gas or vapor is assumed to be a perfect gas, such that the ideal gas law applies.

The four variables of pressure, temperature, density, and velocity (Mach number) at locations 1, 2, and 3 can be solved by four equations of continuity (mass balance), momentum balance, energy balance, and "state" (pressure, temperature, and volume) at any given boundary conditions. Such solutions [10] are shown in Table 1, where the stagnation value (subscript o) is used as the reference condition for isentropic/frictionless flow, while the critical value under sonic/choked conditions (Mach number = 1, denoted by asterisks, whether or not it is actually reached in the pipe exit) is used as a reference condition for adiabatic flow with friction. The stagnation state has a macroscopic flow velocity of zero, which can be an actual state in a pressure vessel, or a hypothetical state in frictional flow where given state conditions are isentropically transformed to a stagnation state. The mass flux can be calculated through Equation [Eq.8a] at any location in the venting line for any given Mach number. Equation [Eq.8b] is the critical mass flux when assuming choked flow at the exit from a pipe or from a nozzle.

**Figure 1. Inlet piping system of safety relief valve**

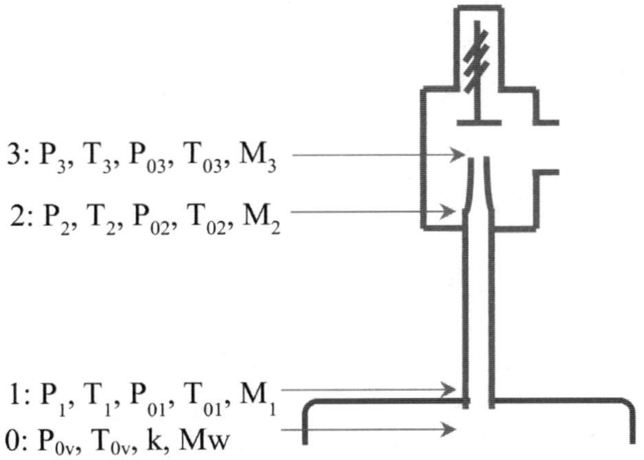


**Table 1. Equations for calculating the state conditions and Mach numbers along the inlet piping of a safety relief valve**

<table>
<tr><td>Variable</td><td colspan="2">Isentropic (frictionless) process</td><td colspan="2">Adiabatic process with friction</td></tr>
<tr><td>Pressure</td><td>$\frac{P_0}{P} = \left(1 + \frac{k-1}{2} M^2\right)^{\frac{k}{k-1}}$</td><td>[Eq.1]</td><td>$\frac{P}{P^*} = \frac{1}{M}\left[\frac{k+1}{2+(k-1)M^2}\right]^{\frac{1}{2}}$</td><td>[Eq.4]</td></tr>
<tr><td>Temperature</td><td>$\frac{T_0}{T} = 1 + \frac{k-1}{2} M^2$</td><td>[Eq.2]</td><td>$\frac{T}{T^*} = \frac{k+1}{2+(k-1)M^2}$</td><td>[Eq.5]</td></tr>
<tr><td>Density</td><td>$\frac{\rho_0}{\rho} = \left(1 + \frac{k-1}{2} M^2\right)^{\frac{1}{k-1}}$</td><td>[Eq.3]</td><td>$\frac{\rho}{\rho^*} = \frac{1}{M}\left[\frac{2+(k-1)M^2}{k+1}\right]^{\frac{1}{2}}$</td><td>[Eq.6]</td></tr>
<tr><td>Mach number relationship</td><td></td><td></td><td>$kN = \frac{1}{{M_1}^2} - \frac{1}{{M_2}^2} + \frac{k+1}{2} ln\left[\frac{2+(k-1){M_2}^2}{2+(k-1){M_1}^2} \cdot \frac{{M_1}^2}{{M_2}^2}\right]$</td><td>[Eq.7]</td></tr>
<tr><td rowspan="2">Mass flux</td><td colspan="3">$G = \rho u = \frac{PMw}{RT} M \sqrt{\frac{kRT}{Mw}} = PM\sqrt{\frac{kMw}{RT}} = P_0 M \sqrt{\frac{kMw}{RT_0}\left(1 + \frac{k-1}{2} M^2\right)^{-\frac{k+1}{k-1}}}$</td><td>[Eq.8a]</td></tr>
<tr><td colspan="3">$G_c = P_0 \sqrt{\frac{kMw}{RT_{0v}}\left(\frac{2}{k+1}\right)^{\frac{k+1}{k-1}}}$ (choked flow)</td><td>[Eq.8b]</td></tr>
</table>

It is worth noting that the actual piping section from location 1 to 2 may be complex – including valves, elbows, tees, and rupture disks, as well as the piping – and not necessarily a straight pipe, as shown in Figure 1. Therefore, the inlet piping velocity head loss, *N*, is calculated from the actual pipe length and fittings as

$$N = f\frac{L}{d} + \sum K_i \qquad \text{[Eq.9]}$$

The Darcy friction coefficient, *f*, and resistance coefficient, $K_i$, of various fittings can be obtained from the often-used Crane Technical Paper 410 [11]. The "state" emergency-relief conditions of pressure and temperature at location 0, i.e., in the vessel, and the required relief capacity, i.e., mass flow rate, $W_r$, in kg/s, are obtained from tests.

Generally, the following assumptions are widely accepted by the industry:

- For adiabatic flow, $T_{01} = T_{02} = T_{03} = T_{0v}$,
- The fluid flow from location 0 to 1, i.e., vessel nozzle, and from 2 to 3, i.e., safety relief valve nozzle, is isentropic,
- The effect of elevation change on pressure drop is assumed to be negligible,
- Both API 520 and ASME code allow an accumulated pressure in the vessel beyond the MAWP during discharge, and is given as a percentage of the MAWP. Therefore, $P_{0v}$ is calculated as

  $$P_{0v} = (1 + x)MAWP + P_{atm} \qquad \text{[Eq.10]}$$

  where *x* is accumulation factor; for example, 10% for single-valve installation if the overpressure scenario is not from fire exposure.

The flow capacity calculation for a safety valve according to API 520 [1] is based on isentropic flow through a nozzle with a correction factor, the so-called discharge coefficient $K_d$, which is the correction factor between the mass flow rate for isentropic flow in a nozzle and the experimentally determined mass flow rate in a safety valve. Section UG131 of the ASME Code [2] requires that the manufacturer-provided coefficient, *K*, be the product of the actual test-result coefficient, $K_d$, and a discounting factor of 0.9. The basic safety relief valve nozzle formula for sonic flow in API 520 is

$$A = \frac{W_r}{CKP_0K_bK_c}\sqrt{\frac{T_0Z}{Mw}} \qquad \text{[Eq.11]}$$

which is the same as [Eq.8b] when Gc = $w_r$/A; when the value of each of the correction factors $K$, $K_b$, $K_c$ is assumed to be 1.0; when the value of the compressibility factor Z is assumed to be 1.0; and when the dimensionless coefficient, C, is given by

$$C = \sqrt{\frac{k}{R}\left(\frac{2}{k+1}\right)^{\frac{k+1}{k-1}}} \qquad \text{[Eq.12]}$$

For conventional and pilot-operated valves, the capacity correction factor, $K_b$, due to backpressure is 1.0. The combination factor, $K_c$, is 1.0 when a rupture disk is not installed. When the compressibility factor Z is less than 1, the pressure loss calculation in which ideal-gas behavior is assumed will yield a larger value than that in which the compressibility factor is considered. Thus, for most common industrial gases, the assumption of an ideal gas will generate conservative results.

## 3. Conventional method to design safety relief valve inlet piping system

The following is the conventional procedure for designing the inlet piping system of a safety relief valve that is summarized from reference [9]:

(1) Determine the required relieving capacity, from all possible sources and causes of overpressure, using a hazard and operability (HAZOP) study, for example. This would include failure of a valve to close, in gas or liquid piping to the protected vessel; pressure or temperature control system failure; fire exposure; failure of a pump in the discharge from the vessel; uncontrolled chemical reaction; and blocked-in vessel or piping subjected to abnormal temperature. The worst case combination of these factors is used to determine the required capacity [12].

(2) Calculate the required area of the internal nozzle or orifice, $A_{3,cal}$, of a safety relief valve from [Eq.11] based on manufacturer-provided discharge coefficient and assuming $P_0 = P_{0v}$. The next-larger standard nozzle size, $A_3$, of a safety relief valve than the calculated size, $A_{3,cal}$, usually is selected. Accordingly, the standard size of inlet, $A_2$, and outlet piping of a safety relief valve can be determined based on API 526 [13] or manufacturer-provided safety relief valve manual.

(3) Calculate the ASME rated flow capacity, $W_{rated}$, based on the actual nozzle size of a safety relief valve from [Eq.11] by assuming $P_0 = P_{0v}$.

(4) Calculate the Mach number in location 2, $M_2$, by combining [Eq.8a] and [Eq.8b] based on a mass balance.

$$W_{actual} = G_2 \cdot A_2 = G_3 \cdot A_3 \cdot (\frac{K}{0.9}) \qquad \text{[Eq.13]}$$

where $G_2$ and $G_3$ are from [Eq.8a] and [Eq.8b], respectively. Because the actual capacity is considered, the ASME rated capacity is divided by 0.9, as in the last term of [Eq.13]. With known $A_2$ and $A_3$, $M_2$ can be calculated by combining [Eq.8a] and [Eq.8b]

$$\frac{0.9A_2}{K \cdot A_3} = \frac{1}{M_2}\left[\frac{2+(k-1){M_2}^2}{k+1}\right]^{\frac{k+1}{k-1}} \qquad \text{[Eq.14]}$$

(5) With a known inlet piping configuration – length, diameter, and fittings – for the safety relief valve system, the velocity head loss, $N$, of the inlet piping can be calculated from [Eq.9].

(6) With the values of $N$ and $M_2$ now known, the Mach number at location 1, $M_1$, can be calculated by solving [Eq.7].

(7) Since $P_{01} = P_{0v}$ and $T_{01} = T_{0v}$, with a known $M_1$ from step (6), all other state variables, $P_1$, $T_1$, and $\rho_1$ can be calculated from [Eq.1], [Eq.2], and [Eq.3].

(8) By applying [Eq.4], [Eq.5], and [Eq.6] to both locations 1 and 2, the state variables, $P_2$, $T_2$, and $\rho_2$ can be calculated based on known $P_1$, $T_1$, and $\rho_1$ from step (7).

(9) With the known state variables $P_2$, $T_2$, and $\rho_2$, and the known Mach number $M_2$ at location 2 from step (8), the stagnation parameters $P_{02}$, $T_{02}$, and $\rho_{02}$ can be calculated from [Eq.1], [Eq.2], and [Eq.3].

(10) With a known $P_{02} = P_{03}$ from step (9), the ASME-creditable flow capacity can be calculated from

$$W_{ASME} = 0.9G_2 \cdot A_2 = G_3 \cdot A_3 \cdot K \qquad \text{[Eq.15]}$$

(11) Now comparing $W_{ASME}$ to $W_r$, and $P_{01}$ to $P_{02}$, if either of the two design criteria

$$\gamma = \frac{W_{ASME}}{W_r} \times 100\% > 100\%, \text{ or} \qquad \text{[Eq.16a]}$$

$$\beta = \frac{P_{01}-P_{02}}{P_{sg}} \times 100\% < 3\% \text{ (3\% rule)} \qquad \text{[Eq.16b]}$$

is not satisfied, it would be necessary to change the inlet piping configuration, usually to decrease the inlet-piping velocity head loss, *N*. Steps (5) to (11) would then be repeated until both [Eq.16a] and [Eq.16b] are satisfied.

The above design calculation method is very tedious and time-consuming. Even with current computer technology, it may still take time and effort to find an inlet-piping configuration that satisfies both [Eq.16a] and [Eq.16b]. The purpose of this article is to propose a calculation method that avoids this trial-and-error process.

## 4. An easy method to design a safety relief valve inlet piping system

An increase of stagnation pressure at location 2, $P_{02}$, would increase flow capacity and reduce unrecoverable pressure loss in the inlet piping system for a safety relief valve. There are two calculation strategies that can result in a minimum stagnation pressure at location 2, $P_{02,min}$, for satisfying either the 3% rule or the flow capacity requirement.

For satisfying the 3% rule, the design of the inlet piping of a safety relief valve system requires that $P_{02}$ cannot be lower than $(P_{0v} - 3\% \cdot P_{sg})$. Therefore,

$$P_{02,min} = P_{0v} - 0.03 \cdot P_{sg} \qquad \text{[Eq.17]}$$

Based on the selected safety relief valve, the ASME-creditable flow capacity, $W_{ASME}$, can be immediately calculated from [Eq.15] with a known nozzle area and $P_{02} = P_{03}$, which is based on the assumption of isentropic flow in the nozzle. Generally, the calculated $W_{ASME}$ will be greater than the required flow capacity, $W_r$. Therefore, both the 3% rule and flow capacity requirement can be satisfied.

In addition, $P_{02,min}$ can also be calculated based on the minimum flow capacity, $W_r$, the required flow,

$$P_{02,min} = P_{03,min} = \frac{W_r}{CKA_3}\sqrt{\frac{T_{03}}{Mw}} \qquad \text{[Eq.18]}$$

Unfortunately, in most cases, substituting the calculated $P_{02,min}$ from [Eq.18] into [Eq.16b] will violate the 3% rule.

Based on the above analysis, an easy method is proposed to design the inlet piping system of a safety relief valve. The procedure involves

(A) The same as step (2) in the conventional method.
(B) The same as step (3) in the conventional method.
(C) The same as step (4) in the conventional method.
(D) Calculating the $P_{02,min}$ from [Eq.17]. With the known $P_{02,min}$, and $M_2$, all state variables in location 2 can be calculated from [Eq.1] to [Eq.3].
(E) Calculating $W_{actual}$ from [Eq.13] from state variables at either location 2 or 3 and the corresponding flow area.
(F) Calculating $M_1$ by solving [Eq.8a] through use of trial-and-error or a spreadsheet solver

$$G = \frac{W_{actual}}{A_1} = P_{o1}M_1\sqrt{\frac{kMw}{RT_{o1}}\left(1+\frac{k-1}{2}{M_1}^2\right)^{-\frac{k+1}{k-1}}} \qquad \text{[Eq.19]}$$

where $W_{actual}$ is calculated from step (E) and $P_{o1} = P_{ov}$, $T_{o1} = T_{ov}$, and $A_1 = A_2$.

(G) Calculating the maximum allowable the inlet piping velocity head loss, $N_{max}$, by solving [Eq.7] through

$$N_{max} = \frac{1}{kM_1^2} - \frac{1}{kM_2^2} + \frac{k+1}{2k} ln\left[\frac{2+(k-1)M_2^2}{2+(k-1)M_1^2} \cdot \frac{M_1^2}{M_2^2}\right] \qquad \text{[Eq.20]}$$

(H) Comparing the actual velocity head loss of the inlet piping system, $N$, (calculated from [Eq.9]) to $N_{max}$, if $N$ is less than or equal to $N_{max}$, the inlet piping system satisfies both the 3% rule and capacity requirement. If $N$ is greater than $N_{max}$, it would be necessary to analyze the inlet piping configuration and determine what practical changes can be made to reduce the velocity head loss of the inlet piping system until $N$ equals or is less than $N_{max}$.

(I) If the inlet piping size needs to be changed, repeating steps from (E) to (G) to find $N_{max}$ of a larger piping size and adjusting the inlet piping configuration till $N < N_{max}$.

Measures to reduce the velocity head loss include decreasing the length of the piping, increasing the diameter of the piping, avoiding high-loss fittings, and avoiding unnecessary changes in flow direction. It is generally economical to determine the smallest standard diameter capable of sustaining an acceptable pressure loss. If a rupture disk is installed upstream of a safety relief valve – for protection against process fluids – it must burst completely, with all fragments remaining attached to the holder. Installing a shutoff valve in the piping to a relief valve is prohibited – with few closely-controlled exceptions – and, if installed, should be selected to contribute the lowest possible pressure drop to the system.

The following case study demonstrates how the proposed method facilitates the design of the inlet piping system of a safety relief valve, as compared to the conventional method.

## 5. Case study

Reference [9] illustrates the design of the inlet piping to a safety relief valve by using the conventional design method. The design requirements are used to compare the conventional method to the proposed simplified method.

A hazard analysis has identified that a pressure vessel should be protected by a safety relief valve, since a process upset could result in a flow of 26,000 kg/h of gaseous nitrogen that must be relieved, to prevent overpressurization. The vessel's MAWP is 50 psig (3.45 bar-g), and a 10% pressure accumulation – above the MAWP – is permitted. The safety relief valve to be installed is 25 meters downstream of the vessel outlet nozzle, and its set pressure is equal to the MAWP in accordance with ASME requirement. Also, a 5% blowdown difference between the opening and seating pressure is allowed. Between the vessel nozzle and the safety relief valve, there would be five long-radius pipe bends and one tee, through which the nitrogen must flow, from one end of the run through the branch. The original design model data are shown in Table 2.

Following steps (1), (2), (3) in the conventional design method, the required nozzle area $A_{3,cal}$ of a safety relief valve is calculated from [Eq.11],

$$A_{3,cal} = \frac{26000/3600}{0.2375 \times 0.95 \times 4.827 \times 10^5}\sqrt{\frac{288.71}{0.02801}} = 67.35\ cm^2$$

where the value of C is obtained from [Eq.12]

$$C = \sqrt{\frac{1.4}{8.314}\left(\frac{2}{1.4+1}\right)^{\frac{1.4+1}{1.4-1}}} = 0.2375\ mol^{\frac{1}{2}} \cdot K^{\frac{1}{2}} \cdot J^{-\frac{1}{2}}$$

This requirement can be met by a spring-loaded safety relief valve, "6Q8"[12], with a "Q" (11.0-square-inch, or 71.29 $cm^2$) orifice and with a 6-inch inlet flange and an 8-inch outlet flange. Therefore, this valve is selected, and $A_3 = 71.29\ cm^2$.

**Table 2. Original design model data**

| | |
|---|---|
| Required flow capacity, Wr, kg/hr | 26,000 |
| Molecular weight, Mw, kg/mol | 0.02801 |
| Specific heat ratio, k | 1.4 |
| Safety relief valve set pressure, bar-g | 3.45 |
| Safety relief valve nozzle coefficient, K | 0.95 |
| Gas stagnation temperature in vessel, $T_{0v}$, K | 288.71 |
| Atmospheric pressure, Patm, bar-a | 1.034 |
| Accumulated pressure, $P_{0v} = 1.1 \times$ MAWP + Patm, bar-a | 4.83 |

With the selected valve and with the 6-inch inlet flange reference [9] describes the trial-and-error calculations and adjustments in the configuration of the inlet piping system to satisfy both the 3% rule and the required flow capacity, as shown in [Eq.16a] and [Eq.16b]. Table 3 shows the four cases of trial calculations that were performed by the author of reference [9], and only Case D satisfied both the 3% rule and the required flow capacity. The detailed calculations and results for each case are also shown in reference [9].

**Table 3. Vent piping setups tried in Reference [9]**

| **Inlet piping system** | **Case A: 5 bends, Flow through branch of tee, 25 m of 6-inch pipe** | **Case B: 2 bends, Flow through run of tee, 25 m of 6-inchpipe** | **Case C: 2 bends, Flow through run of tee, 25 m of 8-inch pipe** | **Case D: 2 bends, Flow through run of tee, 7.9 m of 8-inchpipe** |
|---|---|---|---|---|
| vessel nozzle exit and vent pipe inner diameter, d, cm | 15.41 | 15.41 | 20.27 | 20.27 |
| inlet piping straight length, m | 25.00 | 25.00 | 25.00 | 7.90 |
| Darcy friction factor, f | 0.0149 | 0.0149 | 0.0141 | 0.0141 |
| velocity head loss, Ki | | | | |
| vessel nozzle | 0.5 | 0.5 | 0.5 | 0.5 |
| Piping | 2.42 | 2.42 | 1.74 | 0.548 |
| 90-deg long radius elbows | 1.04 | 0.417 | 0.394 | 0.393 |
| Tee | 0.894 | 0.298 | 0.282 | 0.282 |
| Total velocity head loss, N | 4.85 | 3.62 | 2.92 | 1.72 |

**Table 4. Data and calculation results for piping systems that satisfy the 3% rule or the capacity requirement, with the corresponding maximum allowable velocity head losses**

| Variables | Case D in Reference [9] | Satisfying the 3% rule with 6-inch inlet piping | Satisfying the lowest capacity requirement with 6-inch inlet piping | Satisfying the 3% rule with 8-inch inlet piping | Satisfying the lowest capacity requirement with 8-inch inlet piping |
|---|---|---|---|---|---|
| **Location 1** | | | | | |
| $A_1$, $mm^2$ | 32276.3 | 18640.1 | 18640.1 | 32276.3 | 32276.3 |
| $M_1$ | 0.133 | 0.236 | 0.228 | 0.133 | 0.129 |
| $P_1$, bar-a | 4.767 | 4.643 | 4.656 | 4.767 | 4.771 |
| $P_{01}$, bar-a | 4.827 | 4.827 | 4.827 | 4.827 | 4.827 |
| $T_1$, K | 287.69 | 285.52 | 285.75 | 287.69 | 287.76 |
| $T_{01}$, K | 288.71 | 288.71 | 288.71 | 288.71 | 288.71 |
| $\rho_1$, $kg/m^3$ | 5.581 | 5.476 | 5.487 | 5.581 | 5.584 |
| $u_1$, m/s | 46.03 | 81.43 | 78.46 | 46.15 | 44.52 |
| $W_1$, kg/hr | 29916 | 29925 | 28889 | 29925 | 28889 |
| **Location 2** | | | | | |
| $A_2$, $mm^2$ | 32276.3 | 18640.1 | 18640.1 | 32276.3 | 32276.3 |
| $M_2$ | 0.136 | 0.242 | 0.242 | 0.136 | 0.136 |
| $P_2$, bar-a | 4.661 | 4.535 | 4.378 | 4.662 | 4.501 |
| $P_{02}$, bar-a | 4.722 | 4.723 | 4.560 | 4.723 | 4.560 |
| $T_2$, K | 287.64 | 285.37 | 285.37 | 287.64 | 287.64 |
| $T_{02}$, K | 288.71 | 288.71 | 288.71 | 288.71 | 288.71 |
| $\rho_2$, kg/m3 | 5.457 | 5.352 | 5.167 | 5.459 | 5.270 |
| $u_2$, m/s | 47.18 | 83.33 | 83.33 | 47.18 | 47.18 |
| $W_2$, kg/hr | 29916 | 29925 | 28889 | 29925 | 28889 |
| **Location 3** | | | | | |
| $A_3$, $mm^2$ | 7129.0 | 7129.0 | 7129.0 | 7129.0 | 7129.0 |
| $M_3$ | 1 | 1 | 1 | 1 | 1 |
| $P_3$, bar-a | 2.495 | 2.495 | 2.409 | 2.495 | 2.409 |
| $P_{03}$, bar-a | 4.722 | 4.723 | 4.560 | 4.723 | 4.560 |
| $T_3$, K | 240.59 | 240.59 | 240.59 | 240.59 | 240.59 |
| $T_{03}$, K | 288.71 | 288.71 | 288.71 | 288.71 | 288.71 |
| $\rho_3$, $kg/m^3$ | 3.492 | 3.493 | 3.372 | 3.493 | 3.372 |
| $u_3$, m/s | 316.25 | 316.25 | 316.25 | 316.25 | 316.25 |
| $W_3 = W_{ASME,a}$, kg/hr | 26924 | 26932 | 26000 | 26932 | 26000 |
| **Results** | | | | | |
| $\Delta P_{0,1\text{-}2}$, Bar | 0.105 | 0.103 | 0.267 | 0.103 | 0.267 |
| $\beta$, % | 3.04 | 3.00 | 7.74 | 3.00 | 7.74 |
| $\gamma$, % | 103.6 | 103.6 | 100.0 | 103.6 | 100.0 |
| $N_{max}$ | | 0.54 | 1.48 | 1.70 | 4.63 |

The second column of the following Table 4 shows the calculated results for Case D of Table 3 by the author of this present article, using the conventional method as described above. These results are fairly close to those in reference [9], with an unrecoverable pressure loss of 3.04%, a little bit larger than 3.00% as reported in reference [9].

By using the proposed method in this article, the third and fourth columns of Table 4 show the results of calculations to determine the maximum allowable velocity head loss through the 6-inch-diameter inlet piping that would satisfy the 3% rule alone or the lowest capacity requirement alone, respectively.

As shown in the third column of Table 4, the calculated ASME-creditable capacity – assuming an unrecoverable pressure loss of 3% of the set pressure of the selected safety relief valve, i.e., $\beta = 3\%$ – is larger than the lowest capacity requirement, $W_r$, i.e., $\gamma > 100\%$. However, the maximum allowable velocity head loss is only 0.54 and would be difficult to attain with 6-inch piping.

As shown in the fourth column of Table 4, the calculated unrecoverable pressure loss would be 7.74% - and, thus, greater than 3% - if an e ASME-creditable flow capacity, $W_{ASME,}$, is assumed to be equal to the required flow capacity, $W_r$, i.e., $\gamma = 100\%$. Therefore, both the 3% rule and flow capacity requirement can be satisfied by using $P_{02,min}$ as [Eq.17] to calculate the maximum allowable velocity head loss of inlet piping system. These results are consistent with the analysis in Section 4, above.

If 6-inch-diameter inlet piping is chosen, the maximum allowable velocity head loss of the inlet piping system, $N_{max}$, is 0.54 as shown in the third column of Table 4. Any design of the inlet piping system with a velocity head loss larger than 0.54 will violate the 3% rule. This immediately eliminates Cases A and B of Table 3 because the total velocity head loss for each Case is far greater than 0.54. With a vessel nozzle velocity loss of 0.5, there is only 0.04 velocity head loss left for adjustment, which includes the length of piping length, the 90-deg elbows, and the tee. Thus, it would not be possible to design a practical inlet piping system to satisfy both the 3% rule and the flow capacity requirement for such a small margin of velocity head loss (0.04).. Therefore, any further design calculation must consider enlarging the inlet piping size.

Therefore, a larger inlet piping size of 8-inch-diameter is selected. The fifth and sixth columns of Table 4 show calculation results with corresponding maximum allowable velocity head loss, while satisfying the 3% rule alone and lowest capacity requirement alone, respectively. Both the 3% rule and the flow capacity requirement can be satisfied by using $P_{02,min}$ as given by [Eq.17] to calculate the maximum allowable velocity head loss of the inlet piping system.

With a larger pipe diameter of 8 inches, the maximum allowable velocity head loss of the inlet piping system, $N_{max}$, is 1.70, which leaves more room for evaluating various configurations of the inlet piping system. Thus, a practical inlet piping system could be designed$N < N_{max}$to satisfy both the 3% rule and the flow capacity requirement, while maintaining $N$ less than $N_{max}$. It is not necessary to evaluate Case C in Table 3 because its velocity head loss is 2.92, greater than the maximum allowable velocity head loss of 1.70. Interestingly, Case D in Table 3 as shown in reference [9] has a velocity head loss of 1.72, which is fairly close to the maximum allowable velocity head loss of 1.70. As shown in the second column of Table 4, a velocity head loss of 1.72 gives an unrecoverable pressure loss of the inlet line of 3.04%. To completely comply with the 3% rule and the flow capacity requirement, the straight piping length of Case D in Table 3 should be 7.5 m, rather than 7.9 m.

By comparing the proposed method to the conventional method for the design of the inlet piping system of a safety relief valve, the following advantages are:

- The proposed method is quite straightforward and avoids extensive and time-consuming trial-and-error calculations.
- With a maximum allowable velocity head loss – as calculated by the proposed method – design engineers can quickly identify where or what part of the inlet piping system (including pipe diameter, fittings, piping configuration, etc.) needs to be adjusted or modified to satisfy both the 3% rule and flow capacity requirement.

## 6. Conclusion

An easy method is proposed to facilitate the design of the inlet piping system to a safety relief valve for gas/vapor service. This proposed method allows calculating the maximum allowable velocity head loss, based on both the 3% rule and the required flow capacity. Any practical inlet piping system with a velocity head loss that is larger than the calculated maximum allowable velocity head loss would violate either the 3% rule or the flow capacity requirement. With the known maximum allowable velocity head loss, design engineers can quickly identify where or what part of the inlet piping system needs to be adjusted or modified. This method is quite straightforward and avoids extensive and time-consuming trial-and-error calculations.

## Nomenclature

| | |
|---|---|
| A | Cross-sectional area, $m^2$ |
| C | Thermodynamic coefficient for valve sizing, $mol^{1/2}\cdot K^{1/2}\cdot J^{-1/2}$ |
| D | Pipe internal diameter, m |
| F | Darcy friction factor, dimensionless |
| G | Mass flux, $kg\cdot m^{-2}\cdot s^{-1}$ |
| k | Specific heat ratio, dimensionless |
| K | Manufacturer-provided discharge coefficient, $K = 0.9K_d$, dimensionless |
| $K_b$ | Capacity correction factor due to backpressure, dimensionless |
| $K_c$ | Capacity correction factor due to combination with a rupture disk, dimensionless |
| $K_d$ | Tested discharge coefficient, dimensionless |
| $K_i$ | Resistance coefficient of pipe fitting, dimensionless |
| L | Pipe straight length, m |
| M | Mach number, dimensionless |
| MAWP | Maximum allowable working pressure, bar-g |
| Mw | Molecular weight, $kg\cdot mol^{-1}$ |
| N | Velocity head loss or pipe resistance coefficient, dimensionless |
| P | Pressure, bar-a |
| R | Gas constant, 8.314 $J\cdot mol^{-1}\cdot K^{-1}$ |
| T | Temperature, K |
| u | Velocity, $m\cdot s^{-1}$ |
| W | Mass flow rate, $kg\cdot hr^{-1}$ |
| x | Accumulation factor in percentage |
| Z | Compressibility factor, dimensionless |
| $\beta$ | Ratio of unrecoverable pressure loss to the valve set pressure, dimensionless |
| $\gamma$ | Valve capacity sufficiency ratio, dimensionless |
| $\rho$ | Density, $kg\cdot m^{-3}$ |
| **Subscripts** | |
| 0 | Stagnation condition |
| 1,2,3 | Location of relief line |
| 01,02,03 | Stagnation at location 1, 2, and 3 |
| 0v | Stagnation in vessel |
| actual | Actual |
| ASME | ASME creditable |
| atm | Ambient conditions |
| c | Critical or sonic condition |
| cal | Calculated |
| max | Maximum |
| min | Minimum |
| R | Required |
| Rated | ASME rated |
| Sg | Set gauge pressure |

## References


[1] API STANDARD 520: Sizing, selection and installation of pressure-relieving devices in refineries, 8th Edition, 2008

[2] ASME International, Pressure Vessels – Division I, Boiler and Pressure Vessel Code – Section VIII, ASME, New York, 2010

[3] Smith D., Burgess J., Powers, C., Relief device inlet piping: Beyond the 3 percent rule, Hydrocarbon Processing, November 2011, p 59-66

[4] Cremers J., Friedel L., Pallaks B., Validated sizing rule against chatter of relief valves during gas service, Journal of Loss Prevention in the Process Industries, 2001, V14, p 261-267

[5] Cox O. J., and Weirick M. L., Sizing safety relief valve inlet lines, Chemical Engineering Progress, 1980, V76(11), p 51-54

[6] Kandell P., Program sizes pipe and flare manifolds for compressible flow, Chemical Engineering, 1981, V88 (13), p 89-93

[7] Kern R., How to size piping and components as gas expands at flow conditions, Chemical Engineering, 1975, V82 (22), p 125-132

[8] Van Boskirk, B. A., Sensitivity of relief valves to inlet and outlet line lengths, Chemical Engineering, 1982, V89(17), p 77-82

[9] Westman, M. S., Rapidly design safety relief valve inlet piping system, Chemical Engineering Process, 1997, V93 (3), p 80-88

[10] White F. M., Fluid Mechanics, McGraw-Hill, 2001

[11] Crane Co., Flow of fluids through valves, fittings, and pipe, Technical Paper No. 410, Crane Co., New York, NY., 1982

[12] Crosby Valve Inc., "Crosby Pressure Relief Valve Engineering Handbook", 1997, p 2-6

[13] API STANDARD 526: Flanged steel pressure-relief valves, 6th Edition, 2009

# 348683 Off-Gas Flammability Control for a High Level Nuclear Waste Glass Melter System Based on Process Modeling and Pilot Testing

**Abstract:**

**Off-Gas Flammability Control for**

**A High Level Nuclear Waste Glass Melter System**

**Based on Process Modeling and Pilot Testing**

Alexander S. Choi

Savannah River National Laboratory

Aiken, SC 29808

In February 1996, the Defense Waste Processing Facility (DWPF) at the U.S. Department of Energy's Savannah River Site (SRS) commenced its first radioactive operation to convert over 300 million liters of high level nuclear waste into a stable borosilicate glass matrix for a long-term storage in a geological repository. At the heart of the DWPF operation is the joule-heated melter which converts a 50 wt% slurry feed containing pretreated waste and glass-forming frit into glass and the off-gas treatment system which removes volatile contaminants such as cesium and mercury from the melter exhaust. One of the critical safety requirements for the melter off-gas system is to maintain the concentration of flammable gases in the off-gas downstream of the Quencher below 60% of the lower flammability limit (LFL) under all operating conditions, while keeping the entire system under moderate vacuum.

The original DWPF design for the melter off-gas flammability control was to monitor the concentration of flammable gases using the LFL analyzers and to interlock the feeding off when readings exceeded a preset limit. However, it was shown during cold run tests that even after frequent calibration and maintenance these analyzers still were determined to be not reliable enough to function as a Safety Class instrument. Furthermore, in the event of sudden off-gas surges that could lead to a significant spike in the flammable gas concentration, these LFL analyzers could not provide adequate protection against potential off-gas flammability, since the duration of surges is typically much shorter than the inherent time delay in the response of these analyzers.

As a result, all LFL analyzers and associated interlocks and alarms were removed, and a new strategy was adopted to mitigate potential off-gas flammability that is in compliance with the NFPA Code 69; it requires that all operating variables that can contribute to potential off-gas flammability be monitored, and necessary safety interlocks and operating procedures to control those variables be devised. This paper describes a comprehensive method currently in place at the DWPF to monitor and control these variables using two process models; a steady state model that describes the batch-to-glass conversion chemistry and a transient model that describes the melter off-gas system dynamics and combustion kinetics of flammable gases. Some of the pilot test results that were used to develop and validate the model parameters are also covered briefly.

10TH GLOBAL CONGRESS ON PROCESS SAFETY

# Advancing Process Safety: Operating Risks and Solutions to SOP Compliance

**Samuel H. Solomon**
**Chairman**
**sam@mobideo.com**



Prepared for Presentation at
American Institute of Chemical Engineers
2014 Spring Meeting
10th Global Congress on Process Safety
New Orleans, LA
March 30 – April 2, 2014

UNPUBLISHED

# Advancing Process Safety: Operating Risks and Solutions to SOP Compliance

**Samuel H. Solomon**

**Chairman**
**sam@mobideo.com**

**Keywords:** inspection, compliance, operations risk, mechanical integrity, process safety, EHS, mobile, mobility, software, enterprise software, analytics, procedures, cloud, big data

## Abstract

Advancing process safety requires rethinking technologies necessary to ensure adherence to procedures and effectively monitoring of compliance. With the advent of mobility, cloud and big data analytics, new tools are available to directly impact process safety and its implementation. *Key to successful implementation is the ability to capture and monitor activities at the most granular level of procedural detail – knowledge "one step at a time."* The case study in this paper will be on the inspection process with a specific attention to coating detection and failure as we discuss SOP compliance challenges resolved by this approach.

## 1. Introduction

Significant progress has been made during the past few years to improve the quality of data captured at the point of service in a continuous process facility. This has evolved to improving sensing devices and the data collection from these devices to deliver analytics that reveal the lifecycle of equipment and predicting failures before they occur. This area of machine-to-machine engineering and communications is critical to the long term safety and efficacy of equipment. What has not been addressed is the other side of this issue – the human interaction with this equipment in the form of Standard Operating Procedures, best practices and experience which is, for the most part, still complied with using documents, forms and spreadsheets.

The paper will address these challenges by detailing an information technology approach and its business value impact. The approach will cover just three of the many potential outcomes:

- Increasing process safety while reducing maintenance cycle times
- Increasing asset availability by ensuring latest procedures are followed

- Reducing maintenance costs by aggregating maintenance work and increasing pattern analysis leading to pro-active safety management

As we have said, these procedures are being performed by following checklists, job guides and documentation manuals. And, in every process industry these procedures are typically maintained in hardcopy or an image-equivalent such as PDF. Due to this critical lack of appropriate automation and controls, a number of key business concerns have been uncovered:

- How do we measure risk in people operations?
- How do we ensure compliance with protocols and regulatory standards?
- What do we do as our knowledge and expertise retires at a more rapid rate?
- How do we analyze best practices and evaluate processes?

The answers to these and other important operational questions are hidden in the details of a company's procedures which document exactly "how" a specific activity needs to be executed. These "Standard Operating Procedures" or SOPs, typically cover how every process is supposed to be performed and, yet, for the most part, companies have absolutely no visibility into these activities.

Today, a number of key enablers have come about pointing to new and comprchcnsivc approachcs for technology to address issues of operational integrity, and in particular process safety imbedded in these SOPs.

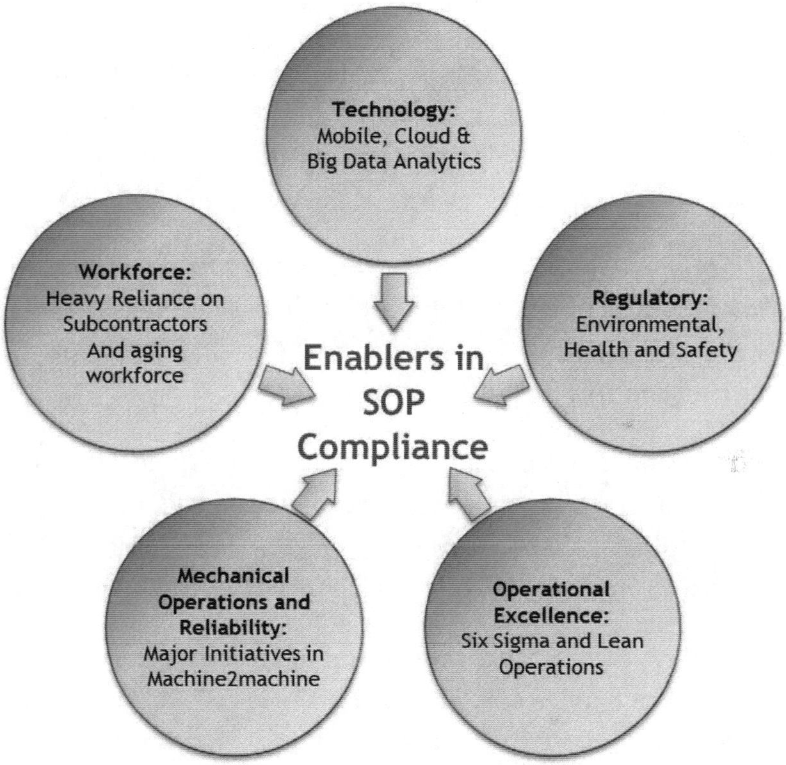


**Figure 1 – Enablers in SOP Compliance**

*Source: Mobideo Technologies LLC*

Let's briefly touch on each of these enablers:

***Technology: Mobile, Cloud and Big Data Analytics*** – The availability and maturity of these technologies has lowered the cost of acquisition and precipitated their convergence. In addition, there is a strong desire within companies today to adopt new solutions that are based on Mobile, Cloud and Big Data Analytics.

***Regulatory: Environmental, Health and Safety*** – Process industry companies now have internal and external compliance requirements. This oversight has grown because of significant risks associated with failure to comply entailing both civil and criminal penalties. SEMS (Safety & Environmental Management System) – a new regulation that is now being imposed on Owner/Operators engaged in off-shore Oil & Gas exploration in the Gulf of Mexico – is just one example of new and more comprehensive compliance requirements being adopted by the industry.

Furthermore, EH&S requirements are becoming critical "KPIs" (Key Performance Indicators) as to the sustainability and viability of companies throughout all process industries, with industry experts and analysts paying increased attention to these indicators. Again, failure to ensure compliance may result in loss of revenue, fines and at-risk employees.

***Workforce: Heavy Reliance on Subcontractors and an Aging Workforce*** – We see this as another critical element in enabling the adoption of workforce process tools, since both the risk of non-compliance is significant, while, simultaneously, the traditional resistance to employee "monitoring" is negated.

For example, a company operating an oil rig must monitor and ensure compliance of its service provider subcontractors (who may represent as much as 90% of the workforce on the platform or in the plant). Liability for subcontractor performance is increasingly being applied not only to the service provider, but also to its customer, the Owner/Operator of the facility.

In addition, companies are constantly looking for ways to solve the "aging workforce" problem – experienced and knowledgeable employees who are retiring from the industry. They are being replaced with a new generation of workers who are technologically savvy and are comfortable working with mobile devices such as an iPAD, as opposed to hard copy checklists or solutions found in a printed reference book. For the new generation of workers, using a mobile device is both natural and expected.

***Mechanical Operations and Reliability: Major Initiatives in Machine2Machine*** – There are major business initiatives by companies such as GE and Emerson to invent and deploy "pervasive sensors" on complex systems, thereby enhancing predictive maintenance and operational KPIs. Monitoring and analysis at the granular level of human interaction with complex systems adds a critical and complementary component to industry trends to better performance measurement and equipment reliability.

***Operational Excellence: Six Sigma and Lean Operations*** – Operational Excellence principles, systems, and tools such as Six Sigma, Lean Manufacturing and Scientific Management are being adopted by companies in order to gain sustainable improvement of key performance metrics. The granular data collected by operators and inspectors provides invaluable information to be used by such Operational Excellence programs.

Our paper will examine key criteria for a successful technological solution to this problem as well as challenges and personnel considerations. Examples will be provided from an actual case study.

## 2. Challenges Facing SOP Compliance

Standard Operating Procedures are a broad description of protocols and procedures which detail the step-by-step guidance necessary to perform a specific task, such as corrosion inspection in a pipe or vessel. They are typically based upon industry standards as are promulgated by organizations including the American Petroleum Institute (API) and would be modified by the company to address actual plant or field conditions or best practices by experienced subject matter experts (SME) at the company. SOPs provide workers with guidance and checklists which detail every step to be taken, whether it is a check, warnings, meter readings and other data gathering requirements. The importance of these procedures in addressing monitoring and compliance is the fact that these SOPs must be followed in a very specific manner – step by step – in a granular fashion in order to adhere to these protocols.

These workflow processes are at the core of a compliance regime, according to the Chevron Richmond Refinery Report on the Aug 6, 2012 refinery rupture and fire, which "were intended to be goal-setting or performance-based (*the emphasis is ours*), in practice they appear to function primarily as reactive and activity-based regulatory schemes that require extensive rulemaking to modify. As a result, the federal and California PSM standards have become static in the face of advancing best practices and technology, with the emphasis placed on the completion of a task or activity rather than achievement of continuous risk reduction."[1] To ensure that goal-setting and performance-based adherence to protocols actually occurs as part of a strategy for continuous risk reduction, one needs to rethink how technology and best practices are implemented to enforce this compliance and to provide the level of monitoring and analytically reporting necessary to comply with emerging standards.

At the center of the issue of adherence to standard procedures and protocols is the question of the origination of errors, and in particular, human errors. The Guidelines for Preventing Human Error in Process Safety by Center for Chemical Process Safety (CCPS)[2] describe opposing approaches to Human Error Reduction and as can be seen from the diagram in Figure 2 below, an appropriate approach requires more than just an understanding of human/organizational/cultural factors – i.e. tools and techniques are on an equal footing. Their suggestion is to focus on a proactive approach that involves five key factors – all of which relate to ensuring compliance to SOP guidelines and procedures. Our focus will be on the tools and techniques related to supporting not just a reactive understanding of what happened but also a proactive approach to error prevention.

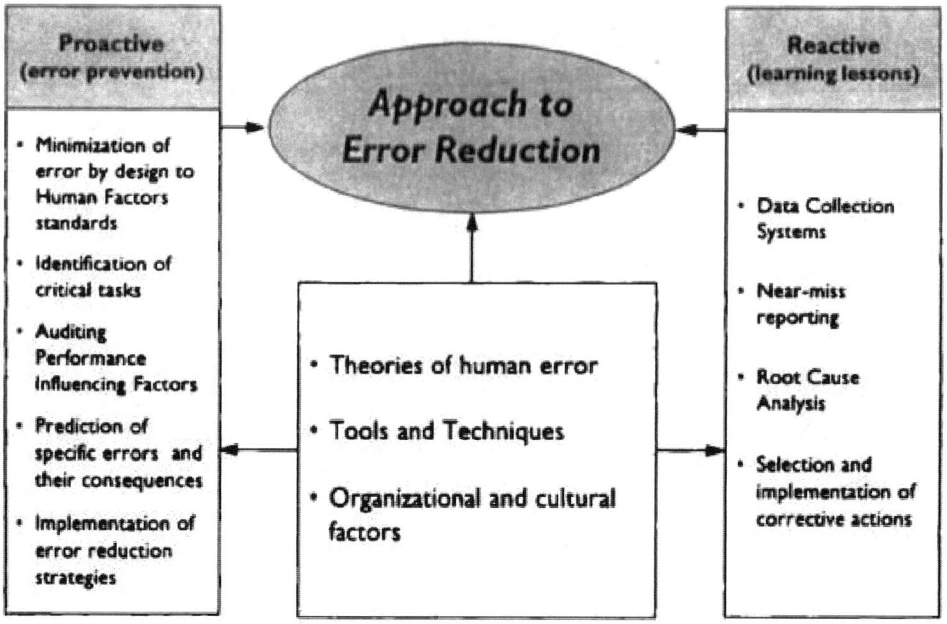


**Figure 2 – General Error Management Structure [2]**

A report from the Institution of Chemical Engineers (IChemE)[3] declares that 20% of equipment failures in chemical process industries are the result of human and organizational errors. The top 5 reasons are all directly related to the administration and deployment of protocols and procedures: poor contractor control, failure to follow procedures, lack of planning, poor management and supervision, and simple misjudgments. David Brown, chief executive of the Institution of Chemical Engineers (IChemE), said: "Accidents caused by equipment failures have many potential causes including contamination, corrosion, poor installation and mechanical failures. But it is worrying to see that human and organizational failures head the list of contributing factors. Everyone employed in the process industries must play their part in managing risk. The recent explosion in West, Texas, which killed 15 people, is another sad reminder of the consequences."

The textbook Human Error – Causes and Control[4] states that in a recent analysis of chemical plant problems, there were five main sources of these problems:

- Maintenance Practices 18%
- Maintenance Materials 7%
- Raw materials 5%
- Design 5%
- Operational discipline 45%

And we quote their conclusion: "The major human error problem was called operational discipline, that is, the failure to follow procedures and to stay within operational envelope. The second human factors problem was maintenance practices, that is, the inspections, repairs and modifications of equipment. Together, these two human error categories account for 63% of the chemical plant problems." They continue to elaborate: "Very often, mechanics do not consult procedures in printed service manuals or those that are available on computer terminals, since they often use the diagnostic challenge of trial-and-error methods to solve problems. In some situations, written instructions are not conveniently available, practical, reasonable or understandable." (*The emphasis is ours*). [4]

When one considers the human error component in compliance to SOPs, it is clear that this major challenge to accountability needs to be addressed at policy, cultural and procedural levels. We will address the procedural level which is really about finding an effective technological solution to SOP compliance given the convergence of mobility, cloud and big-data.

From a technological standpoint, there are a number of challenges to the effective deployment of SOP solutions including:

- Data necessary to carry out the field work is scattered in multiple locations – silo-ed within Enterprise Application Software, inspection, reliability, MOC and other systems for project/content/process management.

- Important data is embedded within disparate fill-in forms and support documents in word/excel/pdf formats.

- Enterprise application software products (such as ERP, EAM, and PLM) and Enterprise middleware (such as BPM and EAI) enable activity management only at higher levels of data resolution (task and sub-task). They fail to provide fine granularity at the "nano" level, necessary to provide deep, detailed and effective business process management and optimization.

- The difficulty of SOP protocol creating, maintenance, distribution and updating for most companies is an expensive nightmare and is embedded within many groups.

SOPs today are still complied with manually, even if the medium of display is an excel spreadsheet or PDF document. And, in many cases, SOPs are followed by adhering to each written step and completing the form with ink and paper. This approach creates severe challenges in addressing compliance and operational risk. These key challenges include:

- Data quality and consistency
- Costly over-maintenance and risk of under-maintenance
- Supervision of sub-contractors
- Defect identification to defect resolution
- Defect identification across equipment and personnel over time

Let us briefly touch on each of these areas:

***Data quality and consistency*** – GIGO is a well-worn acronym in the computer industry: Garbage-in Garbage-out. Current approaches to managing and monitoring SOPs have no effective data validation and consistency systems where information is checked to ensure it is correct the first time. Typically mistakes are only discovered way after the work has already been completed. And, in many cases, the data can never be corrected or made consistent since it cannot be accomplished – sometimes days, weeks or months – after the fact.

***Costly over-maintenance and risk of under-maintenance*** – This is directly related to data availability which drives metrics and decision making. A key part of this decision making process is to contextualize problems, for example, within the overall scope of the maintenance issues and objectives for this part of the plant. Data will come from gauges, sensor and human observation and following SOPs. If the data is unavailable or inconsistent/error prone, then it will be nearly impossible to accurately predict the best maintenance or replacement to support and extend the life cycle and safety of the equipment.

***Supervision of sub-contractors*** – This is probably the most difficult area to ensure compliance to SOPs. On the other hand, they are the easiest to enforce new technology standards. Understanding what service providers are doing, how they are doing it and whether they are in compliance with company/operator standards cannot be accomplished without some sort of real automation at the point of service.

***Defect identification to defect resolution*** – Similar to our comments regarding over and under maintenance, making the identification to resolution process most efficient requires consistent and accurate data at the most granular level of SOP compliance.

***Defect identification across equipment and personnel over time*** – Ultimately, when one starts to contextualize data on human interaction with equipment via SOPs and other procedures, this granular data starts to reveal patterns way beyond the specific piece of equipment and allows for pattern and pier group analysis of the people, processes and equipment to determine errors and failures. We show an example of this in the case study.

## 3. Key Criteria for Competent Technology Solution

Over the past 20 years, Information Technology has successfully automated key back office operations such as Enterprise Resource Management (ERP), Enterprise Asset Management (EAM), Manufacturing Execution Systems (MES), Electronic Document Systems (EDS), Project Lifecycle Management (PLM) and more. A significant gap exists between these back office systems and the work actually being performed by workers in the plant or field.

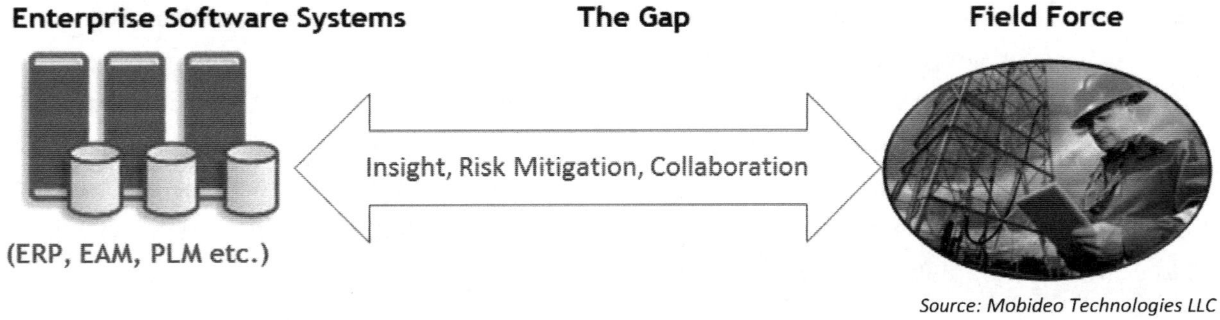


**Figure 3 – Gap Between Back Office and Field Force**

What are key components of a technological solution to address SOP guidelines and their proper deployment in a company?

- Deployment to the workforce with no disruption in the way they currently follow and utilize SOPs.
- Rapid deployment based on inputting existing hardcopy or electronic technical guidelines, which are then converted into confirmation-dependent conditional steps;
- User-defined step-levels for each task, allowing the highlighting of details with whatever degree of granularity is needed to ensure consistent, error free task completion;
- Real-time remote monitoring for supervision and/or collaboration;
- Multi-level guidance sensitive to user experience and expertise allows for significant training and employee mentorship and monitoring;
- Event tracking, information-tagging, real-time analytics, and data mining for continuous improvement programs;
- Cross-platform mobile-device support including: iPAD, Android and Windows;
- Architecture for flexibility and functionality across all deployment scenarios (public/private cloud or on-premises);
- Choice of standalone functionality, or seamless integration with other enterprise application software systems, including: Enterprise Resource Planning (ERP); Enterprise Asset Management (EAM); Enterprise Content Management (ECM); Product Lifecycle Management (PLM); project management and other middleware;
- Allow the system to have full functionality when isolated from the network with automatic updating whenever there is connectivity.

The combination of all factors directly impacts customer bottom line results by reducing operation costs, minimizing potential risks, increasing safety, ensuring quality, improving assets performance and ensuring regulatory compliance.

## 4. Case Study

The solution demonstrated in this case study is based upon Mobideo Technology LLC[5] platform to help companies increase asset capability, performance and situational awareness by ensuring proper execution of document-based activities and best practices across the organization; enabling users to have information where and when they need it. Rapid deployment is made practical by leveraging existing knowledge and expertise in the organization. A key business area impacted by this technology involves adherence and monitoring of SOP procedures and best practices. An important consideration was the need to address the most "granular" or "nano-level" data in an operation.

We have displayed a sample "coating inspection form" within a chemical manufacturing plant. The inspection process occurs on a regular basis as part of a mechanical integrity discipline to ensure the reliability and safety of the plant assets such as vessels, pipes and pressure valves. During the inspection process, an inspector examines the equipment while adhering to the "coating inspection form" and documenting his/her findings. In order to prepare for this pilot we identified the key SOP forms which were incorporated into the Mobideo application. Following this, "metadata" was added, i.e., information useful for validating and making consistent all data entry by the inspector (we used Mobideo's Studio product to accomplish this); we then created links to specific data we will need from the inspection and ERP databases; and finally these forms were distributed to standard (and ruggedized) tablets. We have chosen to implement the coating inspecting report and to focus on just one section being the coating thickness assessment. It took about 3 hours to complete the application presented in this paper.

We found that the manual approach introduced numerous errors by the nature of using a form. Even more important, with automation, we discovered useful and more fine-tuned data that was not possible in a form based approach but become very practical using a mobile solution, such as assessing thickness over a smaller inspection area. In addition, we looked at how mobilization with the right tool impacts the time expended performing the inspection without effecting the trade work itself, such as painting or welding. Finally, we looked at the contextual picture to see how this data, now granular and consistent, would provide a "bigger picture view" of the patterns of the coating failures.

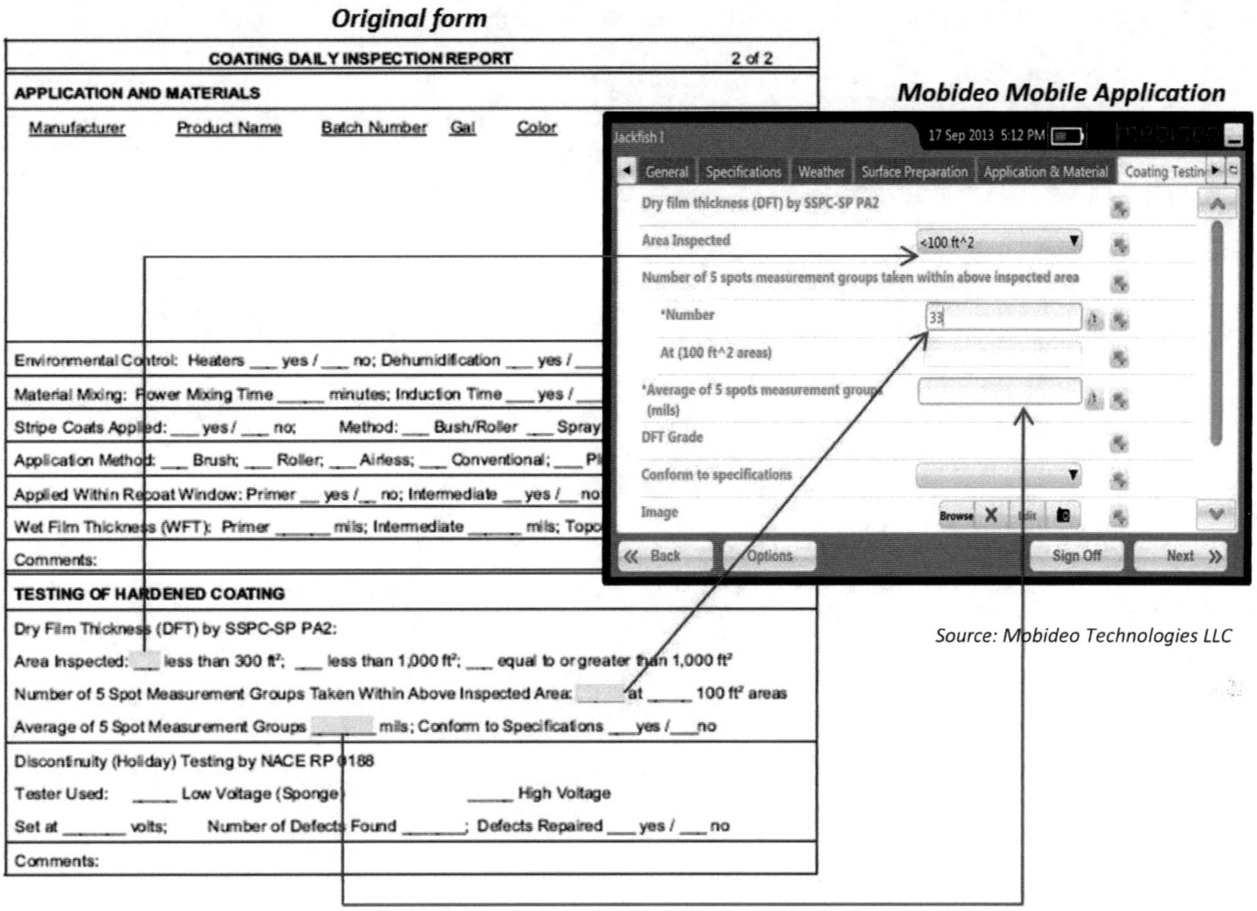


**Figure 4 – Chemical Plant Application**

In the form depicted above, each individual field is a granular step. For example, as highlighted in yellow, one step is to check-off the area inspected, and the next step is to enter the number of 5 spot measurements and finally to enter the average thickness for each of these measurements.

We created the mobile application where each of these granular steps is validated, time and location stamped and sent to the server in real-time (as well as any linked IT systems). The mobile application is built from all the information sources necessary to ensure proper task execution – documents and forms, as well as dynamic content that is retrieved in real time from third party back office systems. These data sources are integrated into a uniform application to be presented to the user. Along with the real time monitoring of the mobile application, a portal is automatically configured to observe, analyze and interact with the people and the data reflecting what they are doing during the SOP process. This flow is depicted in Figure 5.

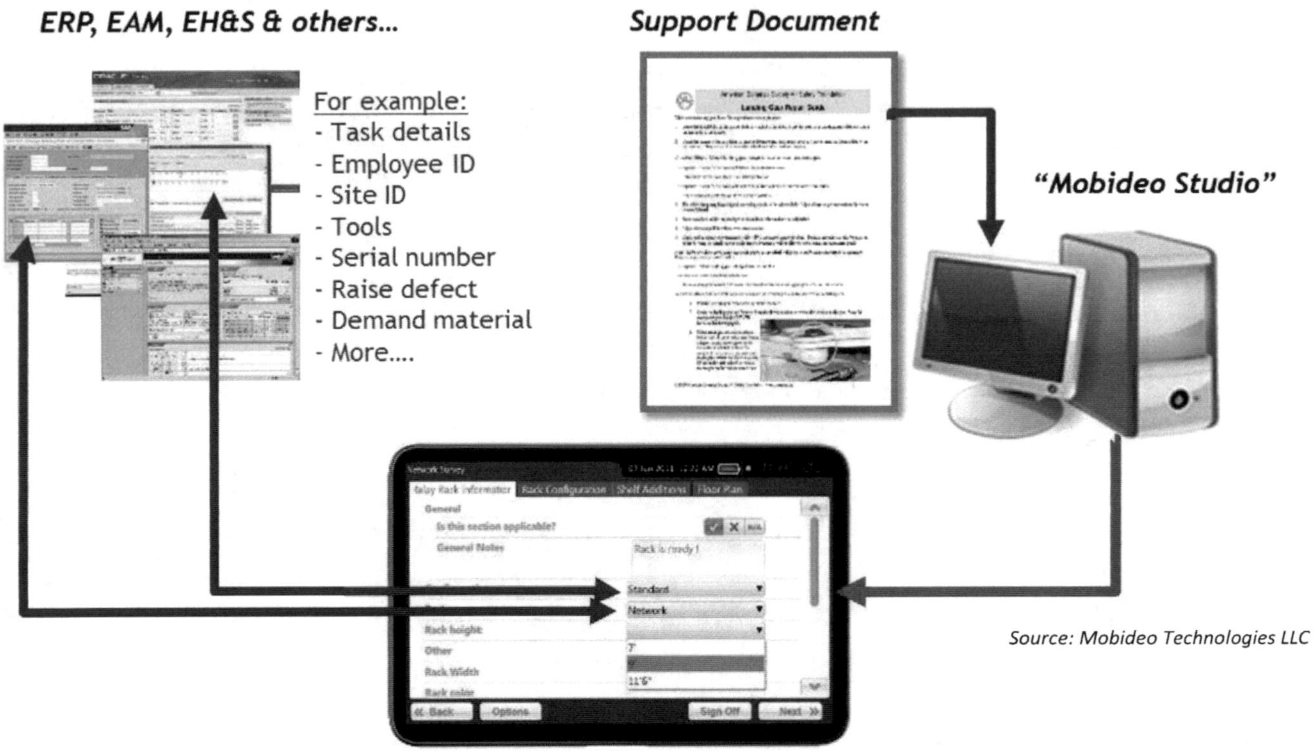


**Figure 5 – Integration of Support Documents with Existing IT Databases**

With the basic design flow in mind, let's look at a few sample images captured from the tablet running the application. The first screen shot (Figure 6) shows how the resulting SOP was incorporated and arranged with meta-data. Along the top there are tabs for each section of the SOP and then the actual SOP details, whether it is a question, selection, notice or display, they are arranged in a similar manner to the SOP therefore requiring little if any training in terms of content. The application itself is based upon simple "big glove" interaction – back, next and options. Everything else is the data within context of how the SOP presents the options. The meta-data would include validation checks, types of data entry (pull-down box, choices, free text, GPS settings, reading from an external meter or device, etc.), warnings, messages and many other data types.

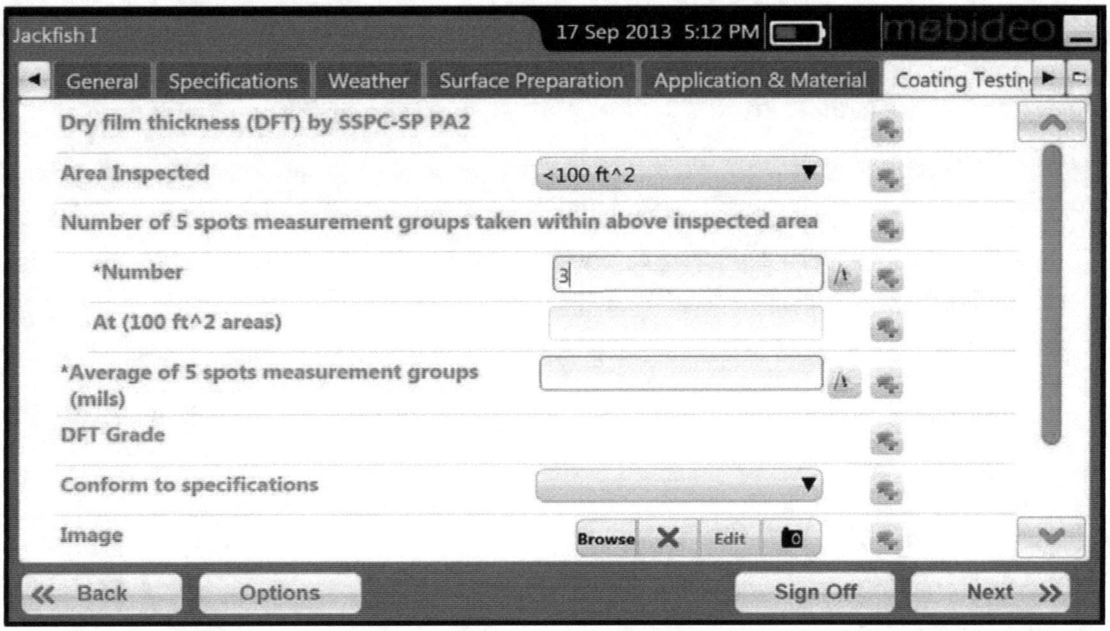


**Figure 6 – Mobideo Form Used During Inspection Process**

Figure 7 shows another example of an interesting input, in this case being a photo that was drawn upon and integrated into the SOP responses. This is an example of extending the impact of the SOP because it is now a mobile application. The other two changes we made by way of Figure is the ability to easily reduce the size of the inspection area (from 100 $ft^2$ to 300 $ft^2$) and allowing a number of spot inspections – something hard to implement and then change and make dynamic in a hardcopy form. In addition, the application makes certain calculations and lookups, such as the DFT Grade, and gives immediate feedback to the operator of the grade assessment.

**Figure 7 – Mobideo Form with Picture Illustrated During Inspection Process**

Figure 8 shows how comments and social collaboration allow for easy capture of issues or best practices at the most granular step level of the process. Each step may be commented upon. Figure 9 shows how metadata enforces validation and consistency.

**Figure 8 – Mobideo Form with Comments Added at the Granular Step Level**

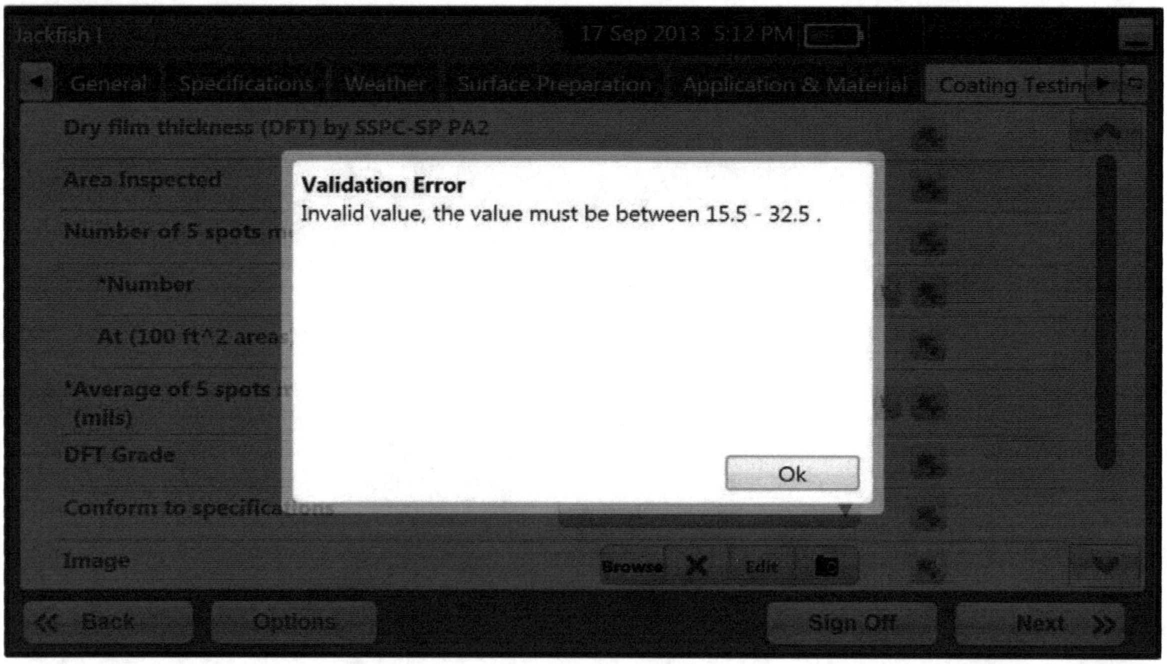


**Figure 9 – Mobideo Form Demonstrating Validation of Data Entry**

Figure 10 is very important. It shows the signature page which ends up becoming a very time consuming and error prone activity since signers may be in different locations and time frames. The system automatically coordinates proper signature signoff and there by enforces oversight and significantly reduces signature coordination time.

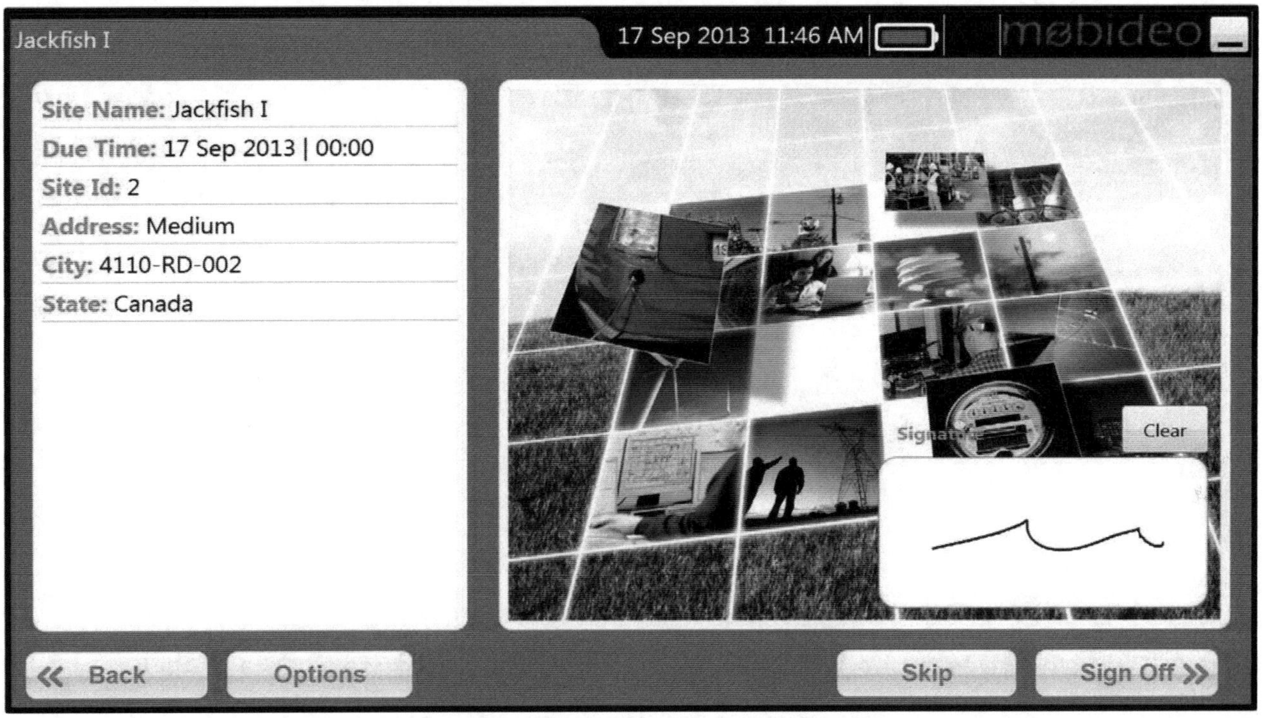


**Figure 10 – Mobideo Form Show Authorization and Signature Coordination**

**Figure 11 – Mobideo Web Portal Status Screen**

Figure 11 is a sample display of the portal. Because every step is being monitored, supervisors have a complete picture of all processes, issues, errors and progress. In addition, traceability of all inputs is assured.

Figures 12 and 13 show an example of the impact the system had on turnaround time of the coating project without attempting to use the system to reduce the time the trades take to do their work. The saving of nearly 30% in time is just related to the logistics of the process, i.e., coordinating sign-offs, generating documentation and facilitating the timing and actions of the various trades. The streamlining of workflow is a direct result of using mobility technology and granular data capture to ensure compliance to the SOP. We now have ways to modify workflow to make it more logical and efficient and we gain productivity just by eliminating the loss due to logistics management.

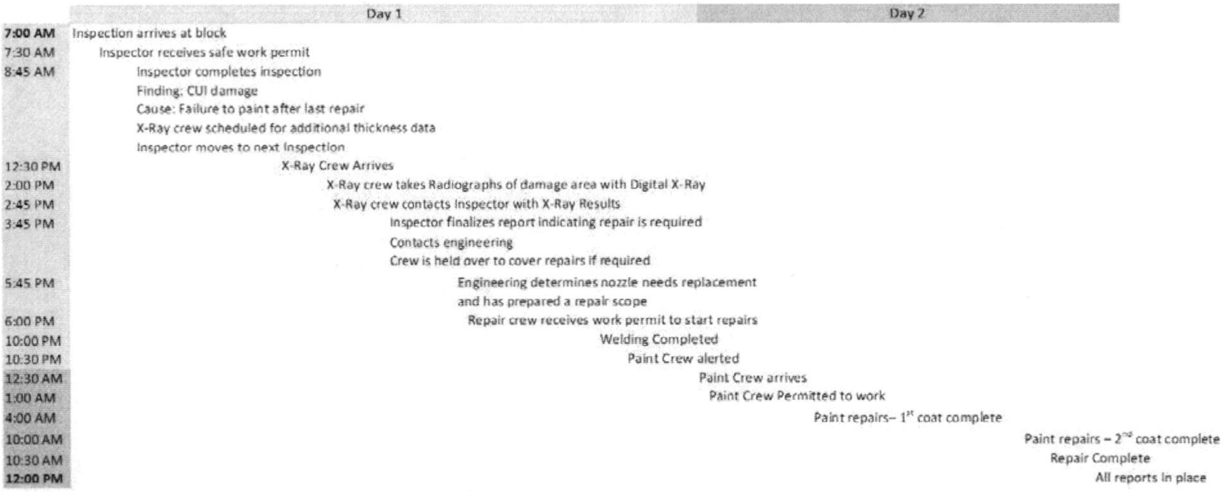


**Figure 12 – Workflow Time Duration Using Standard Manual Approach**

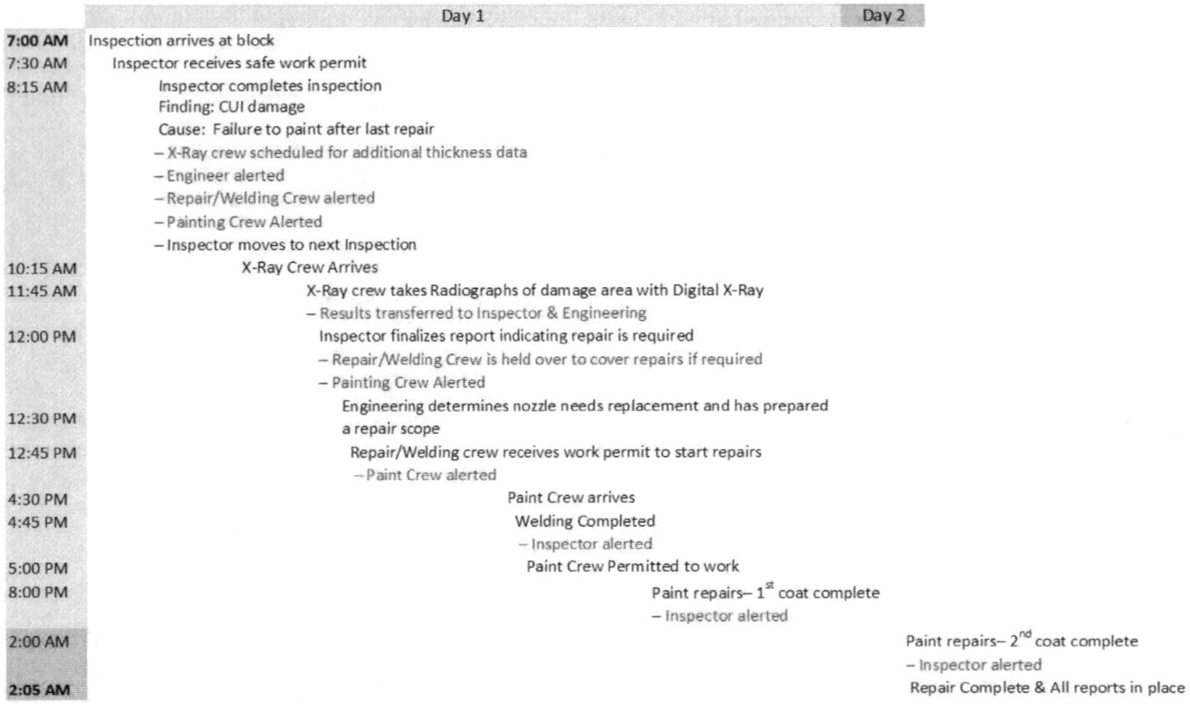


**Figure 13 – Workflow Time Duration Reduced with Mobideo Coordination of Tasks and Trades**

*Source: Mobideo Technologies LLC*

This last Figure 14 demonstrates another business value – the ability to now look across equipment and locations to determine appropriate KPI and intervention strategies. In this case, by looking at the inspection pattern of DFT spot measurements, we can quickly map out an area of corrosion that tells us to potentially redo all the painting as part of a more in-depth inspection process of this area.

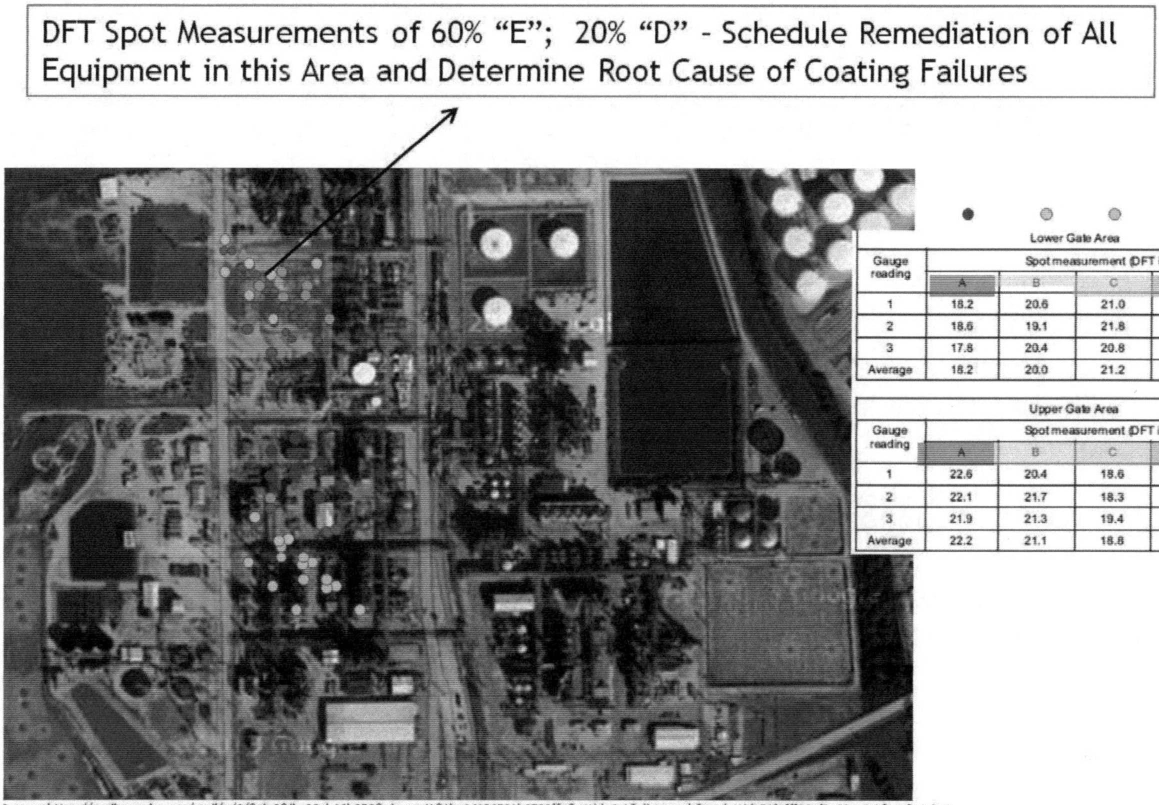


Lower Gate Area

| Gauge reading | Spot measurement (DFT in mils) | | | | |
|---|---|---|---|---|---|
| | A | B | C | D | E |
| 1 | 18.2 | 20.6 | 21.0 | 16.7 | 26.4 |
| 2 | 18.6 | 19.1 | 21.8 | 17.1 | 29.8 |
| 3 | 17.8 | 20.4 | 20.8 | 16.8 | 29.0 |
| Average | 18.2 | 20.0 | 21.2 | 16.9 | 28.4 |

Upper Gate Area

| Gauge reading | Spot measurement (DFT in mils) | | | | |
|---|---|---|---|---|---|
| | A | B | C | D | E |
| 1 | 22.6 | 20.4 | 18.6 | 19.5 | 29.3 |
| 2 | 22.1 | 21.7 | 18.3 | 19.9 | 31.0 |
| 3 | 21.9 | 21.3 | 19.4 | 19.2 | 30.4 |
| Average | 22.2 | 21.1 | 18.8 | 19.5 | 30.2 |

Source: https://mail.google.com/mail/u/0/?ui=2&ik=26ab16b259&view=att&th=14136591b9723ffa&attid=0.1&disp=emb&realattid=51fefff28ef9e23e_0.1&zw&atsh=1

**Figure 14 – Analysis of Corrosion Pattern**

## 5. Issues to Consider

We have experienced a number of customer implementations and it is important to highlight critical success factors outside of an excellent software platform.

The first factor to consider is creating an opportunity to "dip one's toes in the water" before considering a full in-house implementation. With the advent of Software as a Service (SaaS) over the cloud, departments can implement a limited and very defined project to determine design, implementation and organization issues. This also allows for a quick jump to pilot stage without the need of a capital investment based ROI analysis, which could come later but is not necessary to start.

A second success factor is related to training and culture. For the emerging younger workforce, they have little fear of mobile devices and tablets as interactive devices. But the culture of accountability and an understanding that systems make one more effective (and not just big brother) still needs to be addressed. This is where training and cultural awareness by management will be very important.

The third success factor we believe is important is that deployments must be very rapid and the implementations non–technical. In that way, one may push accountability to the teams doing the work as close as possible to the field operations and avoid the loss of knowledge and efficiency as well as cost of implementation in staff functions to support SOPs, as is typically done today.

The forth success factor is the use of existing SOP documents to drive the application process. This is so important since we want to avoid errors and resistance to change because the actual inspection process has been re-engineered. We were able to assure operators and inspectors that they would continue to work and interact with the data exactly the way they had the day before roll-out.

## 6. Conclusion

With the convergence of mobile, cloud and analytics at the point of service, we are ready as an industry to tackle the difficult issues relating to compliance and monitoring of SOP guidelines. This area has essentially remained untouched while the back office and administrative side of the business has been completely automated. ERP, Plant Maintenance System, Inspection database, and Electronic Document Systems, to name a few, have tackled the "what" needs to be done to support operational integrity and safety. The "how" to accomplish this has remained locked in documents, spreadsheets and best practices without any direct way to ensure compliance and to monitor effective deployment from a KPI perspective.

The author believes it is now entirely possible to accomplish this transition of the culture and will be there to make this happen. We also want to keep in mind a number of important ingredients to a successful implementation:

***Converge of all task execution data into a unified application*** – Converge of all task execution information from all sources, whether third party enterprise software such as ERP and EAM, or from documents and forms into a unified, personalized and contextualized mobile application. Integrating ERP, EAM, PLM and other third party enterprise applications enables customers to leverage existing investments in these systems, to extend their reach to the field, to ensure that they are updated in real time, and to increase their value

***Ensure compliance*** – Enable companies to strictly enforce compliance and adherence to standard operating procedures and ensure proper execution of field activities according to company's best practices even in the most remote locations, regardless of whether they are being performed by company's direct employees or sub-contractors.

***Track granular interaction and data entry in real time*** – By offering visibility of critical and complex processes at the most detailed and granular level (nano-step), management has the most comprehensive and detailed insights enabling them to gain situational awareness, perform predictive analytics, manage by exception, assure compliance to protocols and shield employees from mistakes while validating their work.

***Enable rapid deployment driven from the procedures themselves*** – Utilizing the existing knowledge base by incorporating documentation that already exists within the company, as opposed to a top-down approach to a new system, ensures that the deployment is transformative and non-disruptive.

***Technological convergence*** – Take full advantage of mobility with cloud/SaaS for rapid piloting and prototyping, analytics based upon granular data and social collaboration to facilitate the sharing of experience and best practices.

## 7. References

The author wishes to thank Mr. Keith Lapeyrouse, Reliability Manager at Dow Chemical Corporation for his support, insights and counsel.

[1] U.S. Chemical Safety and Hazard Investigation Board, Regulatory Report, Chevron Richmond Refinery Pipe Rupture and Fire, December 2013.

[2] Guidelines for Preventing Human Error in Process Safety, By Center for Chemical Process Safety (CCPS), Page 357, 1994.

[3] http://www.icheme.org/media_centre/news/2013/human-error-risk-for-process-industries.aspx, May 28, 2013

[4] Human Errors – Causes and Control, George Peters and Barbara Peters, CRC Press, page 14, 2006

[5] For more information on Mobideo Technologies LLC please contact yona@mobideo.com

# 349674 “Reducing the Frequency and Lowering the Severity of Human Error: Optimize Performance”

**Abstract:**

**“Reducing the Frequency and Lowering the Severity of Human Error: Optimize Performance”**

Process safety within the chemical, pharmaceutical, and petroleum industries demands extreme rigor where there is little room for human error. However, people are not perfect, and even the best make mistakes. Embracing this truism allows for real safety improvement strategies that are designed with the understanding that human error is inevitable and caused by complex factors. Failure to accept the inevitability of human error will result in continual frustration over adverse events and poor performance. When reactionary measures focus too heavily on the last person(s) involved in the error rather than focusing on the cumulative effects of systemic organizational breakdowns that allowed the event to occur, improvement opportunities are lost. This undue attention on symptoms instead of causes will yield only small gains, whereas accurate analysis and correction of system failures will return great benefits. Human performance will improve once organizational factors are redesigned to promote success and avoid failure. And, as individuals learn and use a set of well-defined, proven tools, common error-traps are mitigated. Thus, a comprehensive, holistic approach is needed that addresses all components of performance – error traps, organizational factors, and performance tools.

Optimize Performance (OP) is the approach that is brilliantly simple, yet profoundly impactful: Simple enough for everyone to learn and use, and impactful because it works. The practical value of Optimize Performance has been realized by more than 9000 individuals over the past 15 months. People embrace the OP brand of human performance because it works for them and their families, on and off the job. When employees embrace error-reduction principles off the job, employers win because of the inseparable connection between safe lifestyles off-the-job and error avoidance on-the-job. Employers reap tremendous benefits as employees put their training to work because they want to, not because they were told they had to.

Optimize Performance methods include:

- A set of well defined “**Tools**” that act as mental PPE to prevent errors and adverse events.
- Recognition of “**Traps**” that create error-likely situations. Both Situational-Factor Traps and Normalized-Drift Traps are identified and mitigated with OP “Tools”.
- An **Error-Review** process that evaluates how OP Tools/Traps can be mined to prevent recurrence and generate concrete and applicable lessons learned.

**OP Traps** increase the risk of error-likely situations. Situational Factors affect individuals at a given point in time, whereas Normalized Drift is a trap in the form of weakness and breakdowns in organizational and personal defenses, which become accepted over time. Both sets of OP Traps result in sub-standard performance and adverse events.

**OP Tools** are a set of thought-provoking defenses that will prevent, predict, or reduce the likelihood of errors and events only after the methods and techniques are fully embraced, learned and practiced.

A simple, but highly effective **Error-Review** process (OPER) offers a more complete understanding of adverse events and is a critical component of the Root Cause Analysis. The OPER helps determine OP Traps that created error-likely situations, and OP Tools that could have helped predict or mitigate the circumstances that resulted in the event.

Regulatory agencies are embracing human performance concepts and have aggressively begun incorporating them into regulations. On many fronts, human performance is emerging as a very effective and valuable performance improvement process.

Once learned, these strategies improve all aspects of operational performance. When applied, Optimize Performance changes culture. Improvements will be widespread and widely recognized. The frequency of human error will be reduced, and the severity of the effects of human error will be lowered. Safety will improve, reliability will be increased, productivity will rise, and people’s lives will be enriched.

**Brochure Description:**

People are not perfect, and even the best make mistakes. Organizational factors and values greatly affect performance and human errors. Optimize Performance is a proven strategy that is brilliantly simple, yet profoundly impactful. Recognize the “Traps” that create error-likely situations and learn how to apply well-defined “Tools” to prevent errors.

# 349820 Management of Process Safety Performance Indicators

**Abstract:**

Those working in the oil and gas industry are reminded with depressing regularity of the importance of process safety management, as post-incident investigation invariably points to one or more deficiencies related to process safety as root causes. Clearly, recent process safety related incidents and the development in process safety regulations have enabled wider awareness and adoption of process safety performance indicators (PSPI).

However, the primary challenge has been related to the selection, collection, analysis, and management of key process safety performance indicators that will eventually contribute to improving process safety in general. A systematic approach to manage these indicators is still evolving.

In this paper, the authors highlight a methodology for selection and measurement of process safety performance indicators which can help drive continuous process safety improvement and manage the performance of critical process safety barriers such as pressure relief devices, instrumentation systems, and mechanical integrity programs. This methodology considers complex set of intra-related factors and parameters that are available to the process operator; however, lack of systematic utilization of such parameters may hinder the realization of valuable process safety trends. In addition, the benefits and drawbacks of manual and automated systems for tracking near real time PSPI's will be presented.

# 349897 Improving Process Safety Performance for Mature Asset By Implementing of the Process Safety Key Performance Indicator

**Abstract:**

A mature asset has its own risk; it especially deals with plant integrity, facility changes, obsolete technology, people culture and stagnant procedure. A mature asset also frequently lay in the area of comfortable zone. Process safety hazard might be forgotten as time goes on and will lead to major accident scenario if it doesn't maintain.

The Offshore North West Java facilities belong to Pertamina Hulu Energy has lots of mature assets. Therefore Pertamina need to measure its process safety performance and be part of the corporate safety performance. The Process Safety Key Performance Indicators (PS KPI) is proposed to be used by Pertamina Hulu Energi Offshore North West Java (PHE ONWJ) to minimize risks from process safety incidents and to improve business performance.

The use of PS KPIs has been championed by the oil and gas industry as follow up lesson learned from several major incidents. It shown eventhough Process Safety incidents have low frequency but it potentially has high consequences. Hence, designing of the PS KPI should be carefully customized, practically implement and provide value to the organization, so that the process safety incident can effectively prevent.

The Pertamina Hulu Energy ONWJ used the combination of published guidance (API 754, CCPS, OGP Report No. 456 and UK HSE HSG254) for designing its PS KPIs. Understanding of the major gaps in facilities integrity, design and personnel culture are the important in designing PHE ONWJ's PS KPI. The effective PS KPI should consider the following criteria:

- The PS KPIs should address operational efficiency, reliability, and avoidance of production disruption.
- At the first phase of deployment, the PS KPIs should leverage the available data from the existing business process.
- The PS KPIs should provide senior management team in the company in the ability to understand the effectiveness of the PS programs
- The leading PS KPIs should be structured and easily tracked.
- Collecting data and tracking PS KPIs should add value to the organization and resources spent on the PS KPIs program should be in line with the value provided.

Pertamina believes that with designing Process Safety KPI that meet the criteria established can improve process safety performance as well as business goals for the company.

# 350057 An Innovative Work Flow for Performing Overpressure Protection Analysis Incorporating Process Simulation, Pressure Relief Valve Sizing, and Flare System Analysis

**Abstract:**

An innovative work flow for performing overpressure protection analysis incorporating process simulation, pressure relief valve sizing, and flare system analysis

By Wilfried Mofor, Aspen Technology Inc.; Nicholas Brownrigg, Aspen Technology Inc.; Dinu Ajikutira Aspen Technology Inc. and Ron Beck, Aspen Technology Inc.

Relief valve sizing is a critical element of the early design of any process facility, both from the point of view of ensuring the feasibility of achieving a safe design but also to understand the capital cost scope for both the relief valves themselves and the associated relief system.

The design of the facility itself is usually performed by a process modeler, employing a simulation system such as Aspen HYSYS or equivalent. Traditionally, the relief valve design has had to be performed in a stand-alone fashion, by an individual charged with that particular design task, employing a spreadsheet-type application that incorporates a particular organization's safety and design rules. Because a project cannot progress in the FEED workflow until the safety elements of the design have been verified, this is often a bottleneck in the process. There are also specific examples (and most organizations can attest to them), where incorrect data was transferred from the process model to the relief sizing programs, leading to incorrect results, sometimes only discovered after the relief valve has been procured and possibly even installed. Additionally, it is important for the safety engineer to rigorously consider all pressure source, routes, and failure potentials in enumerating the scenarios that need to be considered in the sizing analysis. These are usually only able to be derived by examining the process flow schema in the process model (or P&ID).

An innovative solution to this design challenge has recently been introduced by AspenTech, involving integration of an accepted, tested and rigorous relief valve sizing program within the process model workflow. This workflow spans from the equipment source of the pressure through the relief devices into the flare network. And additionally, from the process model, to the valve system design, to the overpressure relief network design and incorporating the documentation needed for the OSHA regulatory process and operator SOPs.

This paper will; (1) describe the new overpressure protection analysis workflow that is enabled by this new technology solution, (2) report on a comparative analysis of a complex overpressure protection analysis design problem as executed with the traditional and new workflow, (3) and report on results to date in industry, both on the E&C and owner operator side, from application of this new workflow.

We expect to be able to report on 3-4 industry case studies by the time this paper is presented, which we are currently in the process of gaining permission to present and to cite the names of the organizations involved.

# 350082 Risk Based Inspection Applied at Aging Chemical Facilities

**Abstract:**

Integrity safety is important to all companies and understanding the asset risk is the major key indicator in this process. Integrity management is underpinned by inspection and maintenance of equipment. Risk based inspection is becoming the major tool in integrity management of equipment at chemical facilities. Process risk management within Mechanical Integrity applying the challenges of using a risk based inspection programme, using a combination of the principles American Petroleun Institute (API) 580, ASME Post Construction Code (PCC-3), and MIL 882 introduced at older chemical plant mixed use environment with a combination of pressure vessels and Atmospheric Storage Tanks.

# 350323 Identifying Early Indicators of Incidents through Near-Misses

**Abstract:**

Post-incident investigations show that there are several near-misses or precursors that occur before undesirable events in different industries (chemical, manufacturing, aerospace, IT, etc.). Near-misses can be categorized into two categories – observable and non-observable near-misses – both of which indicate conditions which had the potential to cause incidents but did not actually result in much impact – due to sheer luck or existence of other protection systems. Observable near-misses refer to events that are readily visible to operating personnel, such as equipment failure, whereas non-observable near-misses refer to events that are not casually visible to the operating team, such as special-causes related to control loops. To improve reliability of systems and processes, it is of utmost importance for the risk management and reliability teams to not only identify and prioritize these near-misses, but also to learn from them. Particularly, cross-industry lessons on near-misses can provide new insights in their identification and management.

The presentation will demonstrate that having a strong near-miss management philosophy, system as well as culture, addressing both observable as well as non-observable events, help companies predict risks at their budding stages and prevent catastrophic events. It will include case studies from different industries on the importance of such events and their timely identification for improvement of reliability, and process safety. New quantitative techniques and methodologies to identify them systematically will also be discussed.

# Methodology to Determine the Minimum Number of PHAs for Projects

**Humbert Joseph Howard**
**Process Risk Specialist**
**Suncor Energy Inc.**
**150 6th Avenue SW**
**Calgary, AB, Canada**
**hjhoward@suncor.com**

Prepared for Presentation at

American Institute of Chemical Engineers

2014 Spring Meeting

10th Global Congress on Process Safety

New Orleans, LA

March 30 – April 2, 2014

# Methodology to Determine the Minimum Number of PHAs for Projects

**Humbert Joseph Howard**
**Process Risk Specialist**
**Suncor Energy Inc.**
**150 6th Avenue SW**
**Calgary, AB, Canada**
**hjhoward@suncor.com**

**Keywords:** Process Hazard Analysis, PSM, Projects

## Abstract

The purpose of this paper is to provide Process Hazard Analysis (PHA) study team leaders and project managers, who drive the need for PHAs during projects, guidance relating to a decision methodology that can be used to assess the minimum number of PHAs to be done for various complexity projects. We will use change complexity analysis to assess the change and use the change complexity factor to determine the number of PHAs required for a project.

Some of the most challenging decisions involve Process Safety risks which often require experience and knowledge in order to be addressed. The use of proper methodology, applied at the right time, can drastically improve the design of new project installations handling hazardous materials.

The determination as to what type of hazard review and the number of PHAs required should be done during the initial engineering phase of the project. This allows recommendations from the PHA(s) to be implemented prior to installation and/or start-up.

## 1. Introduction

In general, the decisions faced by project engineers of process facilities in the chemical and petroleum industries on the number of Process Hazard Analysis reviews are difficult and complex. Companies do many small projects that may only involve a few pieces of equipment and thousands of dollars. Companies also do large projects involving hundreds of pieces of equipment and millions of dollars.

The purpose of this paper is to provide Process Hazard Analysis (PHA) study team leaders and project managers, who drive the need for PHAs during projects, guidance relating to a decision methodology that can be used to assess the minimum number of PHAs to be done for various complexity projects.

Some of the most challenging decisions involve Process Safety risks which often require experience and knowledge in order to be addressed. The use of proper methodology, applied at the right time, can drastically improve the design of new project installations handling hazardous materials.

The determination as to what type of hazard review and the number of PHAs required should be done during the initial engineering phase or front end engineering design (FEED) of the project. This allows recommendations from the PHA(s) to be implemented prior to installation and/or start-up.

This paper is not intended to give guidance on how to conduct the PHA or what methodology should be used. Individual company procedures and practice should be consulted for specific guidance on how to conduct a PHA.

## 2. Methodology

We will use a risk assessment process that I am calling Change Complexity Analysis to assess the change and use the Change Complexity Factor to determine the number of PHAs required for a project.

### 2.1 *Process Boundary for Change Complexity Analysis*

It is very important to determine the process boundary in order to do a proper complexity analysis. The process boundary identifies the area of immediate process impact due to this small project. When the small project change involves equipment that is integral to operation of another piece of equipment the boundary should include both pieces of equipment. This is done to make sure that the true complexity of the change is assessed properly.

**Example 1**

*Replacing a storage tank recirculation pump.*

Boundary for the complexity analysis should include the new pump, the related piping, and the storage tank it recirculates.

**Example 2**

*Replacing a tower overhead condenser.*

Boundary for the complexity analysis should include the new condenser and the tower that the condenser is attached to with all related piping and controls. Intermediate column take-offs should be included with the tower.

**Example 3**

*Replacing a process unit Programmable Logic Controller (PLC).*

Boundary for the complexity analysis should include the new PLC and all the process instrumentation and equipment that is attached to the PLC with all related controls for the equipment.

### 2.2 *Change Complexity Analysis*

Change Complexity Analysis seeks to identify how difficult a change a project will be. The more complex the project, the more carefully the project will need to be analyzed and managed. A project involving water does not need the same level of review as a project using a toxic chemical. A project to purchase and install a bucket conveyor system in alumina service may not need the same level of review as the installation of a new distillation column system in HF service.

We start by identifying the toxicity and flammability of the chemicals involved in the change *without* considering any interaction. Nh is the health factor of a material as defined in NFPA 704 [2] and given in NFPA 49 [1] or in Appendix A in DOW's Fire and Explosion Index 7$^{th}$ edition [3]. Nf is the flammability factor of a material as defined in NFPA 704 [2] and given in NFPA 49 [1] or in Appendix A in DOW's Fire and Explosion Index 7$^{th}$ edition [3]. We use Table 1 to calculate what is called a ***Hazard Material Factor***.

**Table 1 – Hazard Material Factor Multiplier (HMF)**

| Description | Toxic Material (Nh=0) | Toxic Material (Nh=1) | Toxic Material (Nh=2) | Toxic Material (Nh=3) | Toxic Material (Nh=4) |
|---|---|---|---|---|---|
| Flammable material (Nf=0) | **1** | **1** | **2** | **3** | **4** |
| Flammable material (Nf=1) | **1** | **1** | **2** | **3** | **4** |
| Flammable material (Nf=2) | **2** | **2** | **3** | **3** | **4** |
| Flammable material (Nf=3) | **3** | **3** | **3** | **4** | **4** |
| Flammable material (Nf=4) | **4** | **4** | **4** | **4** | **5** |

Nh is the health factor of a material as defined in NFPA 704 [2] and given in NFPA 49 [1] or in Appendix A in DOW's Fire and Explosion Index 7$^{th}$ edition [3].

Nf is the flammability factor of a material as defined in NFPA 704 [2] and given in NFPA 49 [1] or in Appendix A in DOW's Fire and Explosion Index 7$^{th}$ edition [3].

If the fluid you are dealing with is a mixture of chemicals, as is typical in a distillation column, then you should use the chemical (minimum 3% concentration) which results in the highest Hazard Material Factor.

If the fluid you are dealing with is a mixture of mostly water then you should consider the concentration of other chemicals in the water and assess what will be in the vapor space. If the hydrocarbon or toxic will be the dominant material in the vapor space it should be the dominant Hazard Material Factor.

If the fluid in question has special reactivity issues (heat or light sensitivity or not covered in NFPA list), they should be considered but are outside the scope of this paper.

### 2.3 *Correction Factor for Incomplete Design*

Small projects may not have complete information about design when the complexity analysis is performed. The following correction factors should be used to adjust for incomplete design:

| | |
|---|---|
| CFID =4 | PFD drawings only |
| CFID = 2.5 | PFD drawings and P&ID's 20-30% complete |
| CFID = 1.5 | PFD drawings and P&ID's 31-60% complete |
| CFID = 1.0 | PFD drawings and P&ID's 60-100% complete |

Note: All P&ID drawings should comply with ANSI/ISA Standard 5.1 (2009) [4]

### 2.4 *Change Complexity Factor*

To determine the Change Complexity Factor (CCF) for the project change we use the following equation:

$$CCF = HMF *( ET + IT + LT + RD)*CFID \qquad \text{[Eq. 1]}$$

IT: Instrument Total (Non Control)

LT: Total # of Control Loops

ET: Total # of Equipment Pieces (Pumps, Columns, Tanks, etc)

RD: Total # of Relief Devices

HMF: Hazard Material Factor (See Table 1)

CFID: Correction Factor Incomplete Design

2.5 *Change Complexity Factor Score*

| | |
|---|---|
| 0 to 25 | one PHA |
| 25 to 50 | two PHAs |
| 51 and above | three PHAs |

See Table 2 for type of PHA required for each CCF.

**Table 2 - Project PHA Type Required**

| **Change Complexity Factor Score** | **# of PHAs Required** | **Type of PHA required** |
|---|---|---|
| 0 to 25 | 1 PHA to be Done | Detailed PHA only |
| 26 to 50 | 2 PHAs to be Done | FEED and Detailed Design PHAs |
| 51 and above | 3 PHAs to be Done | FEED, Early in Detailed Design and Before start of Construction PHA 's |

## 3. Conclusion

Using Change Complexity Analysis it is possible to determine the complexity of projects. This technique will allow Process Hazard Analysis Study Team Leader and Project Managers to determine the minimum number of PHAs to be done for various complexity projects. The technique will effectively utilize project resources and provide needed input to project teams designing a safe process.

## References


[1] NFPA 49: Hazardous Chemical Data, National Fire Protection Association. Technical Committee on Hazardous Chemicals, American National Standards Institute 1994

[2] NFPA 704: Standard System for the Identification of the Hazards of Materials for Emergency Response, National Fire Protection Association. Technical Committee on Classification and Properties of Hazardous Chemical Data, American National Standards Institute 2012

[3] Dow's Fire & Explosion Index Hazard Classification Guide, 7th Edition ,American Institute of Chemical Engineers (AIChE) 1994

[4] ANSI/ISA 5.1: Instrument Symbols and Identification, International Society of Automation, American National Standards Institute 2009

Additional references not cited:

[5] Robert Johnson, Steven W. Rudy, Stephen D. Unwin, "Essential Practices for Managing Chemical Reactivity Hazards," John Wiley & Sons 2010

10TH GLOBAL CONGRESS ON PROCESS SAFETY

# "People" Means Leadership, NOT Simply Mean People -- 4 New Dimensions of Process Safety Competency

**Steven (Long) Zhang**
**ConocoPhillips China**
**Steven.Zhang@conocophillips.com**
**E4B4, East Wing Binhai Finance Zone, TEDA, P. R. China**



Prepared for Presentation at
American Institute of Chemical Engineers
2014 Spring Meeting
10th Global Congress on Process Safety
New Orleans, LA
March 30 – April 2, 2014

UNPUBLISHED

# "People" Means Leadership, NOT Simply Mean People
# -- 4 New Dimensions of Process Safety Competency

**Steven (Long) Zhang**
**ConocoPhillips China**
Steven.Zhang@conocophillips.com
**E4B4, East Wing Binhai Finance Zone, TEDA, P. R. China**

**"Probably the most difficult point to make and make clearly is that not only is culture imposed upon man but it *is* man in a greatly expanded sense."**

**"By broadening our conception of the forces that make up and control our lives, average people can never again be completely caught in the grip of patterned behavior of which they have no awareness."**

**-- Anthropologist Edward T. Hall *The Silent Language***

**Keywords:** Leadership, Process Safety Competency, Plant Procedure People, Core Value

## Abstract

Within this paper, some attempts to rewrite the CCPS RBPS element *Process Safety Competency* with 4 new dimensions and to characterize the newly upgraded Pillar *Leadership* and new Pillar *Weather/climate* are made so that a *Reflection Model of RBPS Egg* could be crafted, to shed light on the current dead small cycle where the process industry and beyond still might have been trapped till very recently, and to contribute to prospect a future potential alive big cycle which is in fact a real start point of rationale for sustainable cultural change. And furthermore, the implications of the *Reflection Model* are broadened in light both of new RBPS Socio-Technical Systems and of a new map of process safety culture, so that a few of important recommendations regarding sustainable culture change and other important issues rationalized in a strictly anthropological sense.

# 1. Introduction

Within the paper presented at 9th GCPS, the author identified 2 new elements/ universes, i.e. RBPS Core Value and Process Safety Cultural Communication (PSCC), for CCPS RBPS Temple or 4 Pillars and recommended an upgraded model, i.e. RBPS Egg. [1]

And at 1st CCPS APCPS, some benchmarking study of RBPS Egg against *The Art of War* was shared: the element *Process Safety Culture* was absorbed into PSCC; 2 underlying potential pillars *Leadership* & *Weather/climate* were identified from within, and instead of, the original CCPS 1st Pillar *Commit to Process Safety*; the author arrived at a RBPS version of *The Art of War,* a new architecture of 7 mega elements (2 universes + 5 pillars), except with a main gap, i.e. lack of correct *Leadership* or the "people" among 3 "P" not meaning people but leadership in fact. The author recommends that the element *Process Safety Competency* should be revised with the 4 new dimensions. [2]

<table>
<tr><th colspan="2">RBPS Egg</th><th>The Art of War</th><th>Philosophy of Process Safety</th><th></th></tr>
<tr><td colspan="4">RBPS Core Value - Tao - RBPS Worldview</td><td>Worldview</td></tr>
<tr><td rowspan="4">Commit to Process Safety</td><td>Process Safety Competency</td><td rowspan="2">Leadership</td><td rowspan="2">People</td><td rowspan="9">Process Safety Cultural Communication<br>(Process Safety Culture)<br>RBPS Methodology</td></tr>
<tr><td>Workforce Involvement</td></tr>
<tr><td>Compliance with Standards</td><td rowspan="2">Weather/Climate</td><td rowspan="3">Plant</td></tr>
<tr><td>Stakeholder Outreach</td></tr>
<tr><td>Understand Hazards & Risk</td><td></td><td>Terrain</td></tr>
<tr><td>Manage Risk</td><td></td><td>Discipline</td><td>Procedure</td></tr>
<tr><td>Learn from Experience</td><td></td><td>Espionage</td><td>Engineering & Humanities</td></tr>
</table>

Upon the previous work, within this paper, the element *Process Safety Competency* is rewritten with the 4 new dimensions so that the new Pillar *Leadership* could be upgraded as strong as the Sun Tzu's one and RBPS Egg becomes real RBPS version of *The Art of War*. The other new Pillar *Weather/climate* and its relationships with *Leadership* and other mega element/ pillars/ universes would be also examined. And the Reflection Model not sheds light on the current dead small cycle, i.e. failure to *Learn from Experience* or *Espionage,* but also is likely to prospect a potential alive big cycle, upon both of which not *Socio-Technical Systems* for RBPS Egg would be characterized but also *a map of Process Safety Culture* presented to process industry and beyond.

## 2. "People" Means Leadership
## -- 4 New Dimensions of Process Safety Competency

### *2.1 "People" Means Leadership in Internal Communication – 4 New Dimensions of Process Safety Competency*

One of the recommendations the author made of benchmarking RBPS Egg against *The Art of War* is that *"People" means Leadership, not simply mean people*, which might have broad implications, especially to the sphere of original CCPS 1$^{st}$ Pillar *Commit to Process Safety* or the current Leadership (Process Safety Competency + Workforce Involvement) & Weather/Climate (Compliance with Standards + Stakeholder Outreach).

**Leadership in Process Safety Cultural Communication** - Within the paper presented at 9$^{th}$ GCPS, the author went through some of the greatest lessons learned and experience within the cultural traditions of ancient East and West and arrived at conclusion with those *4 ethical and highest strategic imperatives* to process safety practitioners, board of directors, executive management, operations manager, and functions of strategy, legal and HR [3], among which some new concepts, which the author borrowed from Historian Arnold J. Toynbee, have also emerged as "creative minority" and "dominant minority" for process safety leadership. Or more correctly, as the author defined, as the ultimate purpose of PSCC is to convey the message of RBPS Core Value, the correct process safety leadership should be a joint leadership of dominant minority & creative minority in PSCC, allied with corporate functions of strategy, legal & HR.

**A Shorter Chain of Command & Inherently Safer Operational Discipline** - And as the author has commented, RBPS Core Value and PSCC mandate a shorter chain of command directly and primarily from RBPS Core Value rather than from superiors. In another word, the ultimate purpose of the superiors is to support to cultivate and facilitate this new type and shorter chain of command of their subordinates, which is in fact to mentor their subordinates (or their new colleagues) of conveying the message of RBPS Core Value to the latter themselves and others. And some of the subordinates may be likely to become creative minority and ceaselessly remind, communicate or challenge upward the actual status of this desired shorter chain of command, even to the highest level when necessary, till all the superiors' behavior universe sends same messages as tamed as from their verbal universe. And the matches of dominant minority and creative minority might be called inherently safer PSCC leadership, which is in fact in the right direction along PSCC leadership enlightenment roadmap, of which the destination would probably be the *Brotherhood of Bian Que or called "powerful powerless" dominant minority with the "powerless powerful" creative minority*[4].

And within this shorter chain of command and as illustrated within the scheme of the author formally defined Philosophy of Process Safety [5], we also could recognize that RBPS Core Value dictates PSCC, and PSCC dictates PSCC leadership (People), Plant and Procedure. In another word, the role the 2 universes of RBPS Core Value and PSCC jointly play within RBPS Egg is actually of inherently safer operational discipline (ISOD) in a broader sense. And only through RBPS Core Value and PSCC could correct decision making and correct behavior determining be reliably "manufactured" and guiding, organizing and accumulating with other non-ethical building raw materials into the People, Plant & Procedure.

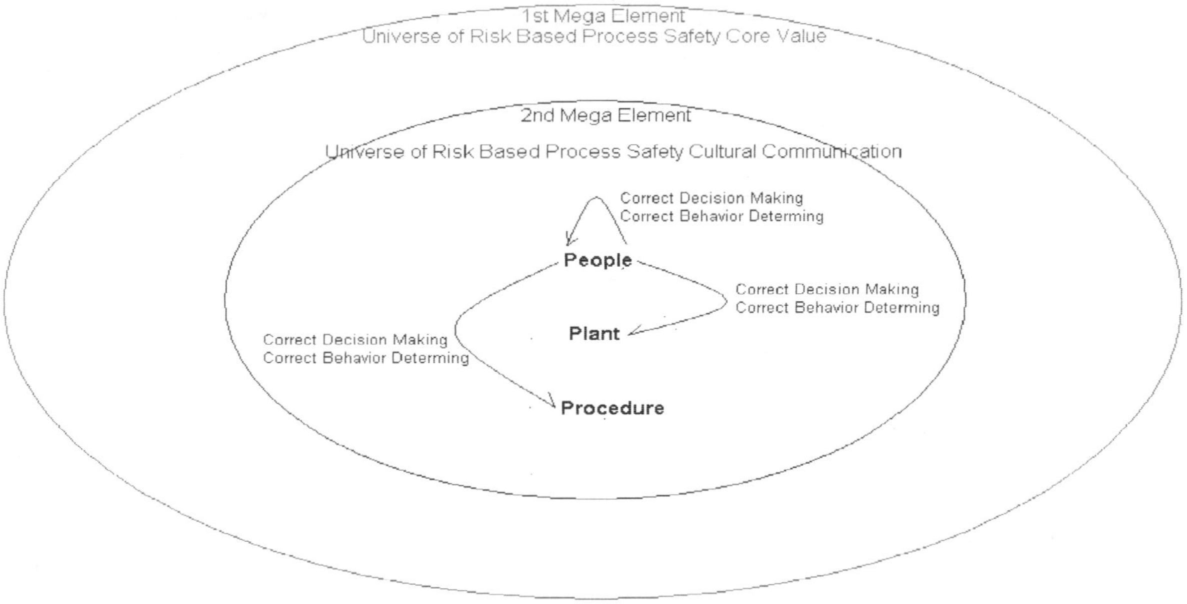


**4 Ethical Imperatives as PSCC Contract** [6] - In retrospect, we might currently find some content within the Baker Report's Recommendation #1 and its commentary not much correct as before. In author's opinion, the Board of Directors, executive management and other members of the corporate management might hardly be able to provide effective leadership or communication leadership until RBPS Core Value could be resiliently documented throughout their collective verbal and behavior universe in light of the new game code or PSCC contract, i.e. those *4 ethical & highest strategic imperatives*. In another word, the PSCC partner "creative minority" among the workforce may get empowered from RBPS Core Value in personal commitment and professional ethics [7] to challenge upward, even to highest level of the organization, the pitfalls and truthfulness of the verbal and behavioral universes of the "dominant minority", who expectedly to be prepared psychologically and rationally for getting empowered to be "powerless", i.e. to obey RBPS Core Value through completely giving up their personal power and authority in deep integrity throughout their verbal and behavioral universes. So some content of the recommendation #1 and its commentary might be revised or added as below [8]:

Process Safety Leadership in fact means PSCC Leadership, not dominant minority but also creative minority;
Dominant minority's leadership could never be effective except to cooperate with creative minority;
*4 ethical and highest strategic imperatives* are the new game code or PSCC contract;
Creative minority gets empowered to upward communicate and dominant minority empowered to obey;
Dominant minority set the "tone at the top" and RBPS Core Value set above the top of the tone;
Dominant minority and creative minority are the inherently safer concepts of PSCC leadership;
……

**4 New Dimensions of Process Safety Competency or PSCC Competency** - Along with the new game code or PSCC contract and the new defined game players, let's examine the required competency of the game players. Sun Tzu defined *Leadership* (Roger T. Ames put it here as *command*) as a matter of wisdom, integrity, humanity, courage and discipline, all of which seems functioning as the military process safety competency of the organization to pursue and convey the message of the Tao. In context of the globalization and global operations, however, Sun Tzu's *Leadership* may be upgraded with the following 4 new dimensions of process safety competency of efficiently facing the challenges of workforce increasing diversity and operational discipline's higher cultural and communication sensitivity [9]:

*Competency in embracing RBPS Core Value;*
*Competency in upward Process Safety Cultural Communication;*
*Competency in downward Process Safety Cultural Communication; and*
*Competency in process safety leadership planning, assessment and intervention*

Theoretically, these 4 new dimensions of process safety competency or PSCC Competency are very correct but in fact less measureable as these 4 ones inherently relate to inner ourselves, the "natural" resources of the professional ethics and personal commitment across the organization, the former of which seems functioning as soil as well as sunshine and the latter seeds.

**"People" means PSCC Leadership & Internal Communication –** Based on above analysis, some following conclusions might be arrived: **Process Safety Leadership is in fact the leadership in Process Safety Cultural Communication; those 4 new dimensions of Process Safety Competency are actually of Process Safety Cultural Communication; RBPS Core Value and Process Safety Cultural Communication mandate a shorter chain of command which substantively functions as an Inherent Safer Operational Discipline (ISOD) in a broader sense; and those 4 ethical & highest strategic imperatives make up Process Safety Cultural Communication Contract.**

### *2.2 Stakeholder Outreach – Extension of Internal Communication*

Another important recommendation the author made of benchmarking RBPS Egg against *The Art of War* should be that "Plant" as small Nature consists of Weather/Climate and Terrain, which seems slightly abstract but is in fact meaningful that our understanding of the "Plant" as well as of "Procedure" have been elevated to a philosophical level, where we might not be able to disagree with currently popular sayings of the contemporary philosophers of technology that our "Plant" is an social artifact rather than pure engineering product[10], i.e. both Weather/Climate and Terrain are of an artifact and social process influenced product. So is "Procedure".

**Without ISOD, What "People, Plant & Procedure" would look like? –** As the scheme within section 2.1 and the form in section 1 jointly illustrate, incorrect Leadership in PSCC would reliably bring about incorrect decision making and incorrect behavior determining across internal communication, external communication, terrain, discipline; and similarly maintaining blindness of humanities espionage[11]. In other word, incorrect Leadership dictates incorrect "People, Plant & Procedure", whereas correct Leadership otherwise with regards "People" (internal communication), Weather/Climate, Terrain, Discipline as well as Espionage.

**External PSCC - Weather/Climate Allied with Terrain & Discipline**

Correct Leadership in PSCC dictates correct "People, Plant & Procedure" through correct decision making and correct behavior determining. And the correct Terrain and Discipline consistently convey the message of RBPS Core Value to Weather/Climate that *there is no power, authority or governance on process safety except on behalf on Process Safety Expertise.* Without receiving correct message from Terrain & Discipline, however, Weather/Climate only could do external communication but not PSCC. And put another way, Weather/Climate only functions as verbal communication, whereas the performance of Terrain and Discipline as the behavior universe, which might convey another message that is very different from the verbal one. In this sense, incorrect Leadership dictates incorrect decision making and behavior determining of Weather/Climate so that Weather/Climate could not be honest enough and eventually failed to convey the message of RBPS Core Value to the public through its verbal universe as well as the behavior universe of Terrain & Discipline.

And simultaneously, Weather/Climate conveys incorrect message back to Terrain and Discipline so that Terrain and Discipline are permanently exposed within the sea of incorrect messages. Marshall McLuhan, one of the greatest thinkers of 20 century, has a popular saying: the medium is the message [12]. The allied media of incorrect Leadership & incorrect Weather/Climate etc completely surrounds, and conveys incorrect messages to, the Terrain & Discipline who inevitably get corrupted and witnessed with one another within the sea of incorrect messages. And the day to day mild Weather/Climate public communication universe would suddenly become fraudulence when the actual message of the wrong behavior of the Terrain and Discipline is perceived by the community. And for the fact that the organization successfully conveyed the message of the wrong core value through the bad behavior of the operations or performance of Terrain and Discipline, the community especially the regulator would "anthropologically" convey message of the community's correct core value to the "unethical citizen" through force measures, i.e. extreme Weather/Climate suddenly arrives.

So effective management of Weather/Climate should ultimately counted on the inherently safer Leadership (correct Leadership in PSCC) and its media the sea of correct messages rather than on Weather/Climate expertise in isolation such as communication initiatives, tools and skills. And interestingly, the author's opinion seems having resonance with some of the catastrophic incident warning signs of CCPS regarding leadership poor behavior on interrelationships between Leadership (internal communication), Weather/Climate (external communication), Terrain and Discipline [13]. Either way, Weather/Climate really does have something critical to do with the status of Leadership's actual belief in RBPS Core Value and of their trust and rationale in PSCC.

In context of those 4 ethical and highest strategic imperatives, the author recommends: external public communication functions should involve into and learn from the internal communication or internal PSCC; and dominant minority should consult with creative minority for the correct decision making and behavior determining throughout the activity universe of Weather/Climate, i.e. Weather/ Climate, allied with Terrain and Discipline, shall be of PSCC. And the failure in PSCC throughout Weather/Climate is the extension of the failure of Leadership (internal communication). Both of them often triggers extreme weather/climate, i.e. the community might have to do their "tribal" PSCC to the bad "citizens" to force them to be ethic enough again. So in the author's opinion, **the ultimate purpose of stakeholder outreach should be to lead the community to the true enlightenment of process safety, i.e. to embrace, and convey the message of, RBPS Core Value, rather than only tactically maintaining stable relationships with stakeholders.**

# 3. Socio-Technical Systems of RBPS Egg -- Leadership Media Universe

## *3.1 RBPS Reflection Model – Dead Small Cycle & Alive Big Cycle*

**Sun Tzu's Legacy of Espionage –** The objects Sun Tzu's Espionage is made to are: Tao (RBPS Core Value), Weather/Climate, Terrain, Leadership and Discipline.[14] And the author added another one i.e. PSCC, which is in fact an advanced form of process safety culture and whose ultimate purpose is to convey the message of RBPS Core Value. So totally, we have 6 strategic assessment criteria/facets, along with Espionage, within the RBPS Egg.

As per Traditional Chinese Medical Methodology, the 6 criteria/facets might be categorized into 2 groups: the group of symptoms of Weather/Climate, Terrain and Discipline, and the group of roots of RBPS Core Value, PSCC and Leadership.

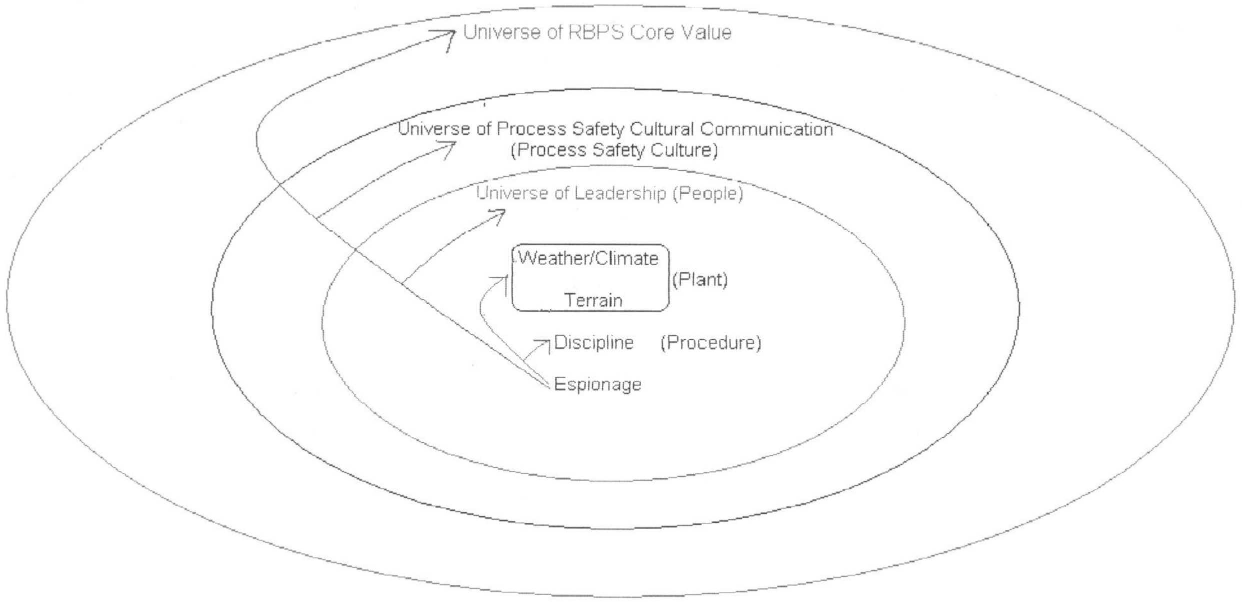


**Symptoms (Syndromes) - Small Dead Cycle Locked within Universe of Leadership -** Whenever creative minority is absent, the dominant minority might probably become 3rd universe which, however, functions as a prison and isolates whatever within from the other 2 universes of RBPS Core Value and PSCC as well as from Leadership itself. And in this manner only could Weather/Climate, Terrain and Discipline be reflected upon in isolation at the level of symptoms so that Weather/Climate is probably not to be deemed as a contributing factor to the catastrophic accidents, but rather, as a post accident remediation measures for the damaged public image and reputation. In another words, in light of the Reflection Small Dead Cycle, Weather/Climate has nothing to do with Espionage and hence its 4 sub-elements i.e. Incident Investigation, Measurements & Metrics, Auditing and Management Review.[15] And the author captures within the form below the inter-relationships mentioned above between Espionage and the symptoms group of Weather/Climate, Terrain and Discipline.

| | **Espionage (Learn from Experience) – Reflection Small Dead Cycle** | | | |
|---|---|---|---|---|
| | **Incident Investigation** | **Measurements & Metrics** | **Auditing** | **Management Review** |
| **Weather/ Climate** | If Weather/Climate is examined as a contributing factor to incident? NO | If Weather/Climate is an item within process safety KPI? NO | If Weather/Climate is an item within auditing? NO | If Weather/Climate is an item within management review? NO |
| **Terrain** | If Terrain is examined as a contributing factor to incident? YES | If Terrain is an item within process safety KPI? YES | If Terrain is an item within auditing? YES | If Terrain is an item within management review? YES |
| **Discipline** | If Discipline is examined as a contributing factor to incident? YES | If Discipline is an item within process safety KPI? YES | If Discipline is an item within auditing? YES | If Discipline is an item within management review? YES |

From the form above, we might recognize that, in context of lack of big alive cycle, the rationale only could be to solely count on the improvement of the current practices and development of new practices within the scope of Terrain and Discipline. In author's opinion, however, Terrain and Discipline function together as "a car", which might get increasingly complicated and complex, whatever extent. But we need to ask other questions: How about the "driver"? Does the driver have a valid "driving license"? Has the driver passed an "alcohol test"?

**Roots (Diseases) - Big Alive Cycle as Espionage through 3 Universes -** And whenever creative minority is active enough in leading PSCC, the wall of the prison of dominant minority would collapse and the organization would succeed in Espionage into the 3 universes as well as subsequently into Weather/Climate, Terrain and Discipline. And due to the nature of Weather/Climate, it might immigrate from small dead cycle to big alive cycle so that Terrain and discipline make up the symptoms group "the car" and RBPS Core Value, PSCC, Leadership and Weather/Climate come together as roots group "the driver". And in this scenario, as captured within the form below, Weather/Climate does have something important to do with Espionage and hence with its 4 sub-elements, and so do RBPS Core Value, PSCC and Leadership. So the rationale becomes of classic Traditional Chinese Medical Methodology, i.e. "both symptoms and roots should be treated, and symptoms only could be eliminated through treatment of roots." In another word, whenever a car knocks upon some people, we need to ask not "what's wrong with the car?", but also "what's wrong with the driver?"

| | Espionage (Learn from Experience) – Reflection Big Alive Cycle | | | |
|---|---|---|---|---|
| | **Incident Investigation** | **Measurements & Metrics** | **Auditing** | **Management Review** |
| **RBPS C.V.** | If RBPS Core Value is examined as a contributing factor to incident? **YES** | If RBPS Core Value is an item within process safety KPI? **YES** | If RBPS Core Value is an item within auditing? **YES** | If RBPS Core Value is an item within management review? **YES** |
| **P.S.C.C. *** | If PSCC is examined as a contributing factor to incident? **YES** | If PSCC is an item within process safety KPI? **YES** | If PSCC is an item within auditing? **YES** | If PSCC is an item within management review? **YES** |
| **Leadership** | If Leadership is examined as a contributing factor to incident? **YES** | If Leadership is an item within process safety KPI? **YES** | If Leadership is an item within auditing? **YES** | If Leadership is an item within management review? **YES** |
| **Weather/Climate** | If Weather/Climate is examined as a contributing factor to incident? **YES** | If Weather/Climate is an item within process safety KPI? **YES** | If Weather/Climate is an item within auditing? **YES** | If Weather/Climate is an item within management review? **YES** |

* P.S.C.C. is the author's addition in light of anthropology and intercultural communication

So the author might say to the process industry and beyond that Sun Tzu's legacy of Espionage has been rediscovered completely through the Reflection Model of RBPS Egg, within which there are 2 symptoms i.e. Terrain and Discipline and 4 roots i.e. RBPS Core value, PSCC, Leadership and Weather/Climate; or otherwise, **there are one symptom with 2 dimensions of Terrain and Discipline and one root with 4 dimensions of RBPS Core value, PSCC, Leadership and Weather/Climate. And Espionage might function as of Traditional Chinese Medical Diagnosis.**

### *3.2 RBPS Socio-Technical Systems – Leadership Media Universe*

Hinted by the topic or perspective of socio-technical systems' failure for catastrophic accidents[16], the author crafts the following RBPS Socio-Technical systems based on RBPS Egg of 7 mega elements (22 sub-elements). And due to the critical role, as the author analyzed above, Leadership the 3rd Universe would play in isolating whatever within from the rest 2 universes as well as from Leadership itself and in closing the big cycle and keeping dead small cycle "work", the author prefers to call the Socio-Technical Systems a Leadership Media Universe. In another word, different Leaderships, or more exactly different Leadership in PSCC, inherently embrace different approaches i.e. either "sole symptom approach" or "symptom-root double approaches". Dominant minority usually favorites on the former, whereas creative minority often holdfasts the latter. And accordingly, they might have polarizing behavioral patterns toward the actual relationships among RBPS Core value, PSCC, Leadership and

Weather/Climate. As we all know, however, RBPS Core Value dictates PSCC; PSCC dictates leadership (dominant minority and creative minority); and leadership dictates Weather/Climate as well as Terrain, Discipline & Espionage.

And obviously, the "sole symptom approach" concerns all the interactions among these 3 mega elements of Espionage, Terrain and Discipline. And all these interactions in fact might have been or are being effectively studied, implemented and leveraged to some extent within and outside the corporate, e.g. with regards Centers of Excellence in asset integrity and in other sub-elements across these 3 mega elements.[17] By and large, the "symptom approach" could be called Process Safety Techno Cultural Communication (PSTCC). And also the "symptom-root double approaches" have actually been defined long earlier by the author, and its name is Process Safety Cultural Communication (PSCC), which consists not only of PSTCC but also of Process Safety Socio Cultural Communication (PSSCC).

Unlike PSTCC which could be acted upon and leveraged effectively to some extent within specific area of media such as Centers of Excellence within corporate, the media for PSSCC is still emerging and to evolve. As the author quoted the popular saying of Marshall McLuhan: "the medium is the message", however, in light of those 4 ethical and highest strategic imperatives, dominant minority such as board of directors, executive management, along with functions of strategy, legal & HR etc, shall explicitly determine what the ultimate purpose /message of their PSSCC media is, or if they are determined to commit faithfully to the PSCC contract.

And it is the ultimate purpose /message of the PSSCC media (or PSCC media) that the board of directors, executive management either explicitly or implicitly testify to, through their verbal and behavior universe organizationally and psychologically aligned with functions of strategy, legal & HR etc, would inherently correlate to the future failure or safe of the RBPS Socio-Technical Systems – the Leadership Media Universe. Or the author might borrow from the concept of the name of McLuhan's masterpiece *Understanding Media: The Extensions of Man* and make saying like this: Leadership Media Universe is to be pursued as the "extensions" of not only dominant minority but creative minority. And in the author's opinion, Leadership Media Universe is in fact an acronym for Process Safety Culture. So the author goes further saying: **Process Safety Culture is not only the "extensions" of dominant minority but also of creative minority. And sure, "performance contract system and other incentive-based compensation plans", probably even the functions of strategy, legal, HR as well as of external communication, remain critical part of the extensions of dominant minority, whereas "decisions about corporate level initiatives, operations, financial performance, resource allocation, capital project, personnel changes etc" the main themes of the "media", either of dominant minority alone or of both.**[18]

And we could say: **the "media" of the 7 mega elements is the "message" of the actual core value, either of RBPS or not at all. And the sustainable change of Leadership Media Universe only initiates when the people at the highest level, probably all board members & executive management, one by one, testify verbally, and subsequently through behavioral universe, to the RBPS Core Value and those 4 ethical and highest strategic imperatives are faithfully committed by all the dominant minority, creative minority and functions of strategy, legal & HR etc.**

| RBPS Socio-Technical Systems - Leadership Media Universe | | | | | | | | | | | | | | |
|---|---|---|---|---|---|---|---|---|---|---|---|---|---|---|
| | **PS Socio Cultural Communication (PSSCC)** Sub-Socio System (7 Elements + Espionage 4 Elements) | | | | | | | | | | | | | |
| | | | | | | | | **PS Techno Cultural Communication (PSTCC)** (Espionage 4 Elements + 11 Elements) Sub-Technical System | | | | | | |
| 22 Sub-elements | RBPS Core Value | Process Safety Cultural Communication | (Process Safety Culture) | Process Safety Competency with 4 new dimensions | Workforce Involvement | Compliance with Standards | Stakeholder Outreach | Incident Investigation | Measurement & Metrics | Auditing | Management Review & Continuous Improvement | Process Knowledge Management | Hazard identification & Risk Analysis | 9 Sub-elements (in light of organizational effectiveness, *inherently safer chain of command* & logistical support) |
| 7 Mega Elements | Tao/ PS Worldview | P.S. Methodology | | Leadership | | Weather/ Climate | | Espionage (RBPS 4th Pillar Learn from Experience) | | | | Terrain (RBPS 2nd Pillar Understand Hazards & Risk) | | Discipline (RBPS 3rd Pillar Manage Risk) |
| CTM* | One Root (4 dimensions) | | | | | | | Traditional Chinese Medical Diagnosis | | | | One Symptom (2 dimensions) | | |
| AIC** | Formal Culture | | | | | | | Informal Culture | | | | Technical Culture | | |

CTM*: Chinese Traditional Medicine
AIC**: Anthropology & Intercultural Communication

### *3.3 Desert & Oasis in Ourselves – A Map of Process Safety Culture*

In order to get more rationale, based on the newly updated pillar Leadership and borrowing anthropologist Edward T. Hall's approach, the author draws a map of process safety culture or called a map of process safety cultural communication, upon which we may be able to estimate how much infra cultural activity of process safety culture still remain desert and how much has already become oasis.

Beside Sun Tzu called "5 strategic assessment criteria" and Clausewitz put it "elements of strategy"[19], the author prefers to treat these 7 mega elements as 7 Primary Message Systems as anthropologist Edward T. Hall did for the culture study so that a similar two-dimensional grid could be constructed as below [20]. And in this manner, we could see there are totally possible 49 infra (strategic) cultural activities resulting from the various combination of the 7 mega elements. And the author defines each infra (strategic) cultural activity as the critical verbal statement & behaviors related to the relationship between mega element A and mega element B. And in special cases, i.e. these 7 ones in the diagonal from upper left to lower right are in fact the critical verbal statement & behaviors related to the definition of the mega element concerned. And for the purpose of simplicity, all the 21 infra (strategic) cultural activities above the diagonal which are exactly the same with the activities collection below the diagonal, along with these 7 ones in the diagonal, would make up the map of process safety culture which the author likes more to call the map of process safety cultural communication.

| | RBPS Core Value 0 | P.S.C.C. (Process Safety Culture) 1 | Leadership 2 | Weather/ Climate 3 | Espionage 4 | Terrain 5 | Discipline 6 |
|---|---|---|---|---|---|---|---|
| RBPS Core Value 0 | 00 | 01 | 02 | 03 | 04 | 05 | 06 |
| P.S.C.C. (Process Safety Culture) 1 | 10 | 11 | 12 | 13 | 14 | 15 | 16 |
| Leadership 2 | 20 | 21 | Leadership with 4 new dimensions 22 | 23 | 24 | 25 | 26 |
| Weather/ Climate 3 | 30 | 31 | 32 | 33 | 34 | 35 | 36 |
| Espionage 4 | 40 | 41 | 42 | 43 | 44 | 45 | 46 |
| Terrain 5 | 50 | 51 | 52 | 53 | 54 | 55 | 56 |
| Discipline 6 | 60 | 61 | 62 | 63 | 64 | 65 | 66 |

A Map of Process Safety Cultural Communication (PSCC)

As we analyzed above, whenever creative minority is absent in PSCC leadership, the leadership would probably become a prison isolating whatever within from the 2 universes of RBPS Core Value and PSCC as well as from Leadership itself so that all the 18 infra (strategic) cultural activities in yellow background would remain dry desert area of the process safety culture. And similarly as we discussed earlier, incorrect leadership also dictates strategic failure bonded Weather/Climate so that those 4 infra (strategic) cultural activities in blue background (33, 34, 35 & 36) remain dry desert of the process safety culture as well. And furthermore, we found the Espionage is solely restricted to the One Symptom with 2 dimensions of Terrain and Discipline. So due to lack of Espionage to the One Root with 4 dimensions, the actual infra (strategic) cultural activity (44) is in fact a half desert of the health area unit. **Totally and approximately, the percentage of the desert area of the current actual industry average process safety culture should be: (18+4+0.5)/28 = 80.4%, whereas the actual oasis 19.6%.**

These statistics of the actual situation of our process safety culture seems a shock and hard to be accepted. Anthropologist Edward T. Hall not only already mentioned this kind of shock and resistance as saying "... people as cultural being are bound by hidden rules and not the masters of their fate ..." within his masterpiece *The Silent Language,* but also reminded that people would become really bound if "they remain ignorant of the nature of the hidden pathways culture provided for them" [21]. And similarly, for our process safety culture, there also have been hidden pathways i.e. RBPS Core Value, PSCC and 4 new dimensions of process safety competency, all of which we could not afford to remain ignorant anymore.

### *3.4 Governance on the Root – Coaching Fishing Better Than Giving Fish*

As anthropologist Edward T. Hall defined, culture has three levels, i.e. formal, informal and technical. Formal culture is always learned through mistake and correction, with a clear right-wrong character. Technical one is also learned through mistake and correction, but being offered with reasons. And informal culture, however, is learned without knowledge of occurrence of the learning process at all.[22] The author innovatively matches the 3 levels of culture with One Root, Diagnosis & One Symptom so that it become manifest that those 4 ethical & highest strategic imperatives would supposedly function as the core principles of the formal culture (One Root with 4 dimensions) and shall be faithfully committed by dominant minority, creative minority and functions of strategy, legal & HR etc. And the rest whole organization psychologically holds the same intentionality as well.

In the author's former opinion, PSCC functions as the impulses from the heart RBPS Core Value. And the impulses consistently transport the nourishment of the highest design principle i.e. RBPS Core Value across the tree of process safety culture so that RBPS Core value solely chairs and dictates correct decision making & correct behavior determining which would guide, organize, and accumulate with, other non-ethical raw materials building and maintaining the rest mega elements i.e. leadership, Weather/Climate, Terrain, Discipline and even Espionage. Put another way, without PSCC, some enemy value or ideology may share the exclusiveness of chairing decision making and behavior determining with RBPS Core Value so that incorrect decision making and incorrect behavior determining are produced and accumulated consistently within last 5 mega elements till, and ever after, the collapse of the "tomb" or what within the 2 universes.[23]

There was an ancient Chinese proverb saying: coaching fishing is better than giving fish. And the author would re-examine in light of the fishing-fish relationship the two accidents CCPS formally used as examples in the 2 elements: asset integrity and readiness.

a. The accident investigation of DC-10 crush on May 25, 1979 concluded that "...the original replacement procedure calls for removal of the engine prior to the removal of engine pylon. To save time and money, the airline instructed its mechanics to remove" them together ...[24]

In the author's opinion, giving up upward PSCC equates, even worse than, giving up fishing the sole livelihood, being bound to starve to death, sooner or later.

b. That major leak of flammable gas and the subsequent fire did occur, and we found that "the startup checks for this equipment had not been performed as a part of overall turnaround , because this particular piece of equipment was not go to be used immediately."[25] The author challenges the logic behind that "because", which was probably a natural mindset and patterned behavior after giving up fishing the sole livelihood or even not attempted to it at all. During that moment of startup check, the accident might have been prevented should the organization concerned have been given with some specific outsourced recommendations toward improvement or reconstruction of the organizational effectiveness and the logistical structure. But could being given with some fish be strategically meaningful for the life of the fishing man who has given up permanently his sole livelihood?

<table>
<tr><th rowspan="2"></th><th rowspan="2"></th><th colspan="3">Sun Tzu's 3 Aspects of Discipline</th><th>RBPS Core Value</th><th>Process Safety Cultural Communication (PSCC)</th><th>Leadership</th><th>Weather / Climate</th></tr>
<tr><th>Organiza-tional Effective-ness</th><th>A Chain of Command</th><th>A Structure for Logistical Support</th><th colspan="4">Formal Culture<br>(4 ethical imperatives as Shorter Chain of Command)</th></tr>
<tr><td rowspan="2">Terrain*</td><td>Knowledge</td><td rowspan="11">Of social construction in context of RBPS Core Value & PSCC (RBPS Core Value dictates correct decision making and correct behavior determining which reliably accumulates within the construction process of Terrain & Discipline)</td><td rowspan="11">To update to Shorter Chain of Command [26], as illustrated at right, primarily and Directly from Tao (RBPS Core Value), rather than from superiors; 4 Ethical & Highest Strategic Imperatives as PSCC contract</td><td rowspan="11">Of social construction in context of RBPS Core Value & PSCC (RBPS Core Value dictates correct decision making and correct behavior determining which reliably accumulates within the construction process of Terrain & Discipline)</td><td rowspan="11">Learn, understand, embrace, testify and communicate of RBPS Core Value through verbal and behavioral universes.</td><td rowspan="11">Learn, understand and communicate that the ultimate purpose of PSCC is to convey the message of RBPS Core Value; what are upward PSCC & downward PSCC; those 4 ethical and highest strategic imperatives</td><td rowspan="11">Learn, understand & communicate of creative minority, dominant minority; Bian Que-Duke Cai, Moses-Pharaoh, Sheba Queen, Solomon and brotherhood of Bia Que; [27]</td><td rowspan="11">Weather/Climate consistent honestly with Terrain & Discipline; ultimately to lead stakeholders and community to the enlightenment of RBPS Core Value</td></tr>
<tr><td>Risk</td></tr>
<tr><td rowspan="9">Discipline (Manage Risk) - 9 Sub-Elements</td><td>Procedure</td></tr>
<tr><td>safe work</td></tr>
<tr><td>asset integrity</td></tr>
<tr><td>Contractors</td></tr>
<tr><td>Training</td></tr>
<tr><td>MOC</td></tr>
<tr><td>Readiness</td></tr>
<tr><td>Operations</td></tr>
<tr><td>Emergency</td></tr>
</table>

*: Sun Tzu's 3 aspects, which were originally specific to Sun Tzu's Discipline, are extended by the author to Terrain.

### *3.5 Recommendations for Sustainable Cultural Change*

To creative minority, dominant minority, functions of strategy, legal and HR etc, the author makes the following recommendations:

a. PSCC is the advanced form of Process Safety Culture, and Process Safety Culture's future is PSCC. And the Map of Process Safety Culture (PSCC) might become a paradigm for Process Safety Culture assessment & sustainable change;

b. The RBPS Reflection Model might prove to be a profound framework for Espionage (incident investigation, metrics, auditing & management review) within which the critical role Leadership (internal communication) would play as the 3rd universe is manifested for both small cycle and big cycle;

c. As RBPS Socio-Technical Systems shows, PSCC consists of both PSTCC (Process Safety Technical cultural Communication) and PSSCC (Process Safety Socio Cultural Communication). And the ultimate purpose of PSCC would be fulfilled mainly through PSSCC. And PSSCC, PSCC and Leadership Media Universe are guided by those 4 ethical & highest strategic imperatives, which shall be faithfully committed by creative minority, dominant minority, and functions of strategy, legal & HR. And Process Safety leadership shall be the leadership in PSCC or PSSCC, and Leadership Media Universe is in fact an acronym for process safety culture;

d. The ultimate purpose of Weather/Climate as part of Leadership Media Universe is to convey the message of RBPS Core Value and lead stakeholders and community to the enlightenment of RBPS Core Value;

e. The health of RBPS Egg, or the actual percentage of desert of the map of process safety culture, is concerned more with messages of actual value transferring among mega elements rather than network and control system. And this concept may be a fundamental challenge to the current popular ones e.g. Bow-Tie & LOPA which inherently have far less relevance with Process Safety Culture or PSCC or RBPS Core Value;

f. Same as culture, process safety culture (Leadership Media Universe) also consists of Formal Culture (One Root with 4 dimensions), Informal Culture (TCM diagnosis) and Technical Culture (One Symptom with 2 dimensions);

g. Couching fishing (PSCC or PSSCC) is better than giving fish (PSTCC). And the true sustainable cultural change only set out the journey with couching fishing (PSCC or PSSCC). PSCC or PSSCC is a growing sphere of communication context, which is governed and dictated by RBPS Core Value as the highest design principle and in which the overall quality of the correct organizational decision making and correct behavior determining could be reliably achieved; and

h. Anthropologist Edward T. Hall reminded us that "the study of institutions and their structures and the study of individuals and their psychological makeup" should be included in study of culture on a higher organizational level. [28] And Hofstede went further with "institutions" within his masterpiece *Cultural Consequences.* [29] The author has to be honest enough to admit that Process Safety Culture is inherently concerned with institutions rather than otherwise. And the content within right side (Formal Culture) of the form within the section 3.4 is in fact a curriculum or competency map for creative minority, dominant minority, functions of strategy, legal and HR etc, to substantively set out a journey of sustainable culture change.

## 4. Conclusion

At the beginning of this paper, the new mega element/ pillar Leadership was upgraded with those 4 new dimensions and newly characterized as leadership in internal communication or in PSCC. And then the mega element Weather/Climate, its interaction with other mega elements as well as it ultimate purpose was explored and examined.

And based on the analysis and characterization and other previous studies, a RBPS Reflection Model was constructed so that we could see how critical the role Leadership (in internal communication or PSCC) actually plays is in locking organizational rationale within the small cycle or otherwise in unlocking rationale through the big cycle.

And this mechanism was further interpreted through not only newly crafted RBPS Socio-Technical Systems but also an unprecedented map of process safety culture or PSCC.

At the end, several important recommendations toward true sustainable culture change were made.

## 5. References


[1] Zhang, Long, *An Upgraded Definition of Risk Based Process Safety Core Value and Its Impact in Light of Anthropology & Intercultural Communication on the Future Evolution of Process Safety Culture & Management*, 2013 AIChE Spring Meeting and 9th Global Congress on Process Safety, San Antonio, April 2013

[2] Zhang, Long, *Benchmarking Risk Based Process Safety Egg (RBPS Temple+2 Universes) against The Art of War*, 1st CCPS Asia-Pacific Conference on Process Safety, Qingdao, Sept 2013

[3] Zhang, *An Upgraded..., ibid.*

[4] Zhang, *An Upgraded..., ibid.*

[5] Zhang, *Benchmarking..., ibid.*

[6] Schramm, Wilbur & Porter, William E., *Men, Women, Messages, and Media: Understanding Human Communication*, Peking University Press, 2007, pp.42-45.

[7] Martin, Mike W., *Meaningful Work: Rethinking Professional Ethics*, Oxford University Press, 2000, pp. vii-xiii.

[8] B.P. U.S. Refineries Independent Safety Review Panel, *The Report of the B.P. U.S. Refineries Independent Safety Review Panel*, 2007, p.244.

[9] Zhang, *Benchmarking..., ibid.*

[10] Bucciarelli, Louis L., *An ethnographic perspective on engineering design*, Design Studies, Volume 9, Issue 3, Jul 1988, pp.159-168.

[11] Zhang, *Benchmarking..., ibid.*

[12] McLuhan, Marshall, *Understanding Media: The Extensions of Man*, MIT Press, 1994

[13] Center for Chemical Process Safety (CCPS), *Recognizing Catastrophic Incident Warning Signs in the Process Industries,* John Wiley & Sons, Inc., 2012, pp.40-41.

[14] Sun-Tzu and Ames, Roger T., *The Art of Warfare,* Ballantine Books, 1993, p.103.

[15] Center for Chemical Process Safety (CCPS), *Guidelines for Risk Based Process Safety,* John Wiley & Sons, Inc., 2007, pp.549-647.

[16] Hurst, Nick W., *Risk Assessment: The Human Dimension*, Royal Society of Chemistry, 1998, pp.25-26.

[17] CCPS, *Guidelines for Risk Based Process Safety,* John Wiley & Sons, Inc., 2007, p.319.

[18] B.P. U.S. Refineries Independent Safety Review Panel, *ibid*, p.244.

[19] Clausewitz, Carl von, *On War*, Wordsworth Editions Limited, 1997, pp.149-150.

[20] Hall, Edward T., *The Silent Language,* Anchor Books, 1990, pp.188-194.

[21] *ibid*, p.117.

[22] *ibid*, pp.67-68.

[23] Zhang, *An Upgraded..., ibid.*

[24] CCPS, *Guidelines for Risk Based Process Safety,* John Wiley & Sons, Inc., 2007, p.317.

[25] CCPS, *Guidelines for Risk Based Process Safety,* John Wiley & Sons, Inc., 2007, p.449.

[26] Zhang, *Benchmarking..., ibid.*

[27] Zhang, *An Upgraded..., ibid.*

[28] Hall, *ibid.* p.186.

[29] Hofstede, Geert, *Culture's Consequences: Comparing Values, Behaviors, Institutions & Organizations Across Nations,* Shanghai Foreign Language Education Press, 2008

# 346632 Evaluating The Need For Depressuring Systems - A Methodology

**Abstract:**

**Evaluating the Need for Depressuring Systems - A Methodology**

A thorough Overpressure Protection Philosophy should consider all options for overpressure protection, and should not be limited to relief devices and rupture disks - they are not the only line of defense.

Another effective means of overpressure protection are depressuring systems. These can be used to mitigate against vessel leaks by reducing inventory or leakage rate; or to mitigate against the effects of external fire.

ioMosaic recently conducted a site-wide overpressure protection study for a large LNG production facility. Part of this scope involved assessing equipment items which did not have a depressuring system installed, and determining if a depressuring system was warranted, in addition to the equipment's other overpressure protection.

This paper outlines the methodology used to determine the need for a depressuring system, and also highlights some of the issues to be aware of when designing a depressuring system.

# 347261 The Capability-Demand Gap In US Refining and Petrochemical Console Operations

**Abstract:**

**The Capability-Demand Gap in US Refining and Petrochemical Console Operations**

By George Dzyacky

Director of Training Services, RSI Simcon, Inc.

Of all the professional, trade and administrative disciplines required to operate a refinery or petrochemical facility, managers, accountants, engineers, electricians, pipefitters, machinists, administrative assistants etc.; console operators who unarguably occupy the most safety-critical position in the plant, are the only ones trained and brought all the way to "qualification" by their coworkers.

**Who is best suited to train operators?**

"We are", say most of the operation training departments in the U.S. RPI , and this has been true for a very long time. The roots of these training practices go back generations, and pre-date the existence of process training companies that now serve our industry. Over the last twenty-five or so years, professional training consultants have developed extensive expertise in the border region between two previously distinct disciplines: unit-specific process operator training and the science of adult learning.

**The Unit Trainer**

The responsibility for operator training (both field operators and console operators) in the U.S. RPI typically resides at the process unit level. This is chiefly due to the fact that for many decades there simply didn't exist external training companies that could tackle the complexities of equipment specificity and the apparent idiosyncratic training requirements encountered in a typical process unit. Conventional wisdom correctly led the industry to conclude that no one was better positioned to train operators, than other operators who had mastered the operational nuances of these complex units.

Traditionally and out of necessity, Operator Trainers are pulled from the ranks of operators who, it should be noted, are very often hired into operating positions without prior specialized training or educational achievement beyond a high school diploma.

**They train what they know**

The operators who become Operator Trainers as I once did, and like the operation supervisor who trained me years before that, train well on the things they know. But their body of unit process knowledge is too often encompassed by the instructions they received from their operator trainer, and the knowledge and skills they acquired as a result of their personal motivation and inquisitiveness. As a result Operator Trainers naturally focus on the essential tasks that made them good at what they do, such as:

- Understanding alarms and how to respond to them
- Knowing where to locate and read instrumentation
- Knowing where to locate and how to make rounds on process equipment, vessels and control valves
- Knowing the purpose and safe use of utilities
- Knowing where to locate and when to utilize procedures
- Describing the basic functionality of technologies in the plant

While these training topics are appropriate for most field operator positions, they fall short of the skills required to be a competent console operator. There still exist in our industry today, many sites where console operator training is accomplished by osmosis. Absent any formal curriculum, operators are assigned to parallel or "shadow" on the console with an experienced, qualified operator for weeks or months until they can replicate the key tasks of their trainer.

**The problem with training by osmosis**

From the Center for Process Safety, American Institute of Chemical Engineers, "Guidelines for Preventing Human Error in Process Safety, on operator training: *"In general, little use is made in the process industry of more sophisticated approaches such as job and task analysis to define the mental and physical skills required for specific types of work, and to tailor the training program accordingly. Instead, informal on-the-job training is common, even in more complex types of work such as control room tasks. Although the necessary skills will eventually be acquired by this process, its inefficiency leads to the need for extended periods of training. In addition, there is the problem that inappropriate or even dangerous practices may become the norm as they are passed from one generation of workers to the next."*

To achieve the level of competency demanded in this job, Console operators in the U.S. RPI must have formal classroom, simulator, and on-the-job training to achieve full capability. Certainly no less rigorous than that of many regulated trades such as, Journeyman Electrician or Machinist. Console operator training must cover topics such as process theory, process control, safety guidelines, safety procedures, math for technicians, instrumentation, pumps, valves, thermodynamics, chemistry, distillation, boilers, furnaces, compressors, reactors, heat exchanges and troubleshooting to name some.

**On human error**

From the Center for Process Safety for the American Institute of Chemical Engineers, the book "Guidelines for Preventing Human Error in Process Safety," defines the cause of human error in the U.S. RPI this way; "… *a systems perspective is taken, which views error as a natural consequence of a mismatch between human capabilities and demands, and an inappropriate organizational culture*."

**In stark contrast**

The State of Texas requires a hair stylist to successfully complete 1500 hours, ten months, of training in a state licensed institution. Only after successful completion of state sanctioned written and demonstration tests, does the Texas State Department of Licensing and Regulation issue a Cosmetology Certificate and allow the graduate to cut hair.

But for console operators, in control of process units that have more instrumentation and greater cost than a Boeing 787 Dreamliner, training and testing is not sanctioned by any state or regulatory body in the U.S.. It is not uncommon for console operators to be in complete, sole control of a process with fast dynamics where hazardous chemicals and hydrocarbons are held under hundreds or thousands of pounds of pressure. Further, it is also not uncommon for these operators to have achieved this level of responsibility with no more than a high school diploma and to have been brought to "qualification" by a coworker over a period of several months.

A cursory review of completed investigations by the US Chemical Safety Board, dramatically illustrates just how dangerous and deadly are these processes. The U.S. RPI conducts business in an era of increased governmental, community and legal scrutiny. Yet despite the known dangers to workers and surrounding communities posed by the variety of refining and chemical processes, there is no regulatory oversight or industry-wide standards that mandate minimum curricula or validate the competency of those in control of these plants.

**Can the industries do it?**

The majority of operator trainers and operator training departments in the U.S. RPI are not prepared to train console operators against a competency framework to appropriate proficiency levels, for reasons of skill, resources and culture. The training interventions (courses) are highly specialized and tend to exceed the expertise and resources of most in-house training departments. The level of training needed to meet the demands of the job is greater in duration and cost than historically presumed appropriate. Lastly, training is not the industry's core business.

**What needs to change?**

We must no longer accept the outdated culture and practices of operators-training-operators. console operators in refining and petrochemical industries should be required to successfully complete comprehensive curricula, based on a competency framework, targeted to teach technical fundamentals, process fundamentals and unit-specifics.

One approach is to have the methods and standards for competency identification, curricula development and assessments managed through an Industrial-Academic Consortium, coordinated by an industry-recognized trade association such as the AFPM or API.

The training must include formal classroom and simulator training covering the entire range of topics from fundamental concepts to unit-specific emergency response.

After trainees demonstrate proficiency in the fundamentals they enter the unit-specific portion of training where they learn to operate their unit. At that point the Operator Trainer comes back into the picture. But this time the trainer is not burdened with the responsibility for designing training programs, that task requires the skill sets of training professionals. The Operator Trainer fills the role in the development of a training program for which they are perfectly suited and qualified – that of Subject Matter Expert (SME). The Operator Trainer works with professional training consultants who specialize in the development and execution of fundamental and unit-specific training material and process simulations, to insure the training is customized for their unit.

After successfully completing the fundamental and unit-specific training courses the console operator would serve an apprenticeship, paralleling on the actual console with a previously qualified operator before the trainee is certified to work the console alone.

As a model for rigorous competency assurance we can look to the nuclear industry, where the knowledge base and performance levels of licensed nuclear board operators parallel closely in depth and breadth to those of console operators in the refining and petrochemical industries, along with salaries.

This type of approach to achieving uniform competency assurance in the refining and petrochemical industries, will not only give plant managers and surrounding communities the innate assurance that all is well in the control room, but will pay for itself many times over, in day-to-day operational benefits and in the timely mitigation of emergencies.

# 348246 Process Safety Management (PSM) In Pilot Plants and Research Laboratories


**Abstract:**

**Process Safety Management (PSM)**

**In Pilot Plants and Research Laboratories**

## Abstract

Pilot Plants are considered to be the important tool for tackling the scale up problems. Even though the quantity of the chemicals used is less compared to a commercial plant, they are capable of creating significant hazards and risks. Hence the hazards and risks posed by pilot plants are critical. The increasing number of accidents in research laboratories and pilot plants all over the world is a testimony to the above fact. Many organizations struggle to determine what guidelines they should apply to their own units. It is essential that the risks at pilot plants and laboratories are to be assessed and properly controlled to protect lives, property and environment.

If the elements of Process Safety Management (PSM) are implemented in a research and development facility accordingly safety can significantly improve. The key strategies suggested include properly conducted preconstruction safety reviews, leak tight design and construction, close attention to area classification, well designed ventilation, well planned storage, properly conducted maintenance, integrity management and strictly controlled procedures for the management of change.


**Paper/Presentation Details**

- Takreer Research Centre (TRC) in brief
- Commit to Process Safety
- Understand Hazards and Risks in Pilot Plants and Research Labs
- Hazard Identification and Risk Analysis
- Manage Risk
- Asset Integrity and Reliability
- Training and Performance Assurance
- Management of Change
- Emergency Management
- Process Safety Performance Measurement Metrics
- Case Study
- PSM implementation in Pilot Plants and Laboratories in TRC

# 348476 Resolving Inherently Safer Design Conflicts with Decision Analysis


**Abstract:**

The Chemical Safety Board (CSB) investigated a fatal explosion at a chemical plant in Institute, West Virginia. The CSB noted that the incident placed at risk a large inventory of methyl isocyanate, the toxic and reactive chemical responsible for the 1984 Bhopal tragedy. The CSB commissioned a study by the National Academy of Sciences (NAS) to investigate the application of inherently safer design (ISD) concepts to the design of the subject chemical plant. One of the findings of the NAS report was the absence of guidance documents to assist the process safety professional when evaluating ISD alternatives. The challenge identified in the NAS report is that process technology alternatives must satisfy multiple objectives, some of which may be in conflict. The NAS panel recognized the potential value of multiattribute utility theory as a basis for evaluating ISD alternatives, but they did not illustrate its use in this report.

In this paper we illustrate the use of multiattribute utility theory (MAUT) as a decision analysis tool for evaluating ISD alternatives. Using the NAS report as the foundation of our study, we show how to use MAUT to evaluate ISD alternatives by formulating utility functions and weights for the decision objectives. A sensitivity analysis is also presented to demonstrate how the final ranking of alternatives varies with the weights. This case study demonstrates that the MAUT technique is an effective tool for resolving ISD conflicts.

# Operating Safely Through Integrated Process Safety Management

**Srinivasan Ramabhadran**
**DuPont Sustainable Solutions**
**#11-01 Harbour Front Tower One, Singapore**
**Ramabhadran.Srinivasan@dupont.com**

**Alfonsius Ariawan**
**DuPont Sustainable Solutions**
**1919 Minnesota Court, Mississauga, Ontario**
**Alfonsius.B.Ariawan@dupont.com**



Prepared for Presentation at
American Institute of Chemical Engineers
2014 Spring Meeting
10th Global Congress on Process Safety
New Orleans, LA
March 30 – April 2, 2014

UNPUBLISHED

# Operating Safely Through Integrated Process Safety Management

**Srinivasan Ramabhadran**
**DuPont Sustainable Solutions**
**#11-01 Harbour Front Tower One, Singapore**
**Ramabhadran.Srinivasan@dupont.com**

**Alfonsius Ariawan**
**DuPont Sustainable Solutions**
**1919 Minnesota Court, Mississauga, Ontario**
**Alfonsius.B.Ariawan@dupont.com**

**Keywords:** Operational Discipline, Felt Leadership, Holistic Approach, Integrated

## Abstract

Recent major incidents across the world have brought the need to operate safely as a top of mind issue for most CEOs. This presentation will focus on how effective implementation of an integrated Process Safety Management system can help achieve safe operation and improve productivity. It will address three key areas: operational discipline - focuses on carrying out each task the right way, every time; integrated approach - a holistic approach to implementing process safety elements that is based on risks rather than implementing individual elements of the system; felt leadership – the critical role of leadership in driving a culture of safety.

## 1. Introduction

Many companies today are beginning or accelerating programs designed to reduce their risks and in particular, prevent potential catastrophic incidents. These efforts are being driven by a number of factors: increased public and regulatory scrutiny of their operational risks, and increased exposure in terms of liability to leaders, and attrition of a competent and skilled workforce. Recent major incidents and the resultant public outcry have also contributed to raising the need to manage operational risks to the top of the agenda. Among other things that are required to effectively implement process safety management, there are three primary factors:

- Operational Discipline
- Integrated Approach
- Felt Leadership

### *1.1 Operational Discipline*

Effective implementation of PSM does not happen only by *designing* good PSM systems; effective implementation requires a continuous, consistent and persistent *application* of these good systems in the day-to-day operations (Figure 1). The principle of "Operational Discipline" is recognized as a key success factor in driving effective PSM implementation. Organizations are increasingly realizing that operational discipline can result in rapid, substantial and sustainable PSM improvements.

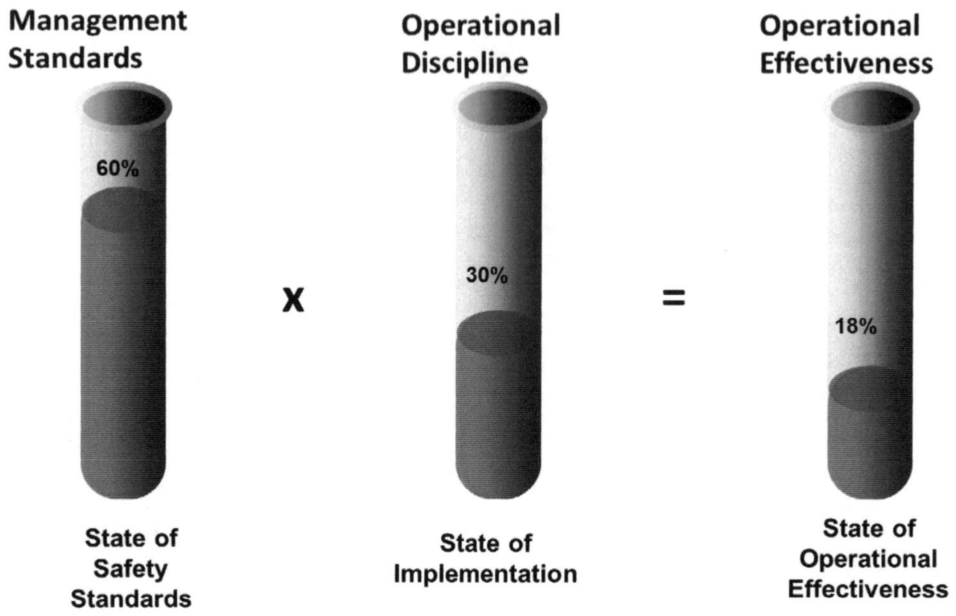


**Figure 1.** Performance is a function of the management system and level of execution

At DuPont, we define operational discipline as "the deeply rooted dedication and commitment by every member of an organization to carry out each task the right way, every time." We believe operational discipline requires change at both the organizational and individual levels in order to be substantial and sustainable.

Based on many years of cumulative working experience internally within DuPont and with clients, we have identified success characteristics that are the foundation of world-class operational discipline. The characteristics cover leadership, culture, processes and standards. A focused effort on improving these characteristics at the organizational and individual levels can lay the foundation for sustainable improvement and change.

### *1.2 Integrated Approach*

Many companies implement the various elements of process safety management, but those elements must work together to achieve success.

DuPont has realized the value of this integrated approach and has defined the process safety management system as a wheel where the hub is the leadership and culture and the spokes are the critical elements of the management system. The way it works is through operational discipline (Figure 2).

**Figure 2.** DuPont PSM Model

### *1.3 Felt Leadership*

What does "felt leadership" have to do with Process Safety Management?

A more descriptive term for what process safety is all about might be "Operational Risk Management." Operational Risk Management implies that:

- It is the operation or the materials processing activities that must be controlled.
- It is the reduction of risk to an acceptable level with the ultimate elimination of significant process incidents that is the goal.
- It is the management of these risks that is one of the organization's key responsibilities and challenges.

The degree of success achieved in the management of operational risks relates to the quality of decision-making by and within the operation's organization. And leadership is all about influencing and improving the quality of decision-making (Figure 3).

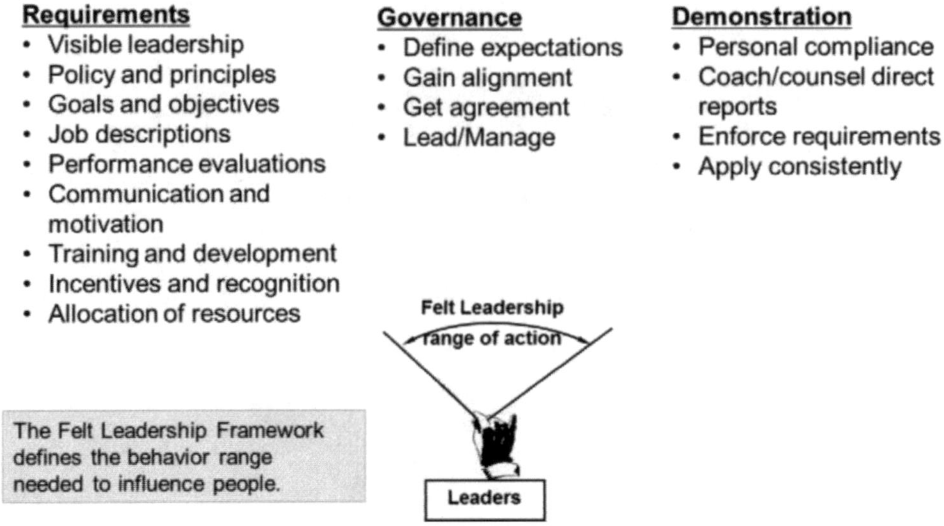


**Figure 3.** Felt Leadeship defined

## 2. Conclusion

- Leaders must have a clear vision of the ultimate goal of the company's process safety management efforts.
- Leaders must set the goals and strategic objectives.
- Leaders must ensure that sufficient, quality resources are dedicated to PSM system requirements.
- Leaders must make line management accountable for PSM implementation and results.
- Leaders must recognize and reward PSM accomplishments.
- Leaders must drive continuous improvement through the use of metrics and audits.

# 349434 Improving PSM Performance through Workforce Culture Assessment

**Abstract:**

Increasing PSM Performance Through Workforce Culture Assessment

By Peter Baker and Carl Green

This paper describes MetaPower's focus on increasing PSM performance by routinely measuring compliance to regulation-based criteria (structure) and simultaneously measuring key dimensions of work groups (culture). This dual approach is built on extensive work process design and implementation experience which convinced us that workforce culture is a key component to successful implementation. So MetaPower went "back to school" for five years, working with scholars who had organizational theory and culture expertise. Our learning and understanding led to the approach presented in this paper.

This presentation will introduce you to the radical difference between structure and culture, to the unique "logic" of each, and to the tools appropriate to cultural issues that can be secured and mastered. You will discover that it is possible, to measure culture and gather clues on how it is impacting PSM performance. You will learn that, because culture is such a powerful force, it is important to change culture only when necessary. You will also be introduced to a method for addressing cultural issues when change is necessary. You will find that when culture is approached with a new understanding, it is not mysterious and forbidding and it can be successfully navigated to achieve consistent performance. Finally, you will learn how structure and culture, integrated into a holistic Work System, can help you achieve your process safety goals.

# 349514 Tracking Instrumentation and Controls Reliability

**Abstract:**

To improve process safety and instrument maintenance decisions, CCPS PERD and the MKOPSC Instrument Reliability Network are developing better reliability tools to provide high quality, valid, and useful data pertaining to the hydrocarbon and petrochemical industries. The documentation required for the validation and maintenance of instrumentation and control devices must be completed properly in accordance with established criteria to demonstrate the appropriateness of credited risk reduction. Maintenance personnel are quite often overwhelmed with the challenge of tracking the testing frequency, inspection procedures, due dates and compliance for the instrumentation and controls in their areas of responsibility.

This presentation will review procedures and work processes that have been incorporated successfully for testing and validation of these devices. Technicians can quickly document compliance, pending due dates, completion dates, corrective actions and comments when an efficient documentation management system is in place. This system can be utilized to track and assess reliability, instrument and compliance data from proof test activities during plant turnarounds, and periodic safety instrumented function validation and routine maintenance.

# PSSR: The Easiest and Sometimes Forgotten PSM Element

**Brian D. Rains**
**DuPont Sustainable Solutions**
Brian.D.Rains@dupont.com

**Srinivasan Ramabhadran**
**DuPont Sustainable Solutions**
Ramabhadran.Srinivasan@dupont.com

Prepared for Presentation at
American Institute of Chemical Engineers
2014 Spring Meeting
10th Global Congress on Process Safety
New Orleans, LA
March 30 – April 2, 2014

UNPUBLISHED

# PSSR: The Easiest and Sometimes Forgotten PSM Element

**Brian D. Rains**
**DuPont Sustainable Solutions**
Brian.D.Rains@dupont.com

**Srinivasan Ramabhadran**
**DuPont Sustainable Solutions**
Ramabhadran.Srinivasan@dupont.com

**Keywords:** PSSR. Pre-start up safety review

## Abstract

Pre-Startup Safety Reviews (PSSR) is to Process Safety what Personal Protective Equipment is to Personal Safety. Often, the PSSR is the last line of defense to ensure that a change or other major plant activity has not introduced an unintended or unmitigated risk. The principles behind the PSSR process are straightforward. The actual PSSR effort is basic and logical. And everyone says they do them and that they are easy. Is there data to support these statements? In this presentation, not only will PSSR best practices be reviewed but common PSSR weaknesses and improvement opportunities will be exposed along with suggested remedies and solutions.

## 1. Importance of PSSR

PSSR is one of the simplest elements of PSM. Does simple translate to less important? The simple answer is NO!

*Figure 1: Where does PSSR come in to play?*

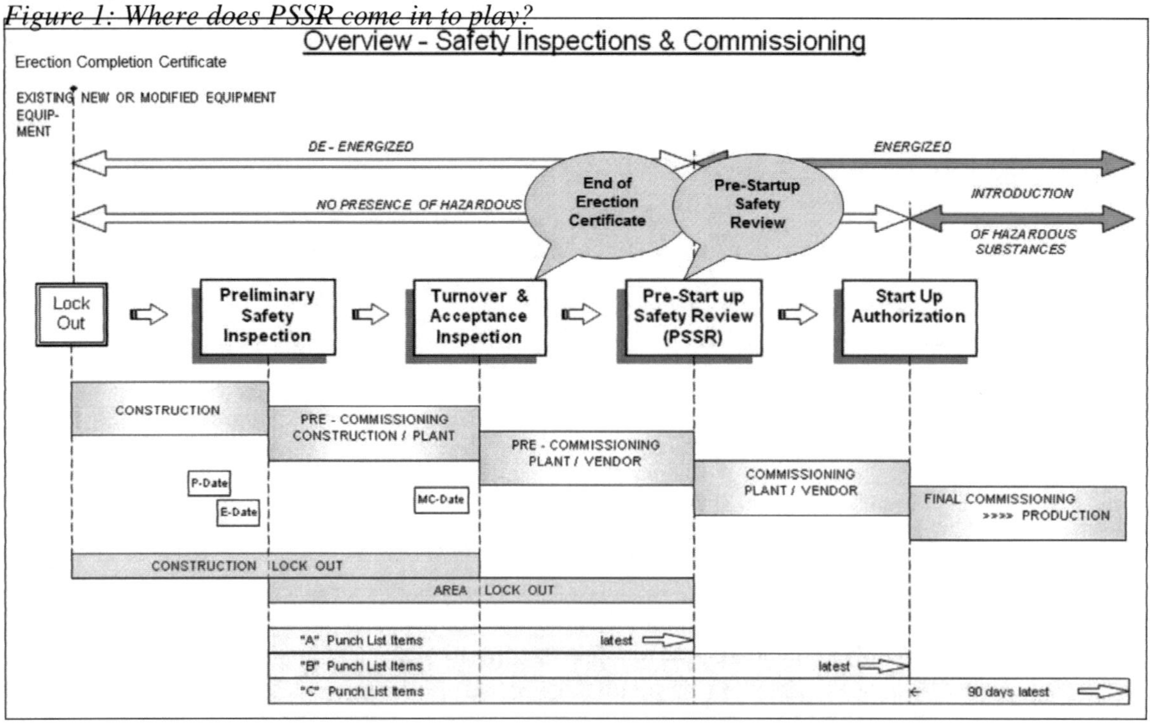

It is usually done after mechanical completion and before commissioning. The "ownership" for PSSR lies with operations as they will be pressing the button to start the operations. The operations leader along with a cross functional team has to satisfy himself that it is safe for him to start up the facility. While mechanical completion will be a responsibility of projects team, PSSR is clearly the responsibility of "operations".

Each element of PSM has a specific purpose and intent and failure to apply even one element in practice can contribute to potential catastrophic incidents. PSSR provides the last line of defense and events that can trigger the need for a PSSR include any change to an existing process or facility or the introduction of a new process or facility.

PSSR is NOT a last minute hazards analysis.

# 2. About PSSR

## 2.1 Principle

PSSR provides a FINAL CHECKPOINT for new and modified equipment to confirm that all appropriate elements of PSM have been addressed satisfactorily and the facility is safe to start up. For capital projects, planning for success with the pre-start-up review begins with assembling a competent project team. Effective project management, including careful planning and coordination of all PSM elements, is important to achieve a smooth start-up.

## 2.2 Essential features

### 2.2.1 Scope—PSSR

Sites shall perform pre-start-up safety reviews for all new (including equipment on hire or lease) and modified facilities for which the modification necessitates a change in the documented process technology.

PSSRs shall be conducted for significant "replacement in kind" equipment capital projects (compressors, heat exchangers, vessels, centrifuges, incinerators,.. ) even when no change has been applied, to help ensure that all relevant elements of PSM have been evaluated (e.g., updating of equipment files, performing QA checks for fabrication and installation, and resetting MI test and inspection values for the new components).

PSSRs may also be conducted for the following circumstances to help ensure that all relevant aspects of process safety have been completed prior to returning to full operations:

- Restarting an idled or mothballed process
- After a major shutdown or turnaround where significant or extended maintenance or repair activities occurred (i.e., emphasizing the field portion of the PSSR process).

Sites may alternatively use process-specific start-up procedures and safety checklists for start-up after these activities.

Prior to the introduction of hazardous substances and associated energies to a facility, the pre-start-up safety review team shall confirm that:

- Construction and equipment are in accordance with design specifications.
- All relevant elements of process safety management have been appropriately addressed (i.e. PSM assessment).
- Basic safety, health, and environmental - and fire-protection items have been checked during a physical inspection of the facility. These inspections are usually guided via a previously pre-prepared "good practice" checklist.
- Installation is in compliance with local regulatory requirements and site and unit operating permit(s), including environmental air permits and approvals.
- The facility is safe to start up. This may be contingent on completion of certain recommendations.

### 2.2.2 Team composition

The pre-start-up safety reviews needs to be conducted by multidisciplinary teams consisting of manufacturing (including operators), technical, design as-needed, maintenance, appropriate safety representatives on an as-needed basis and the Change of Design (COD)/Project originator/leader or the COD Coordinator. Other resources, specialists, or independent or third-party competent persons as specified within various RAGAGEPs (Recognized and Generally Accepted Good Engineering Practices ) may be included as applicable (e.g., electrical, fire safety, instrumentation and control, personnel; safety interlock guardians; ergonomics specialists; or equipment manufacturers). When the new facilities will be used by a resident contractor then a member of that organization must take part in the PSSR.

For very minor changes, a smaller pre-start-up safety review team may be used and could consist of two team members. In this situation, the facility owner shall designate the team members. Examples of minor changes include installation of a new drain valve and instrument transmitter bracket modification.

The team composition is the key to maximizing the value and results of a PSSR. The facility owner must take responsibility for assembling a multidisciplinary team with knowledge and experience appropriate to the task. The Team make up will be predetermined by function and indicated on the Copy of the PSSR circulated with the COD for approval. Team members should include personnel who were not directly involved on the applicable COD/project team.

The following points on team makeup are for guidance only. Advice on this issue can be provided by site or regional/external PSM resources.

- Even for the smallest PSSRs, such as those would be conducted prior to starting up a single piece of equipment (e.g., a pump, heat exchanger or vessel), the team should have three or four representatives from different backgrounds (e.g., manufacturing, maintenance, design/technical, and appropriate safety representatives). This helps create a more searching and valuable review. People who operate or maintain the facility (e.g., an operator and a mechanic) should be included.

- Where the change involves introducing hazards (i.e., process, material or environmental) that is new to the site, the PSSR team should consider including off-site personnel who have relevant knowledge and experience. These could be manufacturing, maintenance, technical or appropriate safety representatives.

- For significant capital projects and for larger projects a formal PSM audit may be recommended. In this case the facility owner must ensure that the team has the skills and capability to carry this out. Consideration should be given to having one or more safety and occupational health consultants on the team.

- The leader of the PSSR team should foster an open atmosphere where the project team members look on the review as an opportunity to upgrade the process safety of the facility and do not become defensive of the design or preparations for start-up.

## 2.3 PSSR Application

The flow chart below represents a recommended process for applying PSSR.

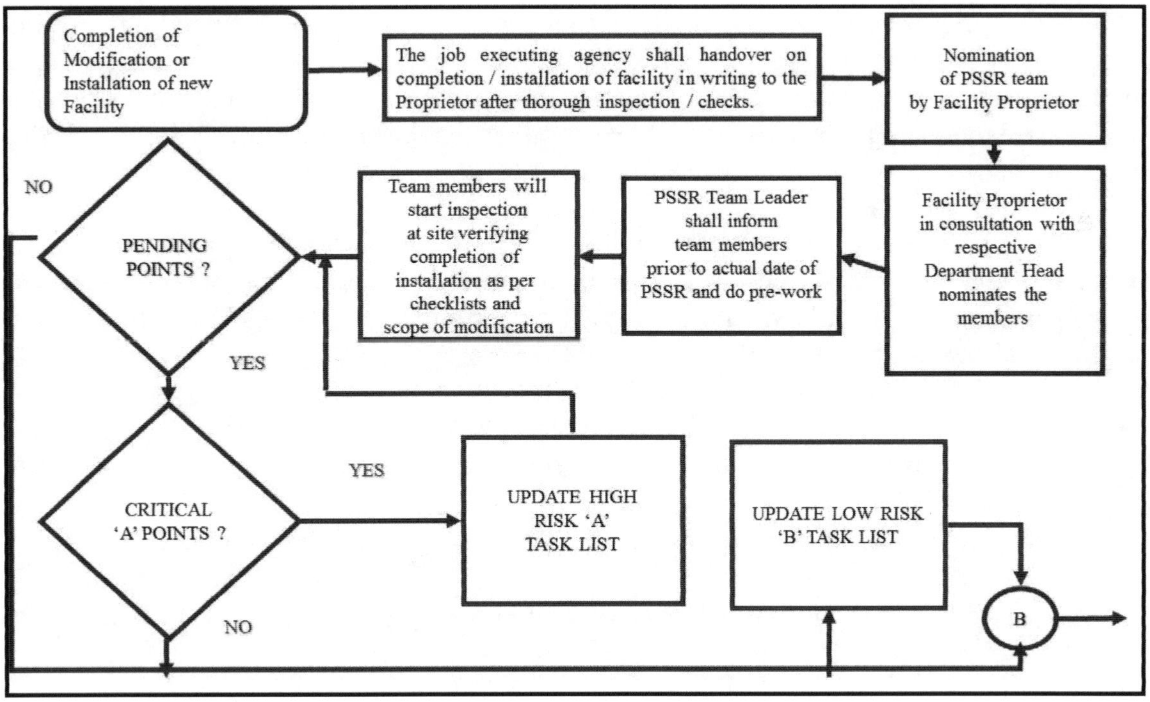


*Figure 2a: PSSR Flow Chart*

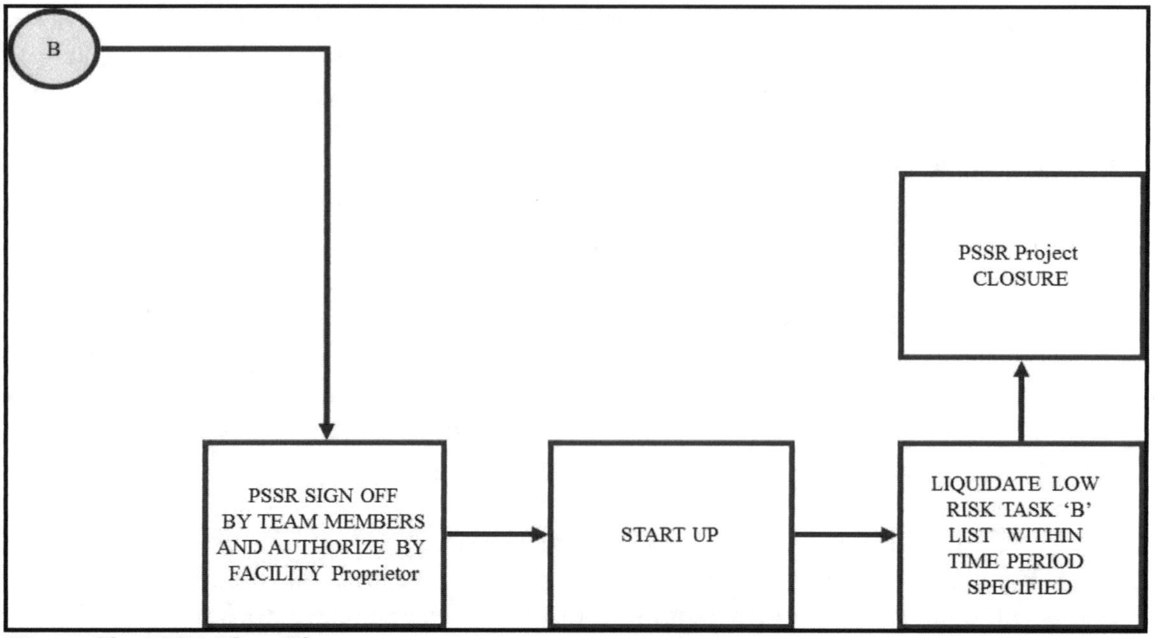


*Figure 2b: PSSR Flow Chart*

### 2.3.1 Significant capital expenditures

Where the change involves significant capital expenditures (typically needing off-site authorization) and one or more complex unit operations with hazardous materials and activities, then it is recommended that the PSM element assessment be conducted by either a first-party or a second-party PSM audit using the appropriate corporate PSM audit checklist(s).

For significant capital expenditures, it may be helpful to carry out a two-part review. The first should take place when design is complete and construction is at an early stage or has not yet begun. This allows operating procedures, emergency planning, and training to be upgraded well in advance of start-up.

The second part of the review can then take place just prior to introduction of hazardous substances and focus on checking that construction and equipment are in accordance with design. For example:

- Prior to charging a substation, instrumentation building (if applicable) for the first time.
- Before introducing hazardous substance / feed in a new, modified, hired or temporary facility

### 2.3.2 Small projects/modifications

For smaller projects/modifications, the PSM element assessment is conducted using a PSSR checklist. The PSSR team should select relevant parts of a PSSR checklist to review, as appropriate, for the project/modification. Although it is not the PSSR team's function to review the process design, its input may affect construction.

- Significant projects involving "replacement in-kind" equipment applications
- Changes to operating facilities
- Temporary changes
- Restarting mothballed process
- Post major turnaround
- Sites may alternatively use specific start-up procedures and safety checklists after these activities

### 2.4 PSSR - PSM element assessment focus

The pre-start-up safety review team shall ensure that all relevant PSM elements have been appropriately addressed and shall be guided by pre-written checklists. Physical inspections should cover:

- Documentation Items
- Vessels, Piping and Valves
- Rotating and Mechanical Equipment
- Electrical systems
- Instruments and control systems
- Structures
- Safety, Health, Security and Fire Protection
- Ergonomics
- Waste, Stream and Environment
- Operation, Inspection and Maintenance and Others
- Fume Hood/Chemical Storage/Gas Cylinder Storage and other Laboratory Area

It also needs to include appropriate checks to confirm that *local regulations* are being followed. The PSM elements to consider included the following:

| | |
|---|---|
| PROCESS SAFETY INFORMATION | Assists in<br>• verifying PT for equipment,<br>• material hazards & technology<br>• Updating in timely fashion. |
| PROCESS HAZARD ANALYSIS (PHA) | • Recommendations have been addressed and appropriate actions associated with s/u have been completed. |
| OPERATING PROCEDURES | • Has the operating/emergency procedures and safe work practices procedures revised to reflect the change? |
| TRAINING AND PERFORMANCE | • Has the operating and maintenance personnel required to be trained for the new facility or change been done? Competency of personnel have been evaluated. |
| CONTRACTORS | • Do contractors require any special training or instructions to be given; are they properly trained? |
| MECHANICAL INTEGRITY and QUALITY ASSURANCE | • Maintenance task procedures are in place?<br>• Equipment needs to be placed in mechanical integrity program?<br>• Installation as per codes, standards and manufacturer's recommendations? |
| MANAGEMENT OF CHANGE | • MOC related activity and its documentation.<br>• 2$^{nd}$ level protection to ensure MOC is working.<br>• Influences complexity of PSSR.<br>• Evaluates MOC scope and intent are met.<br>• Establish criteria that maintain minimum levels of knowledge and experience |
| INCIDENT INVESTIGATION | • PSSR documentation support investigation teams.<br>• LESSONS LEARNED ARE POWERFUL TOOLS OF IMPROVEMENT |
| EMERGENCY PLANNING & RESPONSE | • Verifies that emergency response plan changes are made & affected people are trained. |
| AUDITS | • Following start-up, the new facility is included in the site PSM audit system. |

### 2.4.1 PSSR Attributes

The attributes of a "good practice" checklist are as follows.

- A clear front sheet stating what equipment has been inspected, when and by whom.
- – A statement that the installation is consistent with design specifications.
- – Statements that the PSSR team concludes that the facility is safe to start up after certain recommendations have been satisfactorily resolved.
- – A record of recommendations (if any) with timing and responsibility for completion either before or after start-up.

- A check that the following basic safety and occupational health areas have been appropriately addressed:
- – General Safety
- – Machinery safety
- – Ergonomics
- – Occupational health

- A check that the following PSM elements/topics have been appropriately addressed:
- – Process technology
- – Management of change – technology and facilities
- – Process hazards analysis
- – Quality assurance
- – Mechanical integrity
- – Operating procedures and safe work practices
- – Training and performance
- – Interlocks and alarms
- – Highly toxic materials
- – Contractor safety

- A check that other relevant SHE topics have been addressed:
- – Environmental
- – Community awareness and emergency response
- – Electrical safety
- – Fire protection

- Site should include in their pre-start-up review checklist appropriate checks to ensure that local regulations are being followed.

## 2.5 PSSR documentation, recommendations and closure

The pre-start-up safety reviews shall be documented, signed by each member of the review team, and authorized by the facility owner. The pre-start-up review document shall include a statement as follows:

The PSSR team concludes that the equipment is in compliance with current regulatory requirements and limits stated in the site and/or unit operating permit(s) and is safe to start up after completion of recommendations required before start-up.

A system shall be established and controlled by the owner to ensure review recommendations are resolved (including documentation of resolutions) before hazardous substances and associated energies are introduced to the facility. Start-up authorization shall be issued after verification that all PSSR items to be resolved have been satisfactorily completed. The start-up approval shall be documented on PSSR document. At the site the required for startup recommendations must be signed off as complete before the startup of the facilities is authorized for start up

Similarly, a system shall also be established by the facility owner to ensure resolution of PSSR review recommendations that are to be completed after start-up. At the site those recommendations that are not required for startup are entered in the Site's "Manage IT" system for follow up and tracked to completion.

Periodic reports listing open recommendations should be published at least quarterly and distributed to the site business area management, and all persons assigned to follow up a recommendation, and their supervisors of persons assigned the recommendation. The report should highlight recommendations that are past the due date.

Each item in the Site PSSR Authorization Form is referenced back to this document so that the inspection team may obtain additional clarification as the intent of the items in the Site PSSR Check sheet. It is important to specifically track all and especially punchlist A items for closure and verification before start up. Having a tracking system provides the confidence that all critical recommendations items have been closed and verified and assures that the plant is safe for start-up.

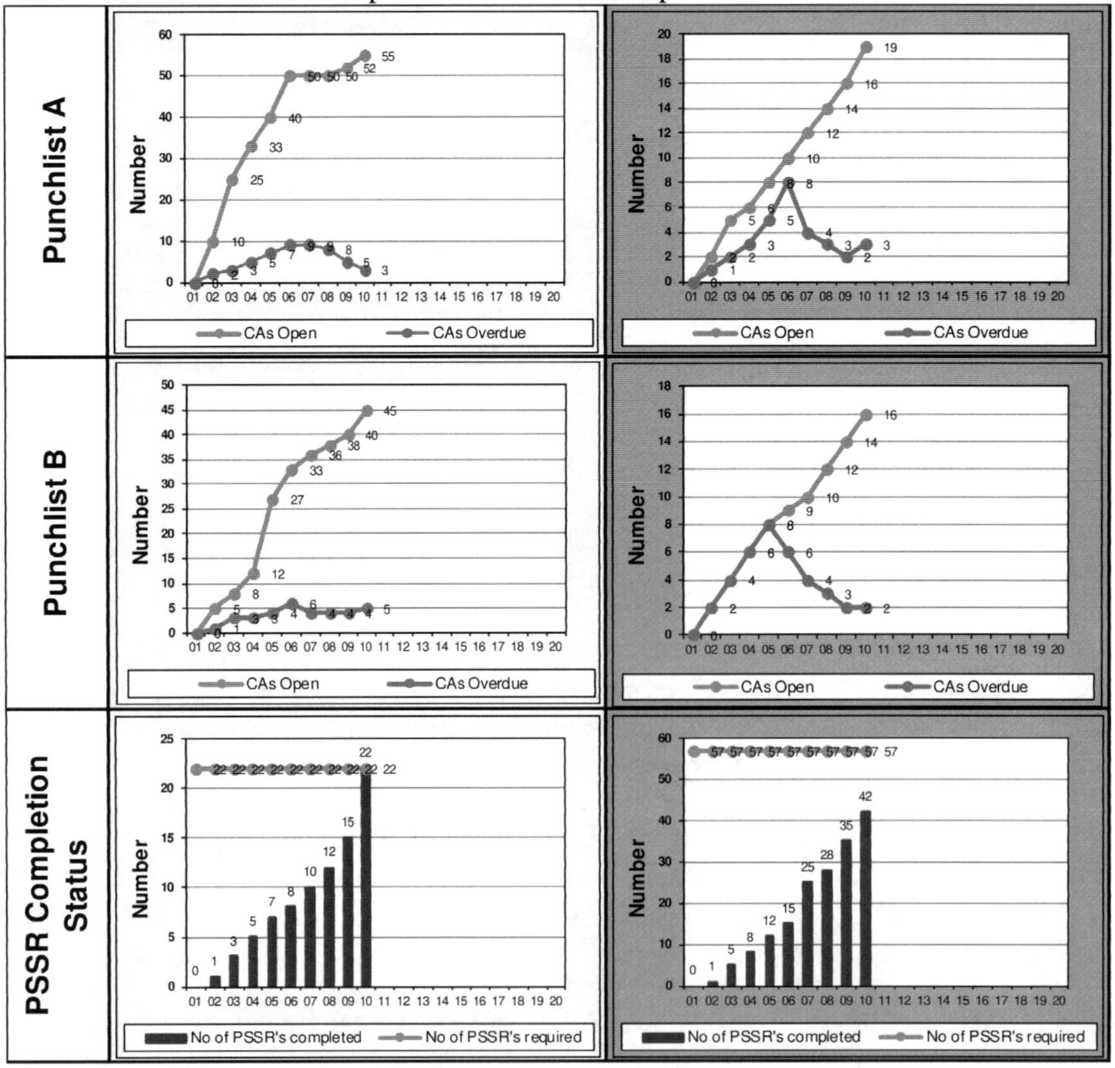


*Figure 3: Tracking PSSR Punchlist Items*

## 2.6 PSSR Responsibilities

Senior Line Supervision: Senior Line Supervision shall authorize the checklist after the team members have signed the checklist and shall ensure that all recommendations required prior to startup are completed and documented prior to the introduction of hazardous materials into the facility.

Review Team Leader: The Review Team Leader is responsible for assembling a multi-discipline team and establishing a review schedule for that team. A member of the safety organization may also be invited to take part in the PSSR inspection. The leader shall guide the team through the review, using the checklist to ensure that all relevant questions have been asked and answered. The leader shall document the answers to the checklist questions and note all deficiencies. The leader will ensure that all deficiencies are assigned to a specific person for correction. The leader will forward the authorized PSSR to the COD Technical Assistant for entering Manage It Central database and filing in the COD folder. Upon completion of all recommendations, the leader will document the date the recommendations completed and initial on the PSSR. The leader is responsible for ensuring all recommendations are completed in a timely manner.

All COD's are entered into a tracking system and their status reviewed by the area once a month.

Review Team Members: These personnel are selected based on their individual skills/knowledge relative to the change being made. They shall use their individual skills and knowledge to ensure that all relevant questions from the checklist are correctly answered and those any and all additional concerns are brought forward for discussion by the team. Members shall sign the checklist, stating that the facility is safe to start up after completion of recommendations required before startup.

Recommendation Assignees: These are the personnel assigned to follow-up on recommendations generated during the safety review. They shall complete their assigned recommendations in a timely manner and document the action taken in Manage It Central database.

COD Technical Assistant: The COD Technical Assistant will be responsible for entering a PSSR and recommendations in Manage It Central database and filing it in the COD folder once received from the Review Team Leader.

Verifier: The area person who is responsible for confirming proper documentation related to the new/modified facilities shall be responsible for checking the PSSR Checklist to ensure all recommendations have either been signed off as complete or are listed on the cover sheet with the respective tracking number from Manage IT prior to signing off for closure of the COD which required the change

## 3 Conclusion

In conclusion, while PSSR is a simple element to understand and deploy, it is an important element and plays a critical role in preventing incidents during start-up. Any process is "more prone" to incidents during start-up due to:

- Involvement of multiple functions and departments
- Understanding of the new technology and/or modified process requirements
- Transfer of ownership from projects to operations

PSSR is a good way to bring all parties together and align on the changes and needs to ensure safe start up.

## 4 References


*[1]* Guidelines for Performing Effective Pre-Startup Safety Reviews, Center for Chemical Process Safety (CCPS), April 2007

[2] DuPont Internal Standards

# 349905 Guidelines for Pressure Relief and Effluent Handling Systems, 2nd Edition


**Abstract:**

The second edition of Guidelines for Pressure Relief and Effluent Handling Systems will be issued later this year. It is the only book to comprehensively cover both design sizing of emergency relief devices and the systems to safely control the effluent after the relief device has activated. It also includes software for designers to size the emergency relief systems, plus worked examples to show how the software can be used for several design scenarios. It is the only book in this field that covers the topic of 2-Phase flow, based on the most up to date research and studies by the Design Institute for Emergency Relief Systems (DIERS). It is also the only book to cover in detail the safe control of the release after the relief device had activated. An overview of the book will be presented.

# 349983 Journey to World Class through Capability Development

**Abstract:**

DuPont has been on a 7 year journey towards world class operational excellence, resulting in significant earnings gains and cost-savings ($6B+) built off a foundation of capability development for individuals and teams.

This strategy has been proven across a wide variety of assets and businesses, both inside and outside of DuPont, deployed and sustained at over 200 DuPont locations.

This framework will be shared/discussed and compared/contrasted with attendee's experiences. Several of the questions to be discussed are:

- How do you intentionally work to build a culture of continuous improvement throughout your organization?
- What specific areas of capability building have you found to be critical (For individuals? For teams? For key leaders?)?
- What cultural elements can be used to engage a multi-generational workforce?

DuPont will share key frameworks and elements of their journey that can be used as criteria for building capability and operational excellence in organizations.

# 349330 Incident Lessons Learned Portal

**Abstract:**

**Incident Lessons Learned Portal**

Marco Vela

Corporate Process Safety Engineer

Celanese

1601 West LBJ Freeway

Dallas, TX 75234

(214) 417-1920

marco.vela@celanese.com.mx

**Abstract**

The Celanese incident lessons learned portal consist of four levels: company lessons learned, technology specific lessons learned, site specific lessons learned and external incidents of interest. This paper will present how we capture knowledge from previous incidents to retell our story at various levels in the organization. It is designed to engrain learnings from past incidents into the corporate process safety culture to prevent repeats.

**Company Lessons Learned**

Company lessons learned use CSB type videos to share the key causes of the major incidents that have occurred in company with all Celanese employees. The system includes the initial delivery method and definition of how the messages are reinforced over time. They describe why we have established a process safety system.

**Technology Specific Lessons Learned**

Technology specific lessons learned are gained either by experience, past incidents, or standard design. These learning's are targeted to hourly workers in each technology, and describe what could go wrong in that technology and the safeguards in place to prevent.

**Site Specific Lessons Learned**

This is a common area on the process safety portal that gives everyone access to one page that contains the information of the most significant incident lessons learned that have happened at each site. Site specific lessons learned are used for toolbox meetings, operator training, new leader process safety certification, PHA preparation, and other process safety activities.

**External Incidents of Interest**

This is a place to share with all the employees the lessons learned from external incidents of interest based on relation that they have with our companies manufacturing operations.

# 349394 A Functional System Approach to Criticality Analysis, Fsca


**Abstract:**

ABSTRACT

A Functional System Approach to Criticality Analysis, FSCA.

This paper will outline and discuss a new and unique approach to physical asset criticality analysis: Functional System Criticality Analysis (FSCA) and illustrate key results with two sample case studies. This approach is applicable to any type of complex facility or process where process security and continuity are paramount. It has consistently identified previously unknown significant risks that were missed by other methods and analyses such as HAZOPS, FMA, process reviews, and can quantify the risk potential in dollars. A functional system based approach (FSCA) is more efficient than current methods through analyzing and ranking system risks via system failure scenarios rather than failure modes at the asset level. It is an invaluable tool and approach when risk mitigation is important to assure overall success balanced against the reality of limited time, capital, maintenance, and staff budgets.

The two case studies featured involve an Envirex ground water remediation facility, and an industrial RO water treatment system within a refinery. Each study had a unique set of evaluation criteria determined by the client's values and the specific performance, contractual, and regulatory issues affecting the facilities. Importantly, in each case study previously unknown significant system risks were identified, as well as highly effective mitigating steps.


Approach:

Several key principles are foundational to the functional system based approach to criticality analysis, FSCA:

- A deep and well structured hierarchy of the asset register, leading to a rigorous parent/child ratio
- Inclusion of all valves and pipes (strive for 100 percent asset listing) in the hierarchy
- Uniform asset levels (every pump is an asset no matter where it is or how big)
- The grouping of assets into single-purpose functional systems

A very clear description of the mission statement/priorities is very important in this type of an analysis. Whether it is defined as a level of service or a more numerical set of KPIs, the defined mission of the physical assets under review will determine the categories evaluated against: safety, environmental, up-time, product quality, maintenance cost, capital replacement, etc. These in turn must also be clearly and uniquely defined from one another on a severity scale ranging from virtually 'no impact' through to 'worst case' in the client specific nomenclature. The next step is to proportion or allocate a contribution ratio for combining categories into overall risk. This will determine, for example, the relative importance or ranking of a worst case scenario compared against a scenario with 2 or 3 instances of a slightly lesser consequence. These can be adjusted to suit a client philosophical approach.

With these principles in place, a risk analysis is then performed at the functional system level derived from the hierarchy, prior to the asset level. Failure scenarios and their resulting consequences are evaluated per the evaluation criteria defined (not failure modes since those are at the asset level). The failure scenarios for each functional system define a composite risk space for each system, and then each system is ranked against all others in terms of risk.

Important to note is that assets within a parent system initially assume the risk ranking of the parent system. Thus, systems with highest risk ranking (mission critical) are then further evaluated at the asset level, employing a suitable mix of evaluation tools (RCM, FMA, etc) to discern which of the contained assets pose the highest risks to the mission objective. Systems with low risk ranking do not have assets that can significantly impact the mission, are thus not mission critical, and are thus relegated to a lesser priority for further evaluation.

In this approach mission critical assets are more rapidly identified with subsequent evaluations more efficiently utilizing complementary tools saving time, staff resources, and money. We have also found that, due to the unique arrangement of the assets within the system level and overall integration, this approach captures risks that might otherwise be missed.

Base Case 1. Envirex Ground water remediation system

In mid 2004 a sequence of unusual, but not extraordinary, events led to an extended shutdown of a key ground water remediation process consisting of extraction wells, a fluidized bed activated media treatment process with filtration, and internal re-use of the treated water. The result of this was that the owner was at considerable risk of consent decree violations. Excellent engineering, standard process reviews, HAZOPs, and an aggressive maintenance program had not correctly identified risk. The need to find the true risks in the system was paramount.

Response:

The owner requested assistance in the evaluation of the overall process to identify the as yet hidden risks and related potential issues in order to prevent future similar and unforeseen occurrences. The Uberlytics criticality analysis approach and an early version of a software tool were brought in and applied in a workshop environment where the owner's system experts, including engineers and operators, were brought together with a criticality analysis expert facilitator.

Inclusion of 100% of the overall process units in this functional area of the refinery, and inclusion of pipes and valves as a separate group asset again proved essential. Successful discovery depended on the aforementioned and what had previously not been done: leveraging the network system approach rather than a discrete assets approach.

Results:

While the owner's focus was primarily on preventing recurrence of the event that had occurred, a full analysis of the system revealed several key aspects that led to a more comprehensive understanding of the process and associated risks, as well as providing specific improvement opportunities that would immensely reduce risk with remarkable little effort and expense. The networks systems approach to criticality analysis pinpointed the issues behind the major occurrence, and also found a number of other previously unidentified risks, including one previously thought to be insignificant. Commensurate simple and cost effective solutions that immensely reduced the risk profile for the entire process in all risk areas were identified.

Base Case 2: Refinery boiler feed pretreatment RO system

Situation

A boiler feed water pretreatment system was due for an expansion and had undergone several design reviews, two HAZOP reviews, and several construction reviews. However, given the importance of this system within the refinery operation, as well as being a key part in the refinery overall water reuse initiative, the refinery elected to commission an analysis on the plant boiler feed water pretreatment system. The decision to perform a criticality analysis was also based, in part, on the success of the previous review methodology. However, since the system had also undergone some annual refurbishments and, after 18 years of operation, was thought to be fully ‘known', it was not expected that any previously unidentified issues would surface. This assumption proved to be false.

Response

As in the previous example, the success of this analysis again hinged on the flexibility of the analysis to include not only areas such as safety, environmental, capacity, and up-time, but also commercial terms, liquidated damages, and contract terms. This proved to be a truly unique analysis in comparison to the many previous ones that had been conducted on the process. The results were reflective of more than just a process view, showing the unique position of this process's risk profile within the larger context of the client operation.

The same steps were conducted as in the previous case, except in here an asset registry had previously been generated as part of a HAZOP review. The database was imported for the analysis. Again, a team comprising various client experts was assembled with a criticality analysis expert facilitator, and again the unique approach proved essential.

Results

While the analysis did clarify and further define the generally known critical areas central to the overall mission, the analysis brought to the surface as yet uncovered critical risks inherent to the system as well as a different than expected risk ranking. One system was a particular surprise to the investigating team as it could shut down the whole operation in a single shift, and two other systems were elevated to a higher criticality ranking than previously thought given the potential for overall system failure. The inclusion of contractual and commercial evaluation parameters highlighted the relationship between process and commercial risks that had previously gone unnoticed. This provided the opportunity to monetize those risks for a more comprehensive business risk evaluation.

Conclusion

A sophisticated yet efficient approach is required to identify all risks, including hidden ones, in a cost effective way that also lends itself to solution identification. Conventional analysis and evaluation often fail to identify some significant mission critical risks because they do not take context into account. The functional systems approach to criticality analysis, FSCA, is a unique and proven approach that consistently meets and exceeds that objective. This approach to criticality analysis reliably identifies previously undiscovered system risks, means of mitigation, does so in a more efficient way than conventional means, and delivers verifiable and reproducible results.

Ignorance of risk it is not absence of risk.

# 349504 Can Black Swans be Red Herrings?


**Abstract:**

At present, our industry faces its biggest challenges due to mismanaged risks and the need to be sustainable. Over a period of time, along with the experience of various major accident events, risk management philosophies and techniques have become more sophisticated. However, there still remains much opportunity to extract better value from risk management activities and investments. This presentation will discuss some of the current weaknesses and highlight specific opportunities for improvement. It is intended to provoke change in how we look at existing risk management processes.

Specific topics will include:

- The potential for "black swan" events is a fallacy. Rather, they are a failing of threat identification, risk assessment and/or mitigation investment decisions.
- Rather than truly exploring the specific causes that might be applicable in that special situation, many threat identification exercises are content with simply listing the generic causes of potential failures.
- Current technology and thinking could enable much more value to be extracted from expensive hazard identification activities that have not evolved since their inception in the 1960s.
- Many risk metrics do not have a practical intrinsic meaning. There is a need to monetize risk metrics so that their intrinsic meaning can be better understood and they can be better related to risk mitigation investments.
- Historically, much risk management effort has gone into demonstrating that the status quo is OK. We need to truly embrace the concept of continuous improvement and focus risk management activities on helping to determine where we can make our best risk mitigation investments.

This will improve our business.

# 349717 Emergency Response Plan: It's Above and Beyond Best Practices

**Abstract:**

Human life is of prime importance in the any industry – offshore, refinery, chemicals, mining, etc., where potential for exposure to hazards can result in harm to people. Although the best way of managing an unexpected event or emergency is through prevention, it is also critical to have measures in place such as a robust Emergency Response Plan (ERP) to manage personnel in emergencies. A well-designed ERP that includes proper escape plans and rescue facilities and procedures can decrease confusion and increases safety of personnel. Among various elements of ERP one of the key decisions to be made is whether or not to follow a Temporary Refuge (TR) philosophy which begins with instructions to assemble in the muster points. Currently, there are regulations and guidelines that require or recommend having a TR and some that do not. This paper explores various escape and rescue philosophies, and the pros and cons of having an evacuation plan. These will be discussed to provide a clearer picture on implications of having one onboard. Chain of command during an emergency is another aspect of ERP that needs clarity. This paper will gather information from major accident investigations and the current ERP regulations and best practices. This information will be combined with expert experience with process safety design and human factors to help determine pros and cons of including a TR in the ERP as well as show the significance of having a specified chain of command for all emergency events.

# 349885 Process Safety Hazard Management Plan : Help You in Sustaining Production and Preventing Lossess


**Abstract:**

## ABSTRACT

*Managing mature assets is a challenge, especially when we have to deal with production target. Sometimes it is not easy to straight on safety as your first priority while in the other hand we need to maintain the production or even increase it. How safety and production can stand together without scarifying each other? Process safety hazard management plan is the key answer to resolve this matter.*

*Process safety practices and formal safety management systems have been in place in some companies for many years, including Pertamina Hulu Energy ONWJ which brought the standards from former company. Process safety hazard management plan is a part of the implementation during the operation stage. Not only assessing safety as a behavior but also assessing the plant and process for reduction in potential major accident risk and an improved plant performance. What can be done on the plant and process area to ensure that hazards are identified, addressed and the consequences are reduced?*

*This paper describes how process safety hazard manament plan can help you in sustaining your production, which includes:*

- *identifying your process hazard and risk*
- *learning your protection layer (Swiss cheese model)*
- *agreeing your acceptable risk*
- *giving commitment on your actions, who is to do what*

*In process safety perspective, all incidents are preventable. The key element is managing risk consistently to face any obstacles, minimize losses and achieve your goals.*

# 349882 Process Risk Assessment and "Safe Area" for a Petro-Chemical Plant in China


**Abstract:**

**Process Risk Assessment and "Safe Area" for a Petro-Chemical Plant in China**



***Jing Yu***, DuPont Engineering, Shanghai, China

***S. Dharmavaram***, DuPont Engineering, Wilmington, DE, USA

***Ji Ming Wang***, SINOPEC, Beijing, China



Many process hazards exist and highly hazardous events are possible in a typical petro-chemical plant. Even though the frequencies of such events occurring are low, the community has limited knowledge of consequences, frequencies, and process risk. Process hazards are like wild animals trapped in cages. Understanding hazards and minimizing risk can still lead to safe operation and contribute to society. Severe objections have been raised recently on building a para-xylene plant in China.

Risk assessment for a petro-chemical plant starts with a thorough identification and understanding of process hazards. Such process safety information is important for conducting process hazards analyses. Practical experience on incidents that have occurred in the past in China and elsewhere the world is also very valuable for determining the frequency of hazardous events.

In this paper/presentation, information about the severity and consequences of hazardous events at a para-xylene plant and associated process risk will be presented. Detailed modeling done for simulating the consequences of accidents and a study of domino effects to determine how fires/explosions propagate within a plant will be discussed. Use of process dynamic modeling tools such as Aspen® to calculate process conditions leading to incidents will be described. Deflagration and detonation tests used as a basis for Aspen® modeling to identify the worst-case scenarios for fires/explosions and toxic releases will also be presented. An understanding of off-site and on-site consequences and frequencies of hazardous events is used ultimately to determine a "safe area" concept for a petro-chemical plant, which will help local government make proper decisions and help the community in understanding process risks.

# IDENTIFYING AND QUANTIFYING MAJOR HAZARD FOR PLATFORMS DECK RAISING USING SYNCHRONOUS HYDRAULIC JACKING SYSTEM

**Authors:**

**Akhmad Tutus Harmantoro**
**PT Pertamina Hulu Energy Offshore North West Java (PT PHEONWJ)**
**Arkadia Office Building, TB Simatupang Street, Kav. 88, South Jakarta, Indonesia, 12740**
**akhmad.harmantoro@pertamina.com**

**Margaretha Thaliharjanti**
**PT Pertamina Hulu Energy Offshore North West Java (PT PHEONWJ)**
**Arkadia Office Building, TB Simatupang Street, Kav. 88, South Jakarta, Indonesia, 12740**
**margaretha.thaliharjanti@pertamina.com**

Prepared for Presentation at
American Institute of Chemical Engineers
2014 Spring Meeting
10th Global Congress on Process Safety
New Orleans, LA
March 30 – April 2, 2014

UNPUBLISHED

# IDENTIFYING AND QUANTIFYING MAJOR HAZARD FOR PLATFORMS DECK RAISING USING SYNCHRONOUS HYDRAULIC JACKING SYSTEM

**Akhmad Tutus Harmantoro**
**PT Pertamina Hulu Energy Offshore North West Java (PT PHEONWJ)**
**Arkadia Office Building, TB Simatupang Street, Kav. 88, South Jakarta, Indonesia, 12740**
**akhmad.harmantoro@pertamina.com**

**Margaretha Thaliharjanti**
**PT Pertamina Hulu Energy Offshore North West Java (PT PHEONWJ)**
**Arkadia Office Building, TB Simatupang Street, Kav. 88, South Jakarta, Indonesia, 12740**
**margaretha.thaliharjanti@pertamina.com**

**Keywords:** Failure Mode, Effect, and Critically Analysis (FMECA), Quantitative Risk Analysis (QRA), subsidence, java sea, deck raising, hydraulic jacking system.

## Abstract

Lima Flow Station is one of the gathering manned stations of Pertamina Hulu Energy Offshore North West Java (PHEONWJ) which is located in Java Sea, Indonesia. The Lima Flow Station consists of four bridge linked platforms and a bridge linked flare structure. Lima Flow Station had experienced subsidence for about 3 meters since it was built in 1973 to early 2013. The subsidence caused significant reduction in the platform air gaps (distance between cellar deck bottom of steel and maximum wave height) which caused Lima Flow Station to not be reliable for safe operation. The synchronous hydraulic jacking system was selected to raise the platform to 4 meters elevation and remedy the air gap for safe operation. Three platforms, Flare tripods, and interconnecting bridges were raised while one platform was not raised as it was isolated and utilized as a muster point and for personnel transfer purposes only. The same technology has been successfully applied in Gulf of Mexico in 2006.

Quantitative Risk Assessment (QRA) study was conducted for the design of a synchronous hydraulic jacking system to measure whether the design provided tolerable risk or not and identify mitigation of any intolerable risk. The assessment was carried out in two main parts: hazard identification in the form of failure modes, effects and criticality analysis (FMECA) to determine major accident events to be carried forward into the QRA; and determination of the impact of safety risk to personnel and assessment of financial risks in the form of delays to resumption of production.

FMECA was used for qualitative and semi-quantitative analysis of engineering systems. The reviewed system was divided into system, subsystem, and equipment. Failure mode, cause, safeguard and effect/consequence for equipment are identified. High consequence was considered as major accident and was further studied in QRA.

QRA consisted of Fault Tree Analysis (FTA) to estimate the probability of a major accident event identified by FMECA, Event Tree Analysis (ETA) to estimate the probability of a major accident leading to the ultimate event, Consequence analysis to estimate the impact of the ultimate event, risk estimation by multiplying probability and impact of ultimate event, risk assessment by comparing estimated risk and risk criteria, and identification of mitigation for any major accident having intolerable risk.

FMECA has identified some major accident events caused by misalignment and structural failure during deck rising. The results of the QRA determined that the individual risk per annum for the most exposed individual was still within tolerable risk of company's criteria as the deck raising was conducted during shutdown and minimum manning level. While the financial risk was high for platform damage due to structural failure for lifted platforms as there was no detection method for structural failure and sudden movement in case of structural failure may cause major damage. The financial loss was mostly contributed by production loss. The major contributor for structural failure was weld failure. Recommendation was raised to minimize the risk by providing extra quality check for welding and material of structural parts on fabrication and installation.

By implementing good project and risk management, Lima Flow Station Deck Raising has completed successfully and started up in 2013 without any accident.

## 1. Introduction

LIMA Flow Station is one of gathering manned station of Pertamina Hulu Energy Offshore North West Java (PHE ONWJ) is located on Java Sea which were installed in 1973. It gathers and produces oil and gas from the LIMA Normally Unmanned Installations (NUIs), with current production rates of 3.000 bopd and 20-30 MMscfd gas and also gathers and produces gas from other stations around 200 MMscfd and oil 4000 bopd. It consists of four bridge linked platforms: LA, LPRO, LCOM, and LSER; and a bridge linked flare structure.

Lima Flow Station has been experiencing subsidence of about 3 meters since it was built in 1973 to early 2013. The subsidence caused significant reduction in the platform air gaps (distance between cellar deck bottom of steel and maximum wave height) which caused Lima Flow Station to not be reliable for safe operation. The synchronous hydraulic jacking system was selected to raise the platform to 4 meters elevation and remedy the air gap for safe operation. LCOM platform, LPRO platform, LSER platform, Flare tripods, and interconnecting bridges were raised while LA platform was not raised as it was isolated and utilized as a muster point and for personnel transfer purposes only. The same technology has been successfully applied in Gulf of Mexico in 2006.

As deck raising activity was a new technology for PHEONWJ, it was decided to conduct a detailed risk assessment on the design of the synchronous hydraulic jacking system proposed by Contractor as part of project risk assessment. It was expected that the risk can be measured quantitatively, whether the design provided tolerable risk or not, the major contributor to risk can be identified, and recommendations can be established to reduce the risk for intolerable risk. The risk assessment will focus on fatality and financial risk.

## 2. Deck Raising Overview

The platforms to be raised were LCOM platform, LPRO platform, LSER platform, including interconnecting bridges and Flare tripods. LA platform was not raised as it was isolated and utilized as a muster point and for personnel transfer purposes only. The deck raising was conducted in two stages due to minimum remaining air gap. On the first stage, the platforms were raised 1 meter and on the second stage the platforms were raised for another 3 meters to give a total of 4 meters elevation.

The deck raising system consisted of 3 main systems: Hydraulic System, Control System, and Structural Parts. The hydraulic system was used to raise the platform by using hydraulic rams. The control system was to ensure the same deck raising elevation and speed to avoid platform tilting during raising. The structural parts consisted of pad eyes for hydraulic jack support, sleeves as the extended deck legs were also used to protect from lateral load during raising, and sleeve pins for temporary securing after completing deck raising.

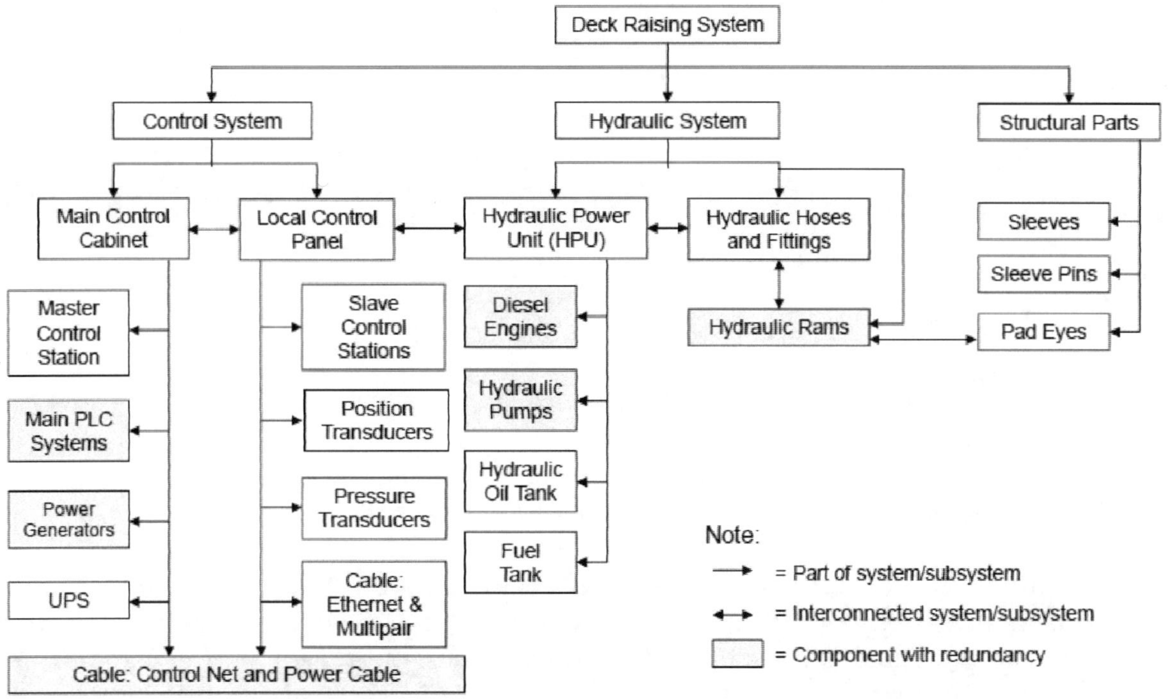


**Figure 1** Simplified Function Diagram For Deck Raising System

# 3. Methodology

The overall methodology for the risk assessment is shown in the flow chart on Figure 2.

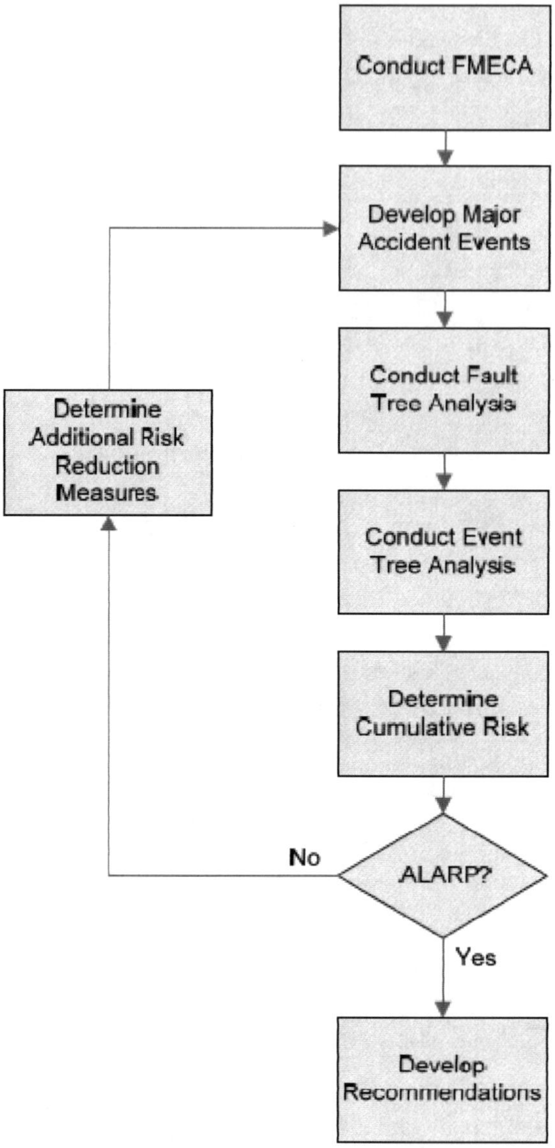


**Figure 2** Flow Chart for Risk Assessment Methodology

### *3.1 Failure Mode, Effect, and Critically Analysis (FMECA)*

Failure Mode, Effect, and Critically Analysis (FMECA) was used for qualitative and semi-quantitative analysis of engineering systems. The review was performed in the meeting with personnel that have good knowledge (designers/operators) of the system. The reviewed system was divided into system, subsystem, and equipment. Failure mode, cause, safeguard and effect/consequence for equipment are identified. Critically Analysis was conducted by using PHE ONWJ risk matrix as reference.

## 3.2 *Development of Major Accident Events (MAE)*

Major Accident Events (MAE) are any unwanted events with potential to cause injury property damage, business loss, and environmental damage. From FMECA result, any event having major and catastrophic consequence based on PHE ONWJ risk matrix was included as MAE and the risk to be further analyzed.

## 3.3 *Fault Tree Analysis (FTA)*

A fault tree analysis is developed based on a fault tree diagram. A fault tree diagram is a structured logic diagram which shows how series and/or combination of failures may lead to MAE. By inputting available failure frequencies into the fault tree diagram, the frequency of MAE can be estimated. Failure frequency was obtained from vendor data, generic failure frequency databases such as OREDA 2009[1] and NPRD 2011[5], and engineering judgments.

## 3.4 *Event Tree Analysis (ETA)*

An event tree analysis which is complementary of fault tree analysis is developed based on an event tree diagram. For developing event tree diagrams, layers of protection were identified. Consecutive events from the initiating event can be identified for each layer by developing two branches for its success and failure. The ultimate consequence can be obtained if review of all layers of protection has been completed. By inputting probability of failure or success to the event tree diagram, the frequency of ultimate consequence was estimated. The probability data was obtained from OGP [4] for ignition probability and engineering judgments for other probabilities, for example the probability of successful escape during structural collapse.

## 3.5 *Consequence Analysis*

The consequences being considered for this risk assessment are uneven lifting of deck leg leading to structural damage and damage to hydrocarbon piping leading to personnel fatality due to flash fire, personnel fatality due to impact of structure collapse, fatality due to personnel over board, and business loss.

## 3.6 *Risk Estimation*

The results of frequency and consequence analysis are combined to obtain the risk for each ultimate consequence. The risk is represented in the form of Individual Risk Per Annum (IRPA) for fatality risk and Probable Loss for Financial Risk. Since the platform raising activity is a short duration activity (3 months), the estimated frequencies will be equivalent 'annualized' frequencies to enable risk comparisons.

### 3.6.1 Individual Risk Per Annum (IRPA)

Individual Risk Per Annum (IRPA) shows the risk exposure to any individual by considering the duration of a specific person on board. To calculate the IRPA, it is required to obtain the fraction of time spent on board and Location Specific Individual Risk (LSIR) which shows the risk exposure to any individual who is present in a particular area for a full one year.

$$LSIR(/year) = \sum_i f_i \times P_{i,impact} \times P_{i,exposure}$$

$$IRPA(/year) = \sum_i LSIR_i \times f_{time}$$

Where :

$fi$ = event frequency (event/year)

$P_{i,impact}$ = probability of harm from individual incident (injury or fatality)

$P_{i,exposure}$ = probability that an individual may be present within the impairment zone

$f_{time}$ = Fraction of time the specified worker group is present at that point

### 3.6.2 Probable Loss for Financial Risk

The financial loss from a major accident event may consider the following costs:

- Asset damage and replacement costs
- Loss of production during the period of investigation, repair and decommissioning
- Cost of investigation

The total probable loss for an associated operation is therefore measured as:

$$Probable\ Loss\ (\$/year) = \sum_i L_i \times f_i$$

Where,

$Li$ = Loss from an accident event ($)

$fi$ = accident event frequency (event/year)

## *3.7 Risk Criteria*

The result of risk estimation is then compared with the risk criteria to identify whether the risk of the deck raising system design is within the tolerable region or not. IRPA Criteria is used for fatality risk while PHE ONWJ risk matrix will be used for financial risk.

3.7.1 Individual Risk Per Annum (IRPA) Criteria

The Individual Risk Per Annum (IRPA) Criteria as shown in the Figure 3 is used to define whether the risk is tolerable or not which refers to Health & Safety Executive in the UK. The risk below $1x10^{-6}$ is considered acceptable risk, while risk above $1x10^{-3}$ is considered unacceptable risk. The risk between $1x10^{-6}$ to $1x10^{-3}$ is considered as tolerable risk. For unacceptable risk, implementation of mitigative action is mandatory to reduce the risk to at least the tolerable region. For tolerable risk, basically the risk can be undertaken and recommendation may be considered to be established.

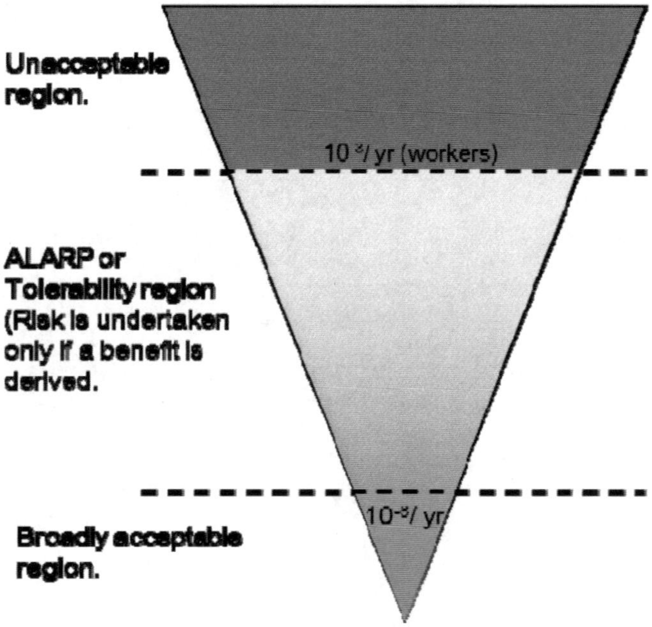


**Figure 3** Individual Risk Per Annum (IRPA) Criteria

3.7.2 Financial Risk Criteria

The PHE ONWJ Risk Matrix as shown in the Figure 4 is used to define whether the risk is tolerable or not. The risk that falls on green cells is considered as low / acceptable risk, risk that falls in yellow cells is considered as medium / tolerable risk, the risk that falls in red cells is considered as high / unacceptable risk.

### *3.8 Develop Recommendation*

For high / unacceptable risk, the additional examination is performed to identify the factor that provides major contribution to the risk. The recommendation is then developed based on the identified major risk contributors. For some cases, residual risk after implementing recommendations is not estimated due to data difficulties.

| | | | Likelihood | | | | |
|---|---|---|---|---|---|---|---|
| | | | **Almost Impossible** $<10^{-6}$ per year | **Very Low** $10^{-6} - 10^{-4}$ per year | **Low** $10^{-4} - 10^{-2}$ per year | **Medium** $10^{-2}$ – 1 per year | **High** > 1 per year |
| Severity | A | **Catastrophic** > $ 200 M | 6 | 7 | 8 | 9 | 10 |
| | B | **Major** $ 100 M - $ 200 M | 5 | 6 | 7 | 8 | 9 |
| | C | **Serious** $ 50 M - $ 100 M | 4 | 5 | 6 | 7 | 8 |
| | D | **Moderate** $ 10 M - $ 50 M | 3 | 4 | 5 | 6 | 7 |
| | E | **Minor** $ 1 M - $ 10 M | 2 | 3 | 4 | 5 | 6 |
| | F | **Slight** ≤ 10 M | 1 | 2 | 3 | 4 | 5 |

**Figure 4** Financial Risk Criteria

# 4. Result

## *4.1 Major Accident Event (MAE)*

Based on Failure Mode, Effect, and Critically Analysis (FMECA), it was identified that uneven raising where there is a deck raising elevation difference relative to other platforms and/or bridges will cause the platform to tilt. The tilting platform and/or bridge will cause the pipe bridge or bridge structure to break down if the stress exceeds the material allowable stress or shifting of the center of gravity that will cause the load to concentrate on a specific platform leg leading to platform collapse. The uneven raising may be caused by misalignment of hydraulic rams during raising or failure of a structural part to withstand the topside load. Table 1 summarizes the identified Major Accident Event (MAE) including its frequency based on Fault Tree Analysis (FTA).

## *4.2 Individual Risk Per Annum (IRPA)*

Based on frequency and consequence analysis, Table 2 and 3 show the resulting Location Individual Specific Risk (LSIR) and Individual Risk Per Annum (IRPA). The risk for misalignment and structural failure is shown to be able to identify the contribution of failure to the risk.

**Table 1** Summary of Major Accident Event (MAE) and Its Probability

| MAE No. | Description | Event probability (per activity) |
|---|---|---|
| 1 | Potential for damage to LCOM – LSER bridge (misalignment) | $5.40 \times 10^{-5}$ |
| 2 | Potential for damage to LPRO – LSER bridge (misalignment) | $5.40 \times 10^{-5}$ |
| 3 | Potential for damage to LPRO – LA bridge and LA lift frame (misalignment) | $6.14 \times 10^{-5}$ |
| 4 | Potential for damage to LPRO – Flare support bridge A and Flare lift frame A (misalignment) | $6.14 \times 10^{-5}$ |
| 5 | Potential for damage to Flare support bridge A – Flare support bridge B and Flare lift frame B (misalignment) | $6.97 \times 10^{-5}$ |
| 6 | Potential for damage to LCOM platform (misalignment) | $8.57 \times 10^{-5}$ |
| 7 | Potential for damage to LSER platform (misalignment) | $8.57 \times 10^{-5}$ |
| 8 | Potential for damage to LPRO platform (misalignment) | $8.57 \times 10^{-5}$ |
| 9 | Potential for damage to LCOM platform (structural) | $7.59 \times 10^{-4}$ |
| 10 | Potential for damage to LSER platform (structural) | $7.59 \times 10^{-4}$ |
| 11 | Potential for damage to LPRO platform (structural) | $7.59 \times 10^{-4}$ |
| 12 | Potential for damage to LPRO – LA bridge and LA lift frame (structural) | $7.22 \times 10^{-4}$ |
| 13 | Potential for damage to LPRO – Flare support bridge A – Flare support bridge B and Flare lift frame A (structural) | $1.23 \times 10^{-4}$ |
| 14 | Potential for damage to Flare support bridge A – Flare support bridge B and Flare lift frame B (structural) | $1.23 \times 10^{-4}$ |

The maximum IRPA is $4.6 \times 10^{-6}$ located on LPRO platform. The risk is considered tolerable as the value is within the range of $1 \times 10^{-6}$ and $1 \times 10^{-3}$. The tolerable fatality risk can be achieved as deck raising was conducted during shutdown and minimum manning level. Structural failure contributes more than 50 % for LSIR and more than 90% for IRPA, except for failure on LPRO-LSER bridge where structural failure is not considered as Major Accident Event. The structural failure provides major contributor as there was not detection method and sudden movement in case of failure may cause major damage.

**Table 2** Location Specific Individual Risk (LSIR) Result

| Location | LSIR (risk / year) | | |
|---|---|---|---|
| | Misalignment | Structural | Total |
| LCOM | 1.1E-05 | 7.6E-05 | 8.7E-05 |
| LSER | 1.4E-05 | 7.6E-05 | 9.0E-05 |
| LPRO | 1.4E-05 | 8.8E-05 | 1.0E-04 |
| HPU (LPRO-LSER) | 2.7E-06 | 0.0E+00 | 2.7E-06 |
| HPU (FSA) | 1.3E-05 | 2.5E-05 | 3.8E-05 |
| HPU (FSB) | 7.0E-06 | 2.5E-05 | 3.2E-05 |
| HPU (LPRO-LA) | 6.1E-06 | 7.2E-05 | 7.8E-05 |
| **Total** | **6.9E-05** | **3.6E-04** | **4.3E-04** |

**Table 3** Individual Risk per Annum (IRPA) Result

| Location | IRPA | | |
|---|---|---|---|
| | Misalignment | Structural | Total |
| LCOM | 6.2E-08 | 3.9E-06 | 3.9E-06 |
| LSER | 7.7E-08 | 3.9E-06 | 3.9E-06 |
| LPRO | 7.9E-08 | 4.5E-06 | 4.6E-06 |
| HPU (LPRO-LSER) | 1.5E-08 | 0.0E+00 | 1.5E-08 |
| HPU (FSA) | 7.2E-08 | 1.3E-06 | 1.3E-06 |
| HPU (FSB) | 3.8E-08 | 1.3E-06 | 1.3E-06 |
| HPU (LPRO-LA) | 3.4E-08 | 3.7E-06 | 3.7E-06 |

### *4.3 Financial Risk*

The result of financial risk is shown in Table 4 which shows the frequency of event, potential financial loss which is detailed to production loss and repair cost, and the risk figure which refers to the PHE ONWJ risk matrix.

Structural failure in LCOM, LSER, and LPRO Platform is considered as high / unacceptable risk as it causes major damage and requires a long period for repair. More than 50% of the potential loss is caused by production loss.

**Table 4** Financial Risk Result

| | MAE No. | Event frequency (per year) | Potential Loss (USD $M per year) | | | Risk Ranking |
|---|---|---|---|---|---|---|
| | | | Production | Repair | Total | |
| Damage due to misalignment during raising | 1 | $5.56 \times 10^{-5}$ | 41.4 | 3.0 | 44.4 | Low |
| | 2 | $5.56 \times 10^{-5}$ | 41.4 | 3.0 | 44.4 | Low |
| | 3 | $6.28 \times 10^{-5}$ | 22.3 | 2.5 | 24.8 | Low |
| | 4 | $6.28 \times 10^{-5}$ | 41.4 | 3.9 | 45.3 | Low |
| | 5 | $7.08 \times 10^{-5}$ | 41.4 | 4.3 | 45.7 | Low |
| | 6 | $8.57 \times 10^{-5}$ | 44.6 | 8.8 | 53.4 | Low |
| | 7 | $8.57 \times 10^{-5}$ | 44.6 | 5.8 | 50.4 | Low |
| | 8 | $8.57 \times 10^{-5}$ | 44.6 | 7.3 | 51.9 | Low |
| Damage due to structural failure during raising | 9 | $7.59 \times 10^{-4}$ | 69.5 | 54.3 | 123.7 | High |
| | 10 | $7.59 \times 10^{-4}$ | 69.5 | 30.3 | 99.7 | High |
| | 11 | $7.59 \times 10^{-4}$ | 69.5 | 42.3 | 111.7 | High |
| | 12 | $7.22 \times 10^{-4}$ | 41.4 | 3.2 | 44.6 | Medium |
| | 13 | $1.23 \times 10^{-4}$ | 41.4 | 5.9 | 47.3 | Medium |
| | 14 | $1.23 \times 10^{-4}$ | 41.4 | 5.9 | 47.3 | Medium |

### *4.4 Recommendation*

The study revealed that structural failure is the major contributor to both fatality and financial risk while production loss is the major contributor to financial risk. Further examinations found that welding failure contributes more than 90 % to structural failure. Therefore, the following recommendations are developed to reduce the risk:

- Conduct robust quality check for welding and material of structural parts on fabrication and installation
- Set-up contingency plans during an accident event to minimized downtime to reduce financial loss, such as installation of a by-pass line.

## 5. Project Implementation

The Lima Flow Station Deck Raising was successfully completed in September 2013. Now the production is operated under normal condition, and is expected to stay in operation until 2026. During the project execution, the recommendations from the risk assessment were consistently implemented. The structural part design was carefully reviewed by a third party and a robust quality check was performed during purchasing, fabrication, and installation. A series of production contingency plans were also set up to minimize loss  in case of failure.

## 6. Conclusion

The potential for overlooked hazards is high for implementation of new technology, such as implementation of the synchronous hydraulic jacking system for Lima Flow Station Deck Raising. Whenever qualitative risk assessment is considered inadequate, , quantitative risk assessment can be conducted by adopting methodology as described on this paper.

The quantitative risk assessment study for Lima Flow Station Deck Raising provided these conclusions:

- Major Accident Events include uneven raising whether it is caused by misalignment of hydraulic rams during raising or failure of a structural part to withstand the topside load. This uneven raising will cause bridge and pipe bridge failure due to excessive stress or platform collapse due to shifting of the center of gravity leading to a concentrated load on a specific leg.
- The maximum IRPA is 4.6x10-6 located on LPRO platform. The risk is considered tolerable as the value is within range of 1x10-6 and 1x10-3.
- Structural failure in LCOM, LSER, and LPRO Platform is considered as high / unacceptable risk as it causes major damage and requires a long period to repair.
- Structural failure contributes more than 50% to fatatility and financial risk while production loss contributes more than 50% to financial risk.
- It was recommended to provide robust quality check for welding and material of structural parts on fabrication and installation and set-up contingency plans during an accident event to minimize downtime to reduce the risk.
- By implementing good project and risk management, Lima Flow Station Deck Raising has completed successfully and started up without any accident.

## 7. References

[1] OREDA, Offshore Reliability Data Handbook, 2009, Vol.I, Topside Equipment, 5th Edition, Oreda participants, Distributed by DNV, Hovik, Norway.

[2] Mannan,S. (Editor), Lees' Loss Prevention in the Process Industries, 3rd Edition, Elsevier Butterworth-Heinemann, 2005

[3] HSE Books, HMSO, London, Reducing Risks, Protecting People: HSE's Decision Making Process, 2001

[4] Ignition Probabilities, OGP Risk Assessment Data Directory Report No. 434 – 6.1, March 2010.

[5] NPRD, Nonelectronic Parts Reliability Data, Reliability Information Analysis Center (RIAC), Utica, NY, 2011.

[6] Safety Equipment Reliability Handbook, Exida, 2005, Exida.com L.L.C, Sellerville, PA, USA.

# 348096 A Creative & Strategic Initiative (Champions Model) In Managing Process Hazard: Process Safety Culture

**Abstract:**

<u>A Creative & Strategic Initiative (Champions Model) In Managing Process Hazard: Process Safety Culture</u>

**Rathore, Sharad**

**Health Safety & Environment**

Honeywell India Technology Centre, UOP IPL, Gurgaon

Abstract

A robust Process Safety Culture within an organization is as important as having a reliable process safety management system; it makes process safety a true part of the day-to-day operations. It develops trust and a sense of importance for process safety, allowing process safety management systems to be effective

Culture is the tendency in all of us - our organization - to want to do the right thing in the right way at the right time, ALL the time – even when no one is looking. Culture is the result of all the actions - and inactions - in institutional/workforce memory. These shape/influence individual behaviors and tendencies. The culture transcends the members of the organization, passing on to new members and enduring over time .Good cultures evolve from common values and attitudes developed as the group seeks to properly solve common problems. It is hard to measure and more difficult to change. Culture will be the "root cause" of the decade.

The significance of Safety Culture is becoming increasingly recognized as a key factor to achieve the safe operation not only of hazardous processes but also in Experimental Pilot plants. This paper describes the systematic approaches that are being practices at HITC R&D facilities to go through in their journey to a world class safety culture and applies this concept to Process Safety Management in experimental pilot plant

The foundation for this journey will be illustrated in a model that describes the strategy for HSE culture strength. The model (called Champions Model) has been calibrated against performance and is based on years of data derived from HSE metrics HLS- ACS facilities. In addition, useful tools, employed by facilities to improve their safety culture and performance, will be presented

Key words: Process Safety Management, Safety Culture

# 348700 Defining Dust Hazard Areas

**Abstract:**

Defining Dust Hazard Areas

Michelle Murphy, Partner

ioMosaic Corporation

As part of the 2013 update, NFPA 654 provides clarification on how to determine if a flash fire or explosion hazard exists, using a hazard assessment. This hazard assessment incorporates dust characteristics, design of the facility and localized accumulations in the analysis. This presentation will present a multi-step approach to conducting these hazard assessments, simplifying the process. Once complete, the hazard assessments define the areas to evaluate during the process hazard analysis and assure that valuable resources are spent appropriately.

# Challenges and achievements in implementing Management of Change system at Binh Son Refinery, Vietnam

**Bong Nguyen Thanh**
**Binh Son Refining and Petrochemical Company Limited**
**208 Hung Vuong Street, Quang Ngai City, Quang Ngai, Vietnam**

**Thuc Do Van**
**Binh Son Refining and Petrochemical Company Limited**
**208 Hung Vuong Street, Quang Ngai City, Quang Ngai, Vietnam**

**Vu Vo Hoang**
**Binh Son Refining and Petrochemical Company Limited**
**208 Hung Vuong Street, Quang Ngai City, Quang Ngai, Vietnam**

Prepared for Presentation at
American Institute of Chemical Engineers
2014 Spring Meeting
10th Global Congress on Process Safety
New Orleans, LA
March 30 – April 2, 2014

UNPUBLISHED

# Challenges and achievements in implementing Management of Change system at Binh Son Refinery, Vietnam

**Bong Nguyen Thanh**
**Binh Son Refining and Petrochemical Company Limited**
**208 Hung Vuong Street, Quang Ngai City, Quang Ngai, Vietnam**
**bongnt@bsr.com.vn**

**Keywords:** Management of Change, Process Safety, Management, MOC, Refinery, implementation

## Abstract

This paper describes the challenges and achievements in implementing a MOC system in BSR, Vietnam, where there is no regulations such as OSHA PSM rule. Binh Son Refining and Petrochemical Co. Ltd, a subsidiary of PetroVietnam commissioned its refinery and polypropylene units located at Dung Quat bay, Quang Ngai Province in the year 2009. This plant was designed and built by a consortium of international process licensors and detailed engineering contractors and has a capacity to handle 6.5 million tons per year of Bach Ho (80 %) and some other imported crude oils. This is the first and only refinery in Vietnam.

This paper describes the evolution of Management of Change system at BSR, the challenges faced and the achievements so far.

## 1. Introduction

Binh Son Refining and Petrochemical Co. Ltd, (BSR) a subsidiary of PetroVietnam commissioned its refinery and polypropylene units located at Dung Quat bay, Quang Ngai Province in the year 2009. This plant was designed and built by a consortium of international process licensors and detailed engineering contractors and has a capacity to handle 6.5 million tons per year of Bach Ho (80 %) and some other imported crude oils. This first and only refinery in Vietnam is a project of national importance, is responsible for the production of petroleum products, and is contributing to energy security and stabilization of oil prices in the domestic market of Vietnam.

The refinery configuration includes 14 technology process units to produce the following products: Polymer Grade Propylene, LPG, Domestic Gasoline, Kerosene and Jet A1, Auto Diesel and Fuel Oil.

**Table 1. Products of Binh Son Refinery**

| Product Name | Tons / year |
|---|---|
| LPG | 400000-420000 |
| A92 MOGAS | 1400000-1800000 |
| A95 MOGAS | 600000-700000 |

| | |
|---|---|
| Kerosene / Jet A1 | 80000-400000 |
| Auto Diesel | 2900000-3200000 |
| Fuel Oil (FO) | 60000-10000 |
| Polypropylene | 130000-150000 |

The refinery also has its own utility and power generation systems. Crude feedstock is unloaded via a single point mooring system and stored in the crude tank farm.An intermediate refinery tankage is provided within the refinery fence, however, finished products are stored in a separate product tank farm. Product loading facilities are provided for both Ocean and Coastal ships. Truck loading is also provided for supplying products locally.

The number of employees at the refinery is around 1100. A very systematic training of employees by the Licensors and EPC contractors, including training in overseas refineries was done before the commissioning of refinery and contributed to the successful commissioning and safe and stable operations thereafter.

## 2. QHSE Management and Process Safety Management

The refinery is certified to Quality Management Standard: ISO 9001:2008, Environmental Management System ISO 14001: 2004, and Management System for Occupational Health and Safety: OHSAS 18001: 2007. Under the QHSE management system, ensuring safe, stable and efficient production is given prime importance. The Process Safety Management (PSM) is part of the HSE Management system and enhances the effectiveness of HSE management system to deal with low frequency high consequence events.

The PSM system of BSR is adopted from the US OSHA PSM elements. References were also made to the UK COMAH Regulations, EU SEVESO Directives. DOE-HDBK-1100-2004, DOE-HDBK-1101-2004; OSHA 3132-2000, OSHA 3133-1994; CCPS; 2007, and Australian Major Hazard Facilities Regulations. The PSM system consists of 14 elements. Process Safety Management implementation started in the year 2010 and one of the initial elements that was put into practice is Management of Change.

Under OSHA PSM rule, Management of Change requires documented procedures aimed at insuring safe system operation as changes to the system occur. The management of change process guides updates or development to process safety information, process hazard analyses, operating procedures, and training to insure all elements of the PSM program are current.

## 3. Evolution of Management of Change system

The objective of Management of Change is to provide a system of administrative controls that ensures:

1. The proper consideration of direct and indirect potential hazards before the change is implemented;
2. That procedures are modified and essential training and communication occurs before the modification is put into service;
3. Enhanced emphasis on the preparation and retention of records of each change; and

4. The ability to audit the changes.

The procedure to follow Management of Change was first issued in 2009. Since then it has undergone many revisions to improve the Management of Change system and its practice. The improvements to the procedure were done based on periodical audits and addressing the gaps in the MOC system. The current MOC procedure is the sixth version from the beginning and put into practice since July 2013.

Key aspects of improvement to MOC procedure and the reason:

| Version | Key aspects of improvement | Reason |
|---|---|---|
| Version 2 | Develop supporting discipline and hazards review checklists for hazard evaluation and design reviews of change | - To assist people to address the important discipline concerns on safety, reliability, operability<br>- Improve the MOC documentation and tracking to ensure it complies with QHSE management system requirements |
| Version 3 | Adopt HAZOP and PSSR procedure to the implementation of MOC | Make links of the MOC procedure with other PSM elements |
| Version 4 | Integrate the MOC procedure with project implementation | To ensure it is in line with the organization and consistency in roles and responsibilities |
| Version 5 | Adopt the Safety Interlock and Protective Systems (SIPS) bypass procedure into MOC system | To manage all bypasses to safety interlocks (e.g. Maintenance Override Switch-MOS) and protective systems (RV, ESD, etc.) |
| Version 6 | Adopt some advanced PHA methods in hazard evaluation of changes such as SIL study, QRA | Improve quality and effectiveness of the process hazard analysis of changes |

The current MOC workflow consists of the following key steps:

1. Change initiation
2. Change screening and hazards evaluation
3. Design review and approval
4. Verification of modification and ready for start-up
5. Post change review

The key role in the success of MOC system implementation is played by MOC Element custodian and the Change Responsible Persons who are responsible for implementing each change.

## 4. Challenges in implementing of MOC

Apart from the most common problems in implementing of MOC that have been seen at our facility as of in worldwide industry, the company has faced some significant challenges which adhere to our local safety and management culture and experiences, which can be addressed as below:

### *4.1. Process Safety Cultural Issues*

In the initial years, a culture that encouraged raising process safety concerns regarding any change proposal was not existing. Corrective actions were not promptly taken when unsafe process safety conditions were brought to management's attention. The awareness of employee about safety was still limited in boundary of occupational safety such as slip, trip or fall. Beside that commitment to Process Safety Management was limited and inconsistent in the context of the production was the first interest.

### *4.2. Employee professionalism and experiences*

As mentioned earlier, Binh Son Refinery is the first refinery in Vietnam. Therefore BSR employees lacked experiences, not only in operating and maintenance but also in management of refinery. It gave a challenge in identification and consideration of suitable and sufficient information to support the MOC process. In addition, the Process Safety Information documents were not sufficient available, e.g., project engineering drawings had not been verified as as-built.

### *4.3. Organizational Issues*

In the early period of the refinery operating, there were several gaps and inconsistencies in defining roles, responsibilities and authorities among facility owner and disciplines. The designing of MOC system therefore became more complex. These gaps in the organization caused a tendency to pass the buck to others whenever a change was proposed.

The employee participation was one of our big challenges during the implementation of MOC and PSM as well. Less people are responsible for process safety task force. Workers at all level of the refinery did not actively participate in hazards evaluations and design review of changes.

One of the other organizational issues is that training and performance systems were not systematically established for effectively providing the knowledge and skill to the MOC implementers.

### *4.4. Other challenges*

The other elements of PSM system had not been implemented fully leading to gaps in MOC activities such as HAZOP study, PSSR, Operating Procedure and PSI availability and updating, etc.

The lack of a Vietnam regulatory in PSM like the OSHA PSM regulation also limited the application of system like MOC in BSR.

## 5. Achievements

As a result of a continuous improvement process, we have accomplished some encouraging achievements in implementation of the MOC system in our refinery which is in line with the implementation of the Process Safety Management in compliance with OSHA's PSM regulation:

1. Good commitment to adhere to MOC procedure at all levels of management for all changes other than replacement in kind.
2. The MOC system covers all disciplines. MOC is even raised for a variety of changes, not having process safety concerns. Examples are: changes such as format change in bill of loading or software upgrade also follow the MOC route and approval process and updating relevant documents.
3. A good training system is in place for maintaining the proper process safety awareness and competency of MOC implementers.
4. An effective compliance audit system which consists of 2 levels (Tier 1 and Tier 2) and continual improvement.
5. Competency of MOC implementers has been much improved and key performance indicators of MOC process have become the job performance evaluation indicators.

## 6. Way forward

The recent PSM internal high level audit has brought out the following gaps, not only with MOC but also with other PSM elements that make links to the MOC implementation:

- The compliance of capital project implementation with MOC requirements.
- Lack of editable engineering documents leading to some difficulties in updating of modification in affected drawings.
- Paper-based MOC document handling leading to cost and schedule over run and gaps on communication and tracking during MOC approval processing and execution

It is proposed to address the above gaps through short-term and long-term action plans for enhancement and consolidation of the MOC implementation within PSM system.

BSR is also developing an electronic-based MOC system to track MOC right from initiation and integrate MOC with work order and purchasing system using our current CMMS (Computerized Maintenance Management System).

Excellence is a never ending journey and BSR is moving towards a MOC system covering all kinds of change and having a simple robust process which is commensurate with the risk introduced by the changes.

## 7. References


[1] Process Safety Management Guidelines for Compliance, https://www.osha.gov/Publications/osha3133.html (accessed Feb. 20, 2014)

[2] Center for Chemical Process Safety, *Guidelines for Management of Change for Process Safety,* 2008

# 349936 Process Safety Culture Applied in Latin America Oil and Gas Industry: Experiences in Ecopetrol Colombia

**Abstract:**

In some installations of oil and gas industry a wide range of risks and hazards are likely to come up together, since there is usually a space shortage. Additionally the complexity of processes has been increased dramatically in petroleum and petrochemical industry, especially in Latin American oil and gas installations. Therefore, there must be a continuous effort to implement strategies that optimize the preventive treatment when hazards like space shortage and process complexity combined by a lack of discipline all mixed with a irresponsible cost reduction policies. This paper will show experiences in Colombia, focused in ECOPETROL improvement in that way.

# 348497 Ammonium Nitrate Condition-Dependent Thermal Decomposition

**Abstract:**

Runaway reactions present a potentially serious threat to the chemical process industry. In general, this research topic will investigate the mechanisms that drive these reactions, develop predictive models, and create appropriate and effective mitigation strategies. Specifically, this project will look at identifying the root causes associated with ammonium nitrate (AN) explosions, so that the risk can be reduced, while maintaining its agricultural benefit. This work focuses on developing reaction stoichiometry under different conditions, thermodynamic parameters and kinetic parameters related to AN thermal decomposition. In this research, Reactive Systems Screening Tool (RSST), Automated Pressure Tracking Adiabatic Calorimeter (APTAC), and Differential Scanning Calorimetry (DSC) will be used.

Ammonium nitrate can explode under favorable conditions, such as the existence of specific additives. Some materials change the chemical reaction conditions, inhibiting the explosion; others act as promoters, increasing the explosion risk. This study will look into the behavior of AN with different types of additives including promoters and inhibitors, combined and in isolation. Apart from that, weathering effect, humidity effect, and volume effect will be examined to find out how the storage condition will affect decomposition of AN. Those results will be compared with the work done in literature and some regulatory issues associated with it.

# 348758 CFD Modeling for Prediction and Prevention of Runaway Reaction

**Abstract:**

Thermal runaway reactions have caused many industrial incidents with dire consequences, including injury and death to humans, negative effects on the environment, and economic losses for companies. The U.S. Chemical Safety and Hazard Investigation Board (CSB) released a report on 167 reactive chemical incidents in the U.S. between 1980 and 2001. About 35% of the incidents were caused by runaway reactions. With the aim to prevent the occurrence of these incidents, it is necessary to have a good understanding of the events that can lead to a runaway reaction.

Mixing operations have an important effect not only in the adequate contact of reactants which affects the performance of the reaction, but also in removing the excess heat generated by the reaction. For exothermic reactions this removal of heat is a crucial operation because when the heat generated exceeds the heat removal, the excess of temperature will increase the heat generation rate, which would produce an auto-acceleration behavior triggering uncontrolled conditions of temperature and pressure inside the reactor.

The mixing inefficiencies lead to non-uniform mass flow distribution which causes inhomogeneous temperature profiles across the reactor. Therefore, it is important to predict this temperature distribution to locate and understand the characteristics of local hot spots that ultimately may provoke a thermal runaway.

Computational Fluid Dynamics (CFD) techniques have been used to identify local and instantaneous values of reactor temperatures and reactant concentrations in mixing tanks to indicate in advance the local hot spots in the reaction mixture. This kind of information is of particular interest to determine situations that can produce a thermal runaway in order to make engineering decisions to avoid them. Specifically, in some cases this information has been used to create an Early Warning Detection System (EWDS) for runaway detection.

The main objective of this work is to discuss the effect of mixing by paddle on the heat accumulation rate, considering the difference between the heat of reaction and the loss of heat by transfer. The results can provide a good understanding of the reaction system using an anchor impeller.

In this work, a time dependent simulation was carried out using a Multiple Reference Frame (MRF) technique; in this method a region away from the impeller is defined as a stationary frame and a region adjacent to the impeller is defined as a rotating frame, but the impeller itself remains stationary with respect to the MRF. Geometry of the tank was created similarly to the Mettler Toledo RC1*e* reaction calorimeter stirred by an anchor impeller. The simulations were carried out with a three dimensional CFD model using the ANSYS FLUENT 14.0 software.

A moderate to high exothermic reaction, the hydrolysis of acetic anhydride, is used as a model reaction in this work. This is a pseudo first-order reaction, ideal for validating the simulation by experimental results on the dynamic response of a calorimetric reactor. This reaction model is a simple mechanism that has been well studied with much experimental work reported in the literature.

The conclusions drawn from this work will have important implications to elaborate future research by developing a method for early warning detection of runaway reaction and extending this idea to more complex mechanisms of highly exothermic reactions.

# 348859 Sensitivity Analysis of Variables Affecting the Runaway Decomposition of Dicumyl Peroxide

**Abstract:**

Dycumil Peroxide (DCP) is widely used as an initiator, hardener and as a cross linking agent in the petrochemical industry. The use of this peroxide poses an intrinsic risk due to its reactivity and the possibility of its highly exothermic self-decomposition and subsequent runaway reaction, which is accompanied by a rapid rate of temperature and pressure increase.

The hazards associated with the process, transport and storage of DCP are clearly reflected by the frequent incidents that have occurred in Asia in the last two decades: Taiwan (2008) [1], Taiwan (2003) [2], Japan (1999) and Taiwan (1988)[1]. Though the accidents associated to DCP have not led to any fatality, it is important to be aware of its decomposition properties so as to prevent further incidents and major consequences.

Emergency relief systems are the most commonly selected measure to protect reactors from thermal explosions (CCPS, 1998[3]). The main advantages of their use are the independence from the control system and passivity (and therefore reliability and robustness). However, the exiting relief sizing methods are consolidated mainly in the case of vapor systems (in which the relief operation is usually enough to cool down the reacting mixture), leading to oversizing in the case of untempered gassy systems (such as the decomposition reaction of DCP[4]).

In the present work, the results of adiabatic experiments using two different calorimeters (PHI TEC I and PHI TEC II) are presented. A sensitivity analysis around the equipment configuration (open and close cell), the concentration of the peroxide (diluted in2,2,4-Trimethyl-1,3-pentanediol diisobutyrate), initial fill level, starting pressure and phi factor on the different runaway parameters (maximum pressure, gas generation, maximum rise of temperature and pressure rates, time to maximum rate and onset temperature) is discussed. Finally vent sizing calculations are performed using different methodologies available in the open literature. Based on the calorimetric data and the vent sizing calculations the limitations of these methodologies are discussed and recommendations for the proper vent sizing and safe scale up of the system studied are made.

**References**

1. Shen S-J, Wu S-H, Chi J-H, Wang Y-W, Shu C-M. Thermal explosion simulation and incompatible reaction of dicumyl peroxide by calorimetric technique. *Journal of Thermal Analysis and Calorimetry*. 2010;102(2):569–577. doi:10.1007/s10973-010-0916-4
2. Wu KW, Hou HY, Shu CM. Thermal phenomena studies for dicumyl peroxide at various concentrations by DSC. *Journal of Thermal Analysis and Calorimetry*. 2006;83(1):41–44. doi:10.1007/s10973-005-6983-2.
3. Center for Chemical Process Safety, 1998, Guidelines for Pressure Relief and Effluent Handling Systems, American Institute of Chemical Engineers
4. Round robin vent sizing exercise on a gassy system: 40% dicumyl peroxide in butyrate solvent. Health and Safety Laboratory ; INERIS Collaboration, 2011. (s)
5. Luc Véchot, Jake Kay, Jill Wilday, Douglas Carson Jean-Pierre Bigot, Runaway reaction of non-tempered chemical systems: Development of a similarity vent-sizing tool at laboratory scale, Journal of Loss Prevention in the Process Industries, Volume 21, Issue 4, July 2008, Pages 359-366, ISSN 0950-4230, http://dx.doi.org/10.1016/j.jlp.2008.01.005

# Beyond the Phi Factor: Correction of Experimental Data for Vaporization in Tempered Reactions, for Pressure-Relief System Design

**Guibing Zhao, Ph.D.**
**Chilworth Technology Inc.**
**113 Campus drive, Princeton, NJ 08540**
**Guibing.Zhao@dekra.com**
**www.chilworth.com**

Prepared for Presentation at
American Institute of Chemical Engineers
2014 Spring Meeting
10th Global Congress on Process Safety
New Orleans, LA
March 30 – April 2, 2014

UNPUBLISHED

# Beyond the Phi Factor: Correction of Experimental Data for Vaporization in Tempered Reactions, for Pressure-Relief System Design

**Guibing Zhao, Ph.D.**
**Chilworth Technology Inc.**
**113 Campus drive, Princeton, NJ 08540**
**Guibing.Zhao@dekra.com**

**Keywords:** Emergency relief system, fill ratio, vaporization, calorimeter, thermo-kinetics

## Abstract

Design of emergency relief systems to accommodate runaway reactions requires using bench-scale adiabatic calorimeters to evaluate thermo-kinetic data of a full-scale reactor. However, the well-known method to correct bench-scale experimental data for "thermal inertia" – the Phi factor – does not consider the effect of vaporization on thermo-kinetic data. Calorimeter volumetric fill ratio is an important parameter directly relating to the vaporization effect and is discussed in this paper. For a 60% fill ratio, a case study of an aqueous reaction system shows that the measured reaction heat would be 7% less than the actual value for a reaction with a heat release of 800 J/g, 5% less than the actual value with a heat release of 600 J/g, and 2% less than the actual value with a heat release of 300 J/g. The self-heating rate is even more sensitive to the fill ratio. A high fill ratio is desirable, to get more-accurate self-heating data for a tempered exothermic reaction. A low fill ratio requires a significant Phi correction and could also give a misleading thermo-kinetic interpretation of a plant-scale reactor. Applying the same fill ratio of a plant-scale reactor to tests in an adiabatic calorimeter – to get thermo-kinetic data for emergency relief sizing – is recommended.

## 1. Introduction

The design of an emergency relief system (that is, a pressure-relief or safety valve or a rupture disk, and the associated piping) for process vessels – in which runaway reactions might occur – requires knowledge of the thermo-kinetics of the reactions involved. The use of adiabatic calorimetry for the quantification of the hazards of exothermic runaway reactions has considerably increased over the last 30 years or so. To a great extent, this has been the result of process hazards analyses that have identified the need for emergency relief systems, and the design of such systems involves calculation methods which require adiabatic data input.

A plant-scale reactor has a smaller surface-to-volume ratio as compared to a laboratory-scale reactor (or to an adiabatic-test cell), since this ratio is inversely proportional to radius for a spherical vessel. As a result, the heat loss to the surrounding environment (which is proportional to the heat transfer/surface area) and the heat loss to the body of the reactor (proportional to the mass of the reactor) per unit mass of reactants loaded are smaller than those characteristics in laboratory equipment. Thus, great efforts have been made throughout the industry to develop an adiabatic calorimeter that can simulate exothermic runaway reactions in a large plant reactor, using a small-scale cell that has similar heat losses to both the environment and to the mass of the cell. Such widely-accepted and commercially-available adiabatic calorimeters include, but are not limited to, the VSP II (Venting Sizing Package II) and the APTAC (Automatic Pressure Tracking Adiabatic Calorimeter). Both of them have adopted automatic pressure-tracking technology to balance the pressures inside the test cell and outside the test cell, and this allows the use of a thin-wall test cell at high pressures. This is critical to reducing the cell mass and, therefore, reducing the heat loss to cell wall. Adiabaticity is accomplished by the surrounding guard heaters which track the cell temperature and thus maintain a near-zero temperature difference between cell wall and surroundings. Three key thermo-kinetic parameters – heat of reaction, temperature-increase rate, and pressure-increase rate – can be measured accurately under the operating conditions of a full-scale reactor. Also, the activation energy, the frequency or pre-exponential factor, and the reaction order can be extracted from the data if a single reaction is involved [1].

The thermal inertia, or Phi factor, is a well-known parameter to describe the heat loss to the test-vessel wall. A higher Phi factor means more energy is absorbed by the testing cell, while a lower Phi factor means more energy is retained in the sample being tested. A Phi factor of unity means that all of the reaction energy is retained in the tested material. The results of testing with a low Phi factor (close to 1) are considered to be representative of the results that would be obtained in a plant-scale reactor and, therefore, could be directly used for designing an emergency relief system. However, the results of testing with a higher Phi factor can be corrected to that with a Phi factor close to 1 if a single reaction is involved [2]. This allows experimentalists to load a smaller quantity of reactants in the test cell (thus, a higher Phi factor, but less likely to rupture the cell) and later correct the results of the test to a Phi factor close to 1. Thus, the results would represent the thermo-kinetic behavior of a plant-scale reactor. The HSE workbook recommends [3] that it may not be necessary to correct self-heat rate data for thermal inertia if the thermal inertia of the calorimeter is less than or equal to that of the full-scale reactor. However, DIERS [2] recommends filling the test cell to 60% or more to obtain the best results, but it does not explain why and what the effect would be with a lower fill ratio (fill ratio means the ratio of sample volume to cell volume). Changes in fill ratio, with corresponding differences in Phi factors, would affect the test results. The real problem is that the corrected results – to obtain the behavior for a Phi factor that would be close to unity – from tests with different fill ratios (with different Phi factors) may not give the same result. For example, a low fill ratio will lead to a high Phi factor and also a high head-space volume, and this would allow more vaporization of volatile materials, such as solvents. The well-known basis for the Phi factor correction [2]Error! Bookmark not defined. only considers heat loss to the cell wall, without considering heat loss due to the endothermic process of vaporization. During the past 30 years of practice with adiabatic test equipment, no reports or publications are found to explore this issue.

For a gassy reaction system, there is no vaporization issue, because little energy is carried from the sample with the generated gas. For a tempered (vapor-generating) system or a hybrid (gas and vapor generating) system, vaporization plays an important role during the testing of exothermic reactions if a low fill ratio is used. The goal of this article is to understand the effects of vaporization on the tested thermo-kinetics of exothermic runaway reactions.

## 2. Modeling of Vaporization in a Closed Test Cell

### *2.1 Reaction Heat*

As shown in Figure 1, consider a closed test cell with a total volume $V_0$ with a mass of $m_0$ of reactants, with the following assumptions:

(1) Liquid specific volume, $v_l$, is far less than vapor specific volume, $v_v$,

(2) Vapor-liquid phase equilibrium is established instantaneously at any temperature, and the relationship between vapor pressure and temperature follows the Antoine equation,

(3) Vapors obey the ideal gas law,

(4) Constant liquid density and constant heat capacity of reaction system,

(5) Constant solvent vaporization latent heat and $\Delta H_{lv} \approx \Delta U_{lv}$.

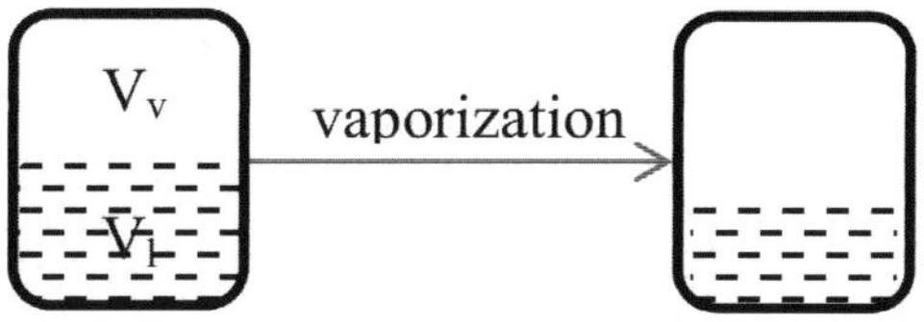


**Figure 1. Liquid vaporization process**

Vapor mass, $m_v$, at temperature $T$, is

$$m_v = \frac{P_v V_v M}{RT} = \frac{P_v M}{RT}\left(V_0 - \frac{m_l}{\rho_l}\right) = \frac{P_v M}{RT}\left(V_0 - \frac{m_0 - m_v}{\rho_l}\right) = \frac{P_v M}{RT}\left(V_0 - \frac{m_0}{\rho_l}\right) + \frac{P_v M}{RT}\frac{m_v}{\rho_l} \qquad \text{[Eq.1]}$$

Solving [Eq.1] gives

$$m_v = \frac{V_0 - \frac{m_0}{\rho_l}}{\frac{RT}{P_v M} - \frac{1}{\rho_l}} \qquad \text{[Eq.2]}$$

Applying Antoine Equation to [Eq.2] gives

$$m_v = \frac{V_0 - \frac{m_0}{\rho_l}}{\frac{RT}{M \cdot \exp\left(A - \frac{B}{T}\right)} - \frac{1}{\rho_l}} \qquad \text{[Eq.3]}$$

From onset temperature, $T_0$, to any temperature, $T$, of a runaway reaction, the vapor mass ratio in the head space is

$$\frac{m_v}{m_{v0}} = \frac{T_0}{T} \cdot exp\left[B\left(\frac{1}{T_0} - \frac{1}{T}\right)\right] \qquad \text{[Eq.4]}$$

From the onset temperature to the maximum temperature of a runaway reaction, the vapor mass change is

$$\Delta m_v = m_{v0}\left\{\frac{T_0}{T_f} \cdot exp\left[B\left(\frac{1}{T_0} - \frac{1}{T_f}\right)\right] - 1\right\} \qquad \text{[Eq.5]}$$

The total energy balance is

$$Q = \phi m_0 C_{vl}\left(T_f - T_0\right) + \Delta m_v \cdot \Delta H_{lv}$$

$$= \phi m_0 C_{vl}\left(T_f - T_0\right) + m_{v0}\Delta H_{lv}\left\{\frac{T_0}{T_f} \cdot exp\left[B\left(\frac{1}{T_0} - \frac{1}{T_f}\right)\right] - 1\right\} \qquad \text{[Eq.6]}$$

where $\phi$ is the Phi factor. If the heat loss to vaporization is not considered, the last term in [Eq.6] is omitted, and the final temperature would be calculated from

$$T_f = T_0 + \frac{Q}{\phi m_0 C_{vl}} \quad \text{[Eq.7]}$$

However, if vaporization is considered, the final temperature, $T_f$, can be solved by "trial-and-error" iteration from [Eq.6].

### 2.2 *Self-heat rate: dT/dt*

From [Eq.7], the heat release rate without considering vaporization is

$$q_{nv} = \frac{dQ}{dt} = \phi m_0 C_{vl} \left(\frac{dT}{dt}\right)_{nv} \quad \text{[Eq.8]}$$

The heat release rate with considering vaporization is, from [Eq.6]

$$q_v = \frac{dQ}{dt} = \frac{dm_v}{dt}\Delta H_{lv} + \phi m_0 C_{vl} \left(\frac{dT}{dt}\right)_v \quad \text{[Eq.9]}$$

From [Eq.2],

$$\frac{dm_v}{dt} = -\frac{V_0 - \frac{m_0}{\rho_l}}{\left(\frac{RT}{P_v M} - \frac{1}{\rho_l}\right)^2} \frac{R}{M} \left(\frac{1}{P_v} - \frac{T}{P_v^2}\frac{dP_v}{dT}\right)\left(\frac{dT}{dt}\right)_v \quad \text{[Eq.10]}$$

Applying the Clapeyron equation

$$\frac{dP_v}{dT} = \frac{\Delta H_{lv}}{T\Delta v} \quad \text{[Eq.11]}$$

to [Eq.10]:

$$\frac{dm_v}{dt} = \frac{V_0 - \frac{m_0}{\rho_l}}{\left(\frac{RT}{P_v M} - \frac{1}{\rho_l}\right)^2} \frac{R}{M P_v^2} \left(\frac{\Delta H_{lv}}{\Delta v} - P_v\right)\left(\frac{dT}{dt}\right)_v \quad \text{[Eq.12]}$$

Applying [Eq.12] to [Eq.9], the heat release rate with considering vaporization can be obtained as

$$q_v = \left[\frac{V_0 - \frac{m_0}{\rho_l}}{\left(\frac{RT}{P_v M} - \frac{1}{\rho_l}\right)^2} \frac{R}{M P_v^2} \left(\frac{\Delta H_{lv}}{\Delta v} - P_v\right)\Delta U_{lv} + \phi m_0 C_{vl}\right]\left(\frac{dT}{dt}\right)_v \quad \text{[Eq.13]}$$

The heat release rate from an exothermic runaway reaction is, using the Arrhenius equation

$$q = r \cdot V_l \cdot \Delta H_r = k \cdot f(C) \cdot V_l \cdot \Delta H_r = A e^{-\frac{E}{RT}} \cdot f(C) \cdot V_l \cdot \Delta H_r \quad \text{[Eq.14]}$$

where f(C) is an "unknown" function of concentration in the expression of reaction rate. Therefore, the heat release rate increases exponentially with an increase in temperature as shown in [Eq.14]. If vaporization is significant, the endothermic vaporization process would exponentially affect the reaction rate, thus the rate of temperature increase, $dT/dt$. For an easy comparison, the ratio of temperature rise rate without considering vaporization and with vaporization considered at the same heat release rate can be deducted from [Eq.8] and [Eq.13],

$$\frac{\left(\frac{dT}{dt}\right)_{nv}}{\left(\frac{dT}{dt}\right)_v} = \frac{V_0 - \frac{m_0}{\rho_l}}{\phi m_0 C_{vl} (\Delta v)^2} \frac{R}{M P_v^2} \left(\frac{\Delta H_{lv}}{\Delta v} - P_v\right)\Delta H_{lv} + 1 \quad \text{[Eq.15]}$$

## 3. Case Study

A case study was conducted to understand how significantly the fill ratio would affect the heat of reaction and the temperature rise rate. A lower fill ratio – corresponding to a higher Phi factor – will have more head space, which allows more solvent to vaporize. Therefore, the effect of fill ratio on the thermo-kinetics of a runaway reaction would be the same as the effect of vaporization on the thermo-kinetics. For an aqueous reaction system, it is not uncommon to find the following physical properties:

$\rho_l = 1200\ kg \cdot m^{-3}$,

$Cv_l = 3000\ J \cdot kg^{-1} \cdot K^{-1}$,

$M = 0.018\ kg \cdot mol^{-1}$.

The Antoine constants for water vapor, with pressures in Pascals and absolute temperatures

$A = 24.24$

$B = 4736$

$\Delta H_{lv} = 2188\ kJ \cdot kg^{-1}$

The error of the calculated vapor pressure based on the above Antoine constants and the reported water vapor pressure data [4] is about ±3%. An adiabatic VSP cell material made of Hastelloy C has a $Cp = 420\ J \cdot kg^{-1} \cdot K^{-1}$, an empty cell weight of 41.5 g (including stirrer), and a volume of 110 ml.

Assuming the onset temperature of a runaway reaction to be $T_0 = 100°C$, i.e., the normal boiling temperature of water, the final temperature, $T_f$, can be calculated based on the reaction heat, $\Delta H_r$, in J/g by solving [Eq.6] or [Eq.7]. The total energy released, $Q$, can be calculated by the multiplication of mass of reactant loaded and the reaction heat,

$$Q = m_0 \cdot \Delta H_r \qquad \text{[Eq.16]}$$

The result of the energy release is: increased temperature of the sample, increased temperature of the cell wall, and increased vaporization of water from the sample. The heat loss to cell and sample is expressed in the first term of [Eq.6]. The heat loss to vaporization is expressed in the second term of [Eq.6]. The numerical iteration is conducted to calculate the final temperature, $T_f$, for different fill ratios from [Eq.6]. The fraction of heat loss to vaporization ($x_v$), test cell ($x_c$), and sample ($x_s$) can further be calculated as follows

$$x_v = \frac{m_{v0}\,\Delta H_{lv}\left\{\frac{T_0}{T_f}\cdot exp\left[B\left(\frac{1}{T_0}-\frac{1}{T_f}\right)\right]-1\right\}}{m_0 \cdot \Delta H_r} \qquad \text{[Eq.17a]}$$

$$x_c = \frac{(\phi-1)C_{vl}\left(T_f - T_0\right)}{\Delta H_r} \qquad \text{[Eq.17b]}$$

$$x_s = \frac{C_{vl}\left(T_f - T_0\right)}{\Delta H_r} \qquad \text{[Eq.17c]}$$

As an example, the calculated results for a reaction heat of 800 J/g are shown in Table 1, for a 110-cubic-centimeter test vessel. For different scenarios, similar calculations can be made.

**Table 1. The calculated results for a reaction heat of 800 J/g for an aqueous system**

| $m_0$, g | $V_l$, ml | fill ratio | Phi factor | $(T_f)_{nv}$, C | $m_{v0}$, g | $(T_f)_v$, C | Pv, bar | $m_v$ at $T_f$, g | $\Delta m_v$,g | $x_v$ | $x_c$ | $x_s$ |
|---|---|---|---|---|---|---|---|---|---|---|---|---|
| 1.00 | 0.830 | 0.00758 | 6.81 | 139 | 6.56E-02 | 130 | 2.63 | 1.55E-01 | 8.91E-02 | 0.243 | 0.646 | 0.111 |
| 5.00 | 4.17 | 0.0379 | 2.16 | 223 | 6.36E-02 | 188 | 11.7 | 5.86E-01 | 5.22E-01 | 0.284 | 0.385 | 0.331 |
| 10.0 | 8.33 | 0.0758 | 1.58 | 269 | 6.11E-02 | 223 | 23.9 | 1.07E+00 | 1.01E+00 | 0.273 | 0.267 | 0.460 |
| 20.0 | 16. 7 | 0.152 | 1.29 | 307 | 5.61E-02 | 259 | 45.9 | 1.77E+00 | 1.71E+00 | 0.231 | 0.173 | 0.596 |
| 30.0 | 25.0 | 0.227 | 1.19 | 323 | 5.11E-02 | 280 | 64.9 | 2.20E+00 | 2.15E+00 | 0.192 | 0.131 | 0.677 |
| 40.0 | 33.3 | 0.303 | 1.15 | 333 | 4.61E-02 | 296 | 81.6 | 2.44E+00 | 2.40E+00 | 0.160 | 0.107 | 0.734 |
| 50.0 | 41.7 | 0.379 | 1.12 | 339 | 4.11E-02 | 307 | 96.4 | 2.53E+00 | 2.49E+00 | 0.132 | 0.090 | 0.777 |
| 60.0 | 50.0 | 0.455 | 1.10 | 343 | 3.61E-02 | 317 | 110 | 2.50E+00 | 2.47E+00 | 0.109 | 0.079 | 0.813 |
| 70.0 | 58.3 | 0.530 | 1.08 | 346 | 3.11E-02 | 325 | 122 | 2.37E+00 | 2.34E+00 | 0.088 | 0.070 | 0.842 |
| 80.0 | 66.7 | 0.606 | 1.07 | 349 | 2.61E-02 | 331 | 133 | 2.15E+00 | 2.13E+00 | 0.070 | 0.063 | 0.867 |
| 90.0 | 75.0 | 0.682 | 1.06 | 350 | 2.10E-02 | 337 | 143 | 1.86E+00 | 1.84E+00 | 0.053 | 0.057 | 0.889 |
| 100 | 83.3 | 0.758 | 1.06 | 352 | 1.60E-02 | 342 | 153 | 1.50E+00 | 1.49E+00 | 0.039 | 0.053 | 0.908 |
| 110 | 91.7 | 0.833 | 1.05 | 353 | 1.10E-02 | 347 | 162 | 1.09E+00 | 1.08E+00 | 0.026 | 0.049 | 0.926 |
| 120 | 100 | 0.909 | 1.05 | 354 | 6.01E-03 | 351 | 170 | 6.22E-01 | 6.16E-01 | 0.013 | 0.046 | 0.941 |
| 130 | 108 | 0.985 | 1.04 | 355 | 1.00E-03 | 355 | 178 | 1.08E-01 | 1.07E-01 | 0.002 | 0.043 | 0.955 |

### *3.1 Reaction Heat*

Figure 2 shows the calculated results with vaporization considered, by solving [Eq.6], and without considering vaporization, by solving [Eq.7], at reaction heats of 300 J/g, 600 J/g, and 800 J/g. For the same fill ratio, the higher reaction heats lead to higher final temperatures, whether vaporization is considered or not. With increasing fill ratios, the Phi factor decreases more than proportionally, as shown in Table 1. For the same reaction heat, $\Delta H_r$, the final temperature, $T_f$, increases with an increase in fill ratio. An increase in fill ratio leads to (1) a lower Phi factor, with a smaller fraction of energy absorbed by the cell wall; (2) a smaller head-space volume, and less energy absorbed by vaporization. Therefore, an increase in fill ratio leads to an increase in the final temperature. For the same reaction heat, the temperature difference between vaporization considered and without considering vaporization is small at a low fill ratio, gradually increasing to a maximum, and then decreasing at further increases in fill ratio. This type of trend is further shown in Figure 3.

**Figure 2. The final temperature and Phi factor as a function of fill ratio**

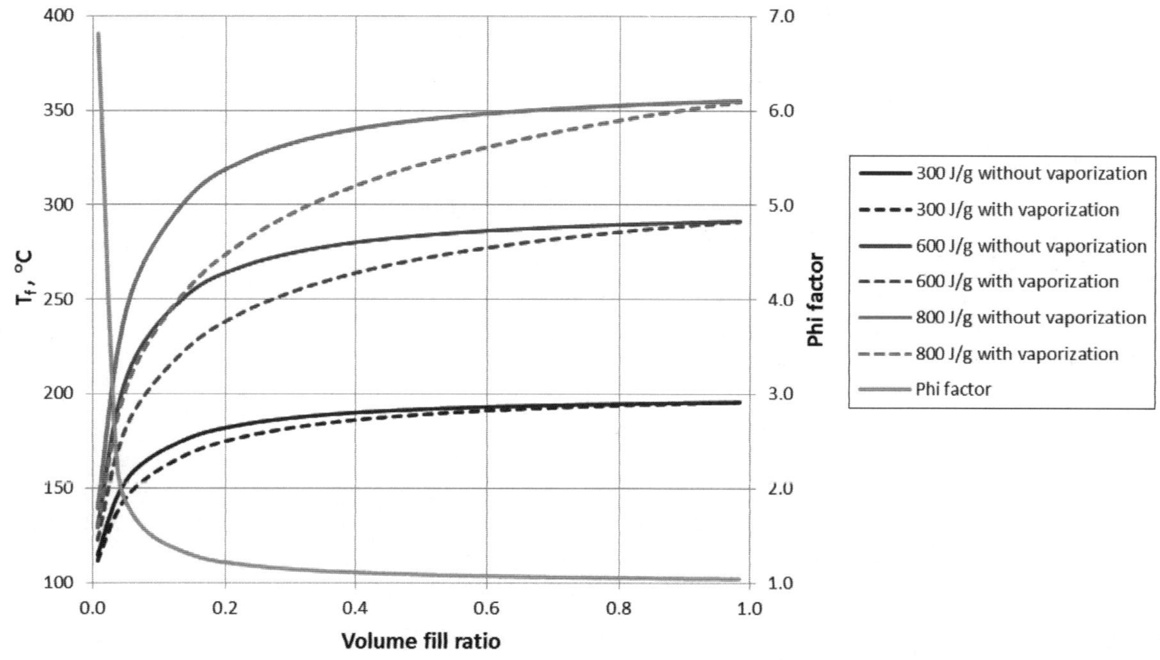


At very low fill ratios, the Phi factor is high. Most of energy that is released from the reaction contributes to heating the cell wall, and the energy contributing to vaporization is low. That is the reason that the temperature difference between vaporization considered and without considering vaporization is small at very low fill ratios, as shown in Figure 3.

With slight increases in fill ratio, more energy is released. Although the head space is reduced (which leads to less volume available for vapor), more energy is contributed to vaporization, since more energy is available from a larger reaction mass. This behavior continues until a maximum difference is attained.

With further increases in fill ratio, more energy is available for heating both the cell wall and the sample mass, while less volume is available for vaporization. This leads to a decrease in the temperature difference between vaporization considered and without considering vaporization. For a fill ratio of unity (100% full), there would be no head space available for vaporization.
As shown in Figure 3, the maximum temperature difference – when vaporization considered and when vaporization is not considered – is attained at higher fill ratios with higher reaction mass and energy. The temperature difference is smaller for a lower reaction mass (fill ratio) and energy. This is consistent with [Eq.17b] and [Eq.17c].

**Figure 3. Final temperature difference between considering vaporization and without considering vaporization, and Phi factor, as a function of fill ratio**

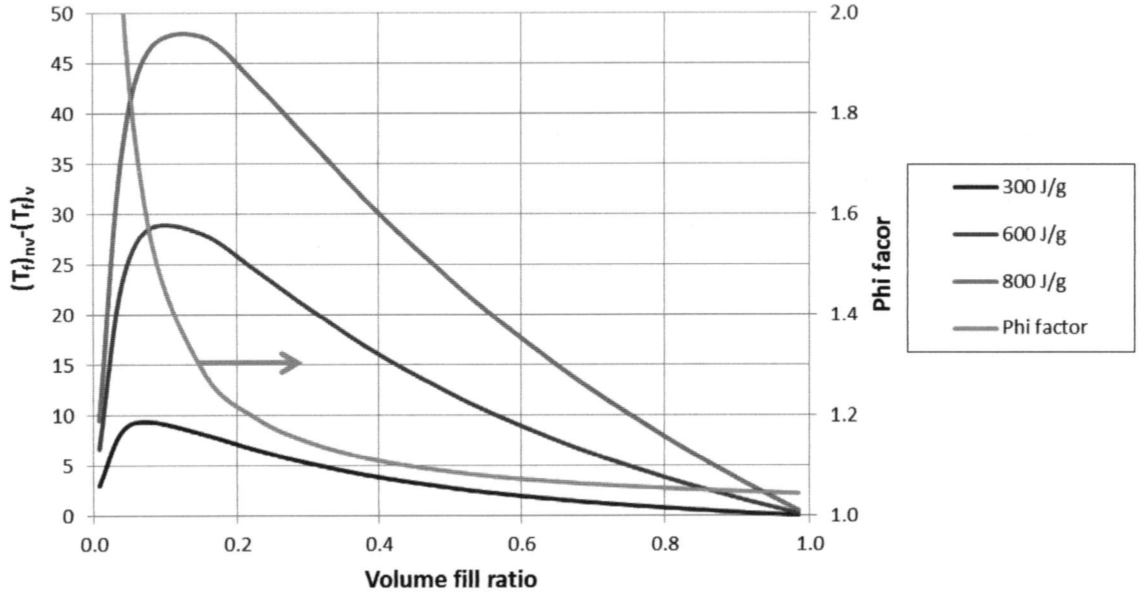


With a known final temperature, the fractions of heat losses to vaporization, the cell, and the sample can be calculated from [Eq.17a-c]. Figure 4 shows the calculated fractions of these heat losses as a function of fill ratio, for reaction energies of 300 J/g, 600 J/g, and 800 J/g. For the same fill ratio, the heat losses to the test cell and the sample decrease, while heat losses to vaporization increase, with an increase in the reaction heat. As would be expected, more-energetic reactions have more energy to expend in vaporization. For a lower reaction heat – such as 300 J/g – the fraction of heat loss to vaporization monotonously drops with an increase in fill ratio. However, for higher reaction energies – such as 600 J/g and 800 J/g, the fraction of heat loss to vaporization initially increases, then arrives at a maximum, and then decreases with fill ratio. These results are consistent with the results as shown in Figure 3. At a lower reaction heat – such as 300 J/g – the heat losses to both the test cell and the sample is higher, and this limits the heat loss to vaporization in the whole range of fill ratios. However, at a higher reaction heat – such as 600 J/g and 800 J/g – the heat losses to vaporization would increase with the fill ratio at lower ranges – due to more energy expended in vaporization, but the heat loss to vaporization decreases at the higher fill ratios, due to less space available for vaporization, and more energy contributes to increasing the temperature of the sample.

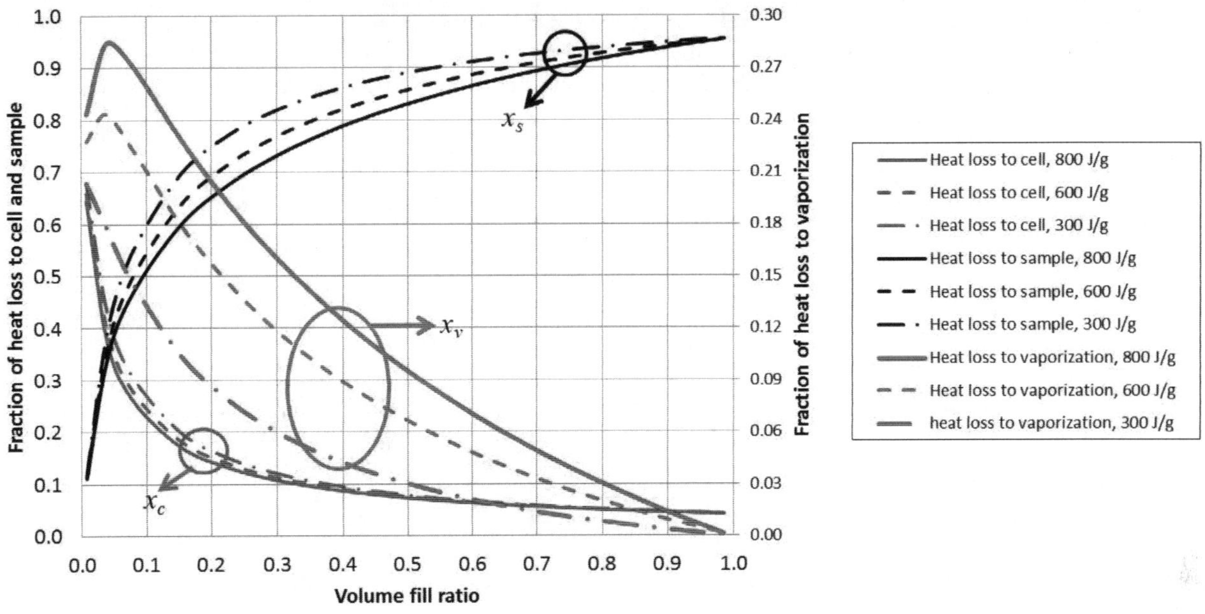


**Figure 4. The fraction of heat loss to vaporization, cell, and sample as a function of fill ratio.**

The results in Figure 4 indicate that, for a fill ratio of 60% – as recommended as a low fill-ratio limit by DIERS [2] – the measured reaction heat would be 7% less than the actual value for a reaction with a reaction heat of 800 J/g, 5% less than the actual value with a reaction heat of 600 J/g, and 2% less than the actual value with a reaction heat of 300 J/g, if vaporization is not considered in the energy calculations.

### *3.2 Self-heating Rate: dT/dt*

Figure 5 shows that the effect of vaporization on the temperature-increase rate, dT/dt, as a function of fill ratio, at temperatures of 150 °C, 200 °C, 250 °C, and 300 °C. The ratios of temperature-increase rate without considering vaporization and with vaporization considered are calculated from [Eq.15]. The reaction heats calculated from [Eq.6] and [Eq.16] are also shown in Figure 5. At the same fill ratio, a higher temperature indicates that more energy is generated by the reaction system, which would cause more vaporization. Therefore, the ratio of temperature-increase rate without considering vaporization to the temperature-increase rate with vaporization considered increases with an increase in temperature. At a given temperature, combining [Eq.2], [Eq.6], and [Eq.16] gives the reaction energy

$$\Delta H_r = \phi C_{vl}\left(T_f - T_0\right) + \frac{\frac{V_0}{m_0} - \frac{1}{\rho_l}}{\frac{RT_0}{P_{v0}M} - \frac{1}{\rho_l}} \Delta H_{lv}\left\{\frac{T_0}{T_f} \cdot exp\left[B\left(\frac{1}{T_0} - \frac{1}{T_f}\right)\right] - 1\right\} \qquad \text{[Eq.18]}$$

Therefore, as the fill ratio increases – for a given volume – the Phi factor decreases and the sample mass increases, and this leads to decreases in both terms in [Eq.18]. Therefore, the measured reaction heat decreases, as shown in Figure 5. At the same reaction temperature, the ratio of the temperature-increase rate without considering vaporization to the temperature-increase rate with vaporization considered decreases with an increase in fill ratio. As the fill ratio approaches 1.0 – where no head space available for vaporization – the temperature-increase rates become equal. As would be expected, the effect of vaporization on the temperature-increase rate becomes more significant with lower fill ratios and more energetic reactions.

**Figure 5. Effect of vaporization on temperature-increase rate, dT/dt, as a function of fill ratio and the ratio of dT/dt without consideration of vaporization to dT/dt with vaporization considered.**

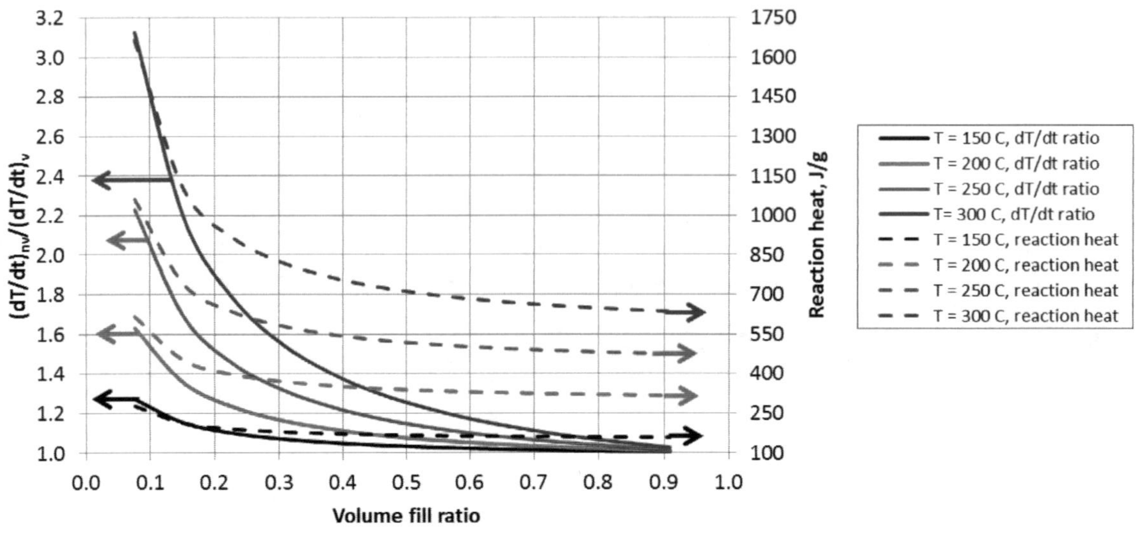


Figure 6 shows the effect of vaporization on temperature-increase rate, dT/dt, as a function of temperature at different fill ratios and Phi factors. At a given fill ratio, the ratio of temperature-increase rate without considering vaporization and with vaporization considered increases with an increase of temperature. A lower fill ratio causes a higher ratio of temperature-increase rate without considering vaporization to the temperature-increase with vaporization considered.

**Figure 6. Effect of vaporization on temperature rise rate, dT/dt, as a function of temperature at different fill ratios and Phi factors.**

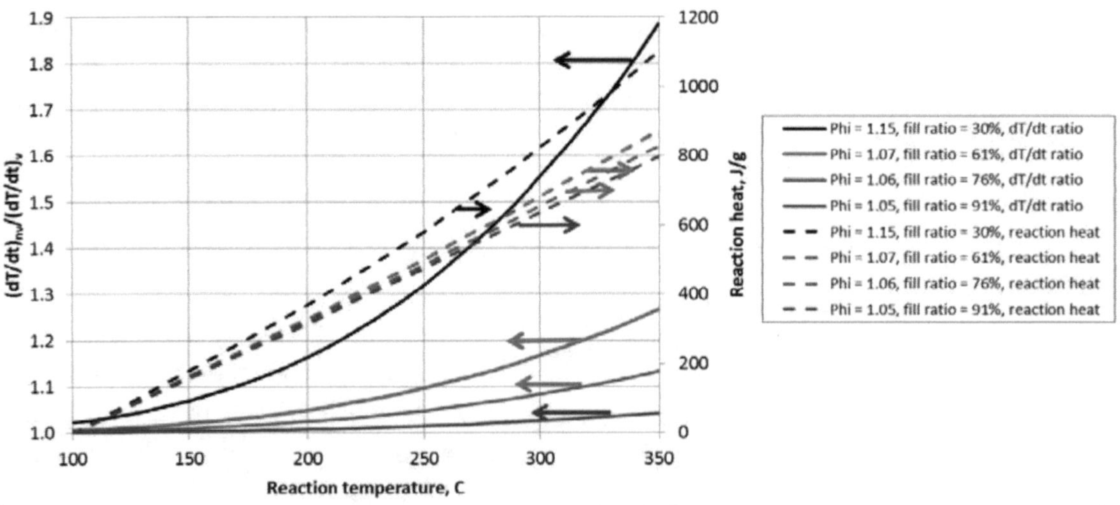

Figure 7 shows the ratio of temperature-increase rate without considering vaporization to the temperature-increase rate with vaporization considered, as a function of reaction heat at different fill ratios and Phi factors. For a given fill ratio, the ratio of these temperature-increase rates increases as the reaction energy increases. For a given heat of reaction, the ratio of temperature-increase rates increases as the fill ratio decreases. For a fill ratio of 60% – which is the lower limit of fill ratio that is recommended by DIERS [2] – the measured temperature-increase rate would be 22% lower than the actual value at a reaction energy of 800 J/g, 13% less than the actual value at a reaction energy of 600 J/g, and 5% less than the actual value at a reaction energy of 300 J/g, if vaporization is not considered in the calculations. For a lower fill ratio of 30%, the measured temperature-increase rate would be 52% less than the actual value at a reaction energy of 800 J/g, 34% less than the actual value at a reaction energy of 600 J/g, and 13% less than the actual value at a reaction energy of 300 J/g, if vaporization is not considered in the calculations.

**Figure 7. Ratio of temperature-increase rate without considering vaporization to the temperature-increase rate with vaporization considered, as a function of reaction heat at different fill ratios and Phi factors.**

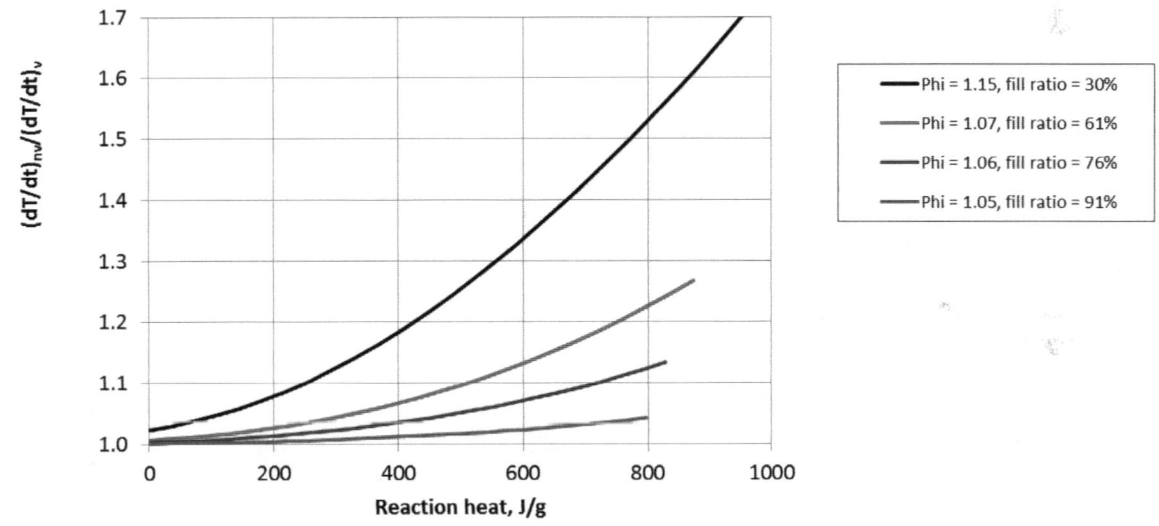


## 4. Conclusion

(1) A case study of an aqueous reaction system shows that reaction heat with a fill ratio of 60% would be 7% less than the theoretical value for a reaction heat of 800 J/g, 5% less than the theoretical value for a reaction heat of 600 J/g, and 2% less than the theoretical value for a reaction heat of 300 J/g, if vaporization is not considered in the evaluation of the test results.

(2) The self-heating rate is sensitive to the fill ratio. For a fill ratio of 60%, the measured temperature-increase rate would be 22% less than the actual value at a reaction heat of 800 J/g, 13% less than the actual value at a reaction heat of 600 J/g, and 5% less than the actual value at a reaction heat of 300 J/g, if vaporization is not considered in the evaluation of the test results.

(3) A high fill ratio is required to obtain more-reliable data for a high-energy exothermic reaction that involves a volatile component in a tempered system.
(4) A low fill ratio would give misleading thermo-kinetic data for the design of an overpressure-protection system on a full-scale reactor that involves one or more volatile components.

## 5. Recommendations.

This investigation of adiabatic calorimetry results in the following recommendations:

(1) For emergency relief sizing, use the same fill ratio of the full-scale reactor in the test vessel of the adiabatic calorimeter, to obtain valid thermo-kinetic data, rather than just considering thermal inertias.
(2) Generally speaking, the fill ratio of 60% as recommended by DIERS should be a minimum charging ratio for adiabatic calorimeters, for obtaining the valid thermo-kinetic data for emergency relief sizing.
(3) Considering the heat transfer limitations of full-scale reactors, Phi correction of the measured self-heat rates from adiabatic calorimeters is always necessary.

## Nomenclature

| | |
|---|---|
| $\Delta H_{lv}$ | Vaporization latent heat at constant pressure, $J \cdot kg^{-1}$ |
| $\Delta U_{lv}$ | Vaporization latent heat at constant volume, $J \cdot kg^{-1}$ |
| $\Delta H_r$ | Reaction heat, $J \cdot kg^{-1}$ |
| $\Delta v$ | Specific volume difference between vapor phase and liquid phase, $m^3 \cdot kg^{-1}$ |
| A | Antoine constant [natural log], or frequency factor |
| B | Antoine constant, for temperatures in $^oK$ |
| C | Concentration, $mol \cdot m^{-3}$ |
| $C_p, C_v$ | Heat capacity, $J \cdot kg^{-1} \cdot K^{-1}$ |
| E | Reaction activation energy, $J \cdot mol^{-1}$ |
| k | Reaction rate constant |
| m | mass, kg |
| M | Molecular weight, $kg \cdot mol^{-1}$ |
| P | Pressure, Pa |
| Q | Total energy released, J |
| q | Heat release rate, $J \cdot s^{-1}$ |
| R | Gas constant, 8.314 $J \cdot mol^{-1} \cdot K^{-1}$ |
| r | Reaction rate, $mol \cdot m^{-3} \cdot s^{-1}$ |
| T | Temperature, K |
| t | Time, s |
| V | Volume, $m^3$ |
| ρ | Density, $kg \cdot m^{-3}$ |
| $\phi$ | Phi factor, $\phi = 1 + \frac{m_c C_{pc}}{m_0 C_{pl}}$ |
| Subscripts | |
| 0 | Initial, or onset |

| | |
|---|---|
| c | Cell |
| f | Final or end |
| l | Liquid |
| nv | Without considering vaporization |
| s | Sample |
| v | Vapor, or with considering vaporization |

## References


[1] Townsend D. I. and Tou J. C., "Thermal Hazard Evaluation by an Accelerating Rate Calorimeter", Thermochimica Acta, Vol. 37, Pages 1 to 30 (1980).

[2] Fisher H. G., et al., "Emergency Relief System Design Using DIERS Technology", Experimental Emergency Relief System Design – Some Do and Do Not Recommendations, American Institute of Chemical Engineers, New York, Pages 382 to 391 (1992)

[3] Etchells J., Wilday J., "Workbook for Chemical Reactor Relief System Sizing", Contract Research Report 136, HSE books, Page 142 (1998).

[4] Green D. W., and Perry R. H., "Perry's Chemical Engineers' Handbook", $8^{th}$ Edition, The McGraw-Hill Company, Inc., Page 2-413 (2008).

# 349001 The Effect of Non-Uniform Distribution of Obstacles on Deflagration-to-Detonation Transition (DDT)

**Abstract:**

**The effect of non-uniform distribution of obstacles on**

**Deflagration-to-Detonation Transition (DDT)**


Camilo Rosas[1, 2], Hao Chen[1, 2], Eric Petersen [3], and M. Sam Mannan[1, 2]*

[1] Mary Kay O'Connor Process Safety Center

[2] Artie McFerrin Department of Chemical Engineering

[3] Department of Mechanical Engineering

Texas A&M University, College Station, Texas 77843-3122

*Corresponding author: mannan@tamu.edu

The mechanisms for explosions are usually classified as two different types. The first mechanism is deflagration, a sub-sonic combustion wave with respect to the unburned gas ahead of the flame, during which the flame front and the shock front can be distinguished. The second type is detonation, which is a supersonic combustion wave propagating at 1500-2000m/s in fuel-air and can produce overpressures up to around 2 MPa. To obtain a detonation there are two different approaches. The first way is by direct initiation with a high energy source, and the second one is by inducing a transition from deflagration to detonation (DDT) with a low energy source and turbulence enhancers, which are also known as obstacles.

From previous research, the obstacles' shape, distribution, and blockage ratio along the tube were always designed to be uniform. However, these designs deviate from real processes. For instance, different layouts of obstacles (*e.g.* pipe racks, turbines, and vessels) are substantially present in any facility, which could either accelerate and lead to DDT or quench the combustion process. Thus, in this work, the obstacles were designed to be distributed non-uniformly along the detonation tube to determine the effect of the randomness in the obstacles on the distance necessary to obtain DDT. Details of these experiments and the results are presented in the paper.

Computational Fluid Dynamics (CFD) models were also applied in this work to better understand the mechanisms. Even though CFD models do not represent DDT, various models are still able to predict high pressure values which are typically in the range of detonation pressures. Moreover, a CFD model capable of predicting the explosion behavior is of particular interest for the situation that the flame front captures the pressure front by the spatial pressure gradients across the flame front, which is one of the main mechanisms that may initiate a DDT.

# Dispersion Modeling of a Cloud Generated by the Depressurization of a Flashing Multi-Component Liquid System

**Laurent Nouailhetas**
**ERM, America Risk Practice,**
**1776 I Street Northwest, Suite 200**
**Washington D.C., 20036, USA**
**laurent.nouailhetas@erm.com**

**Ralph Mancik**
**ERM, America Risk Practice,**
**840 West Sam Houston Parkway North, Suite 600**
**Houston, 77024, USA**
**ralph.mancik@erm.com**

Prepared for Presentation at
American Institute of Chemical Engineers
2014 Spring Meeting
10th Global Congress on Process Safety
New Orleans, LA
March 30 – April 2, 2014

UNPUBLISHED

# Dispersion Modeling of a Cloud Generated by the Depressurization of a Flashing Multi-Component Liquid System

**Laurent Nouailhetas**
**ERM**
**1776 I Street Northwest, Suite 200**
**Washington D.C., 20036, USA**
**laurent.nouailhetas@erm.com**

**Ralph Mancik**
**ERM**
**ralph.mancik@erm.com**

**Keywords:** Depressurization, multi-component mixture, flashing liquid, dispersion, modeling

## Abstract

Depressurization is usually a process operation during which the gas contained in a pressurized system is sent either to a flare or to a cold vent. However, depressurization can also be the consequence of an unexpected event as, for example, the accidental opening of a vent or an unintended leak of a pressurized system. One of the immediate outcomes is the release and dispersion of a cloud into the atmosphere that creates a hazard when the cloud is flammable and/or toxic. The prediction of the harmful extent of such dispersing clouds is a common task when assessing the risk associated with such events.

Depressurization events are of several types which differ in the number and complexity of the mechanisms involved. In general, there is either a system containing a single compressed vapor phase (referred to in this paper as a "gas system"), or a system containing a flashing liquid phase that is in thermodynamic equilibrium with a vapor phase (referred to in this paper as a "flashing liquid system"). The flashing liquid is either a pure component or a multi-component mixture.

With a multi-component flashing liquid system, the temperature and the composition of the released vapor will change over time. The resulting dispersing cloud could transition from a buoyant vapor mixture early in the depressurization event to a dense vapor mixture later in the depressurization event. One immediate consequence is that in stable atmospheric conditions, a dispersing cloud could ascend initially, but after a few minutes it could descend and disperse at ground level. Not accounting for the changing characteristics of the system could result in a misleading assessment.

With a good consequence modeling program, modeling the time-varying discharge and dispersion of a depressurizing gas system is relatively easy to handle. However, the same problem for a system containing a flashing liquid under pressure (LNG, LPG, etc.) is more complex even with a good commercial consequence modeling program. Adding an appropriate accounting of the changing characteristics of a multi-component system further complicates predicting the harmful extent of a dispersing cloud.

The objective of this paper is to present a few methods to appropriately model such a complex dispersion problem. After comparing the merits and demerits of each method, ERM proposes the best method for the modeling of a time-varying multi-component flashing liquid release.

# 1. Introduction and context

Depressurization is a common process operation during which the gas contained in a pressurized system is sent to a flare or to a cold vent. However, depressurization can also be the consequence of an unexpected event as in the accidental opening of a vent or an unintended leak of a pressurized system. One of the immediate outcomes is the release and dispersion of a cloud into the atmosphere. This creates a hazard when the cloud is flammable and/or toxic. A depressurization through a flare would also generate a hazardous cloud in the event of a flame out. Figure 1-1 illustrates common situation for controlled and uncontrolled depressurizations.

**Figure 1-1: Vessel depressurization and gas discharge generating a hazardous cloud**

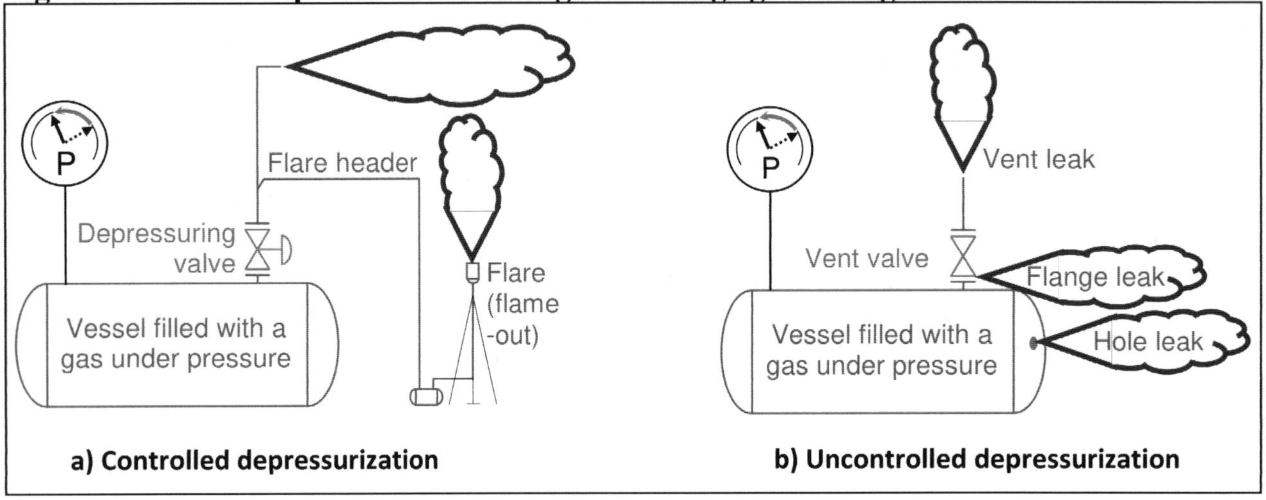


Depressurization events are of several types which differ in the number and complexity of the mechanisms involved. In general, there is either a system containing a single compressed vapor phase (referred to in this paper as a "gas system"), or a system containing a flashing liquid phase that is in thermodynamic equilibrium with a vapor phase (referred to in this paper as a "flashing liquid system"). The gas or the flashing liquid is either a pure component or a multi-component mixture. Figure 1-2 illustrates a typical depressurization of a gas system and a flashing liquid system.

**Figure 1-2: Depressurization through a vent: two accidental configurations**

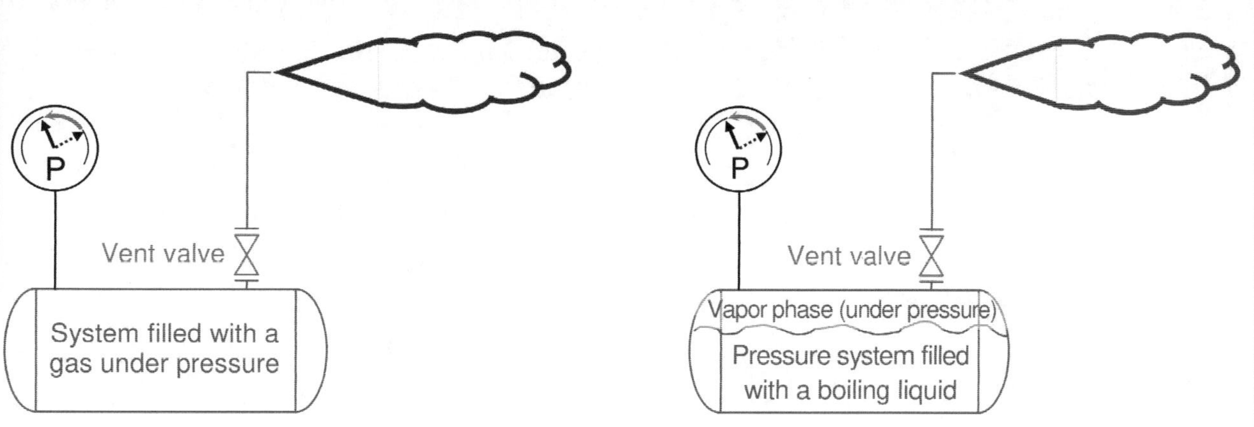


**a) Depressurization of a gas system** **b) Depressurization of a flashing liquid system**

Other variations of these include systems containing a liquid and a vapor phase that are not in equilibrium, systems with a vapor phase in presence with multiple liquid phases, and others. However, this paper will focus on the "flashing liquid system" with a multi-component composition, and will make references to the "gas system" for comparison purposes.

## 2. Characteristics of the discharged gas

### *2.1 Gas system*

During the depressurization of a gas system, the characteristics of the gas will evolve over time as follow:

- The mass inventory will decrease as gas inventory (and gas volume) is withdrawn from the gas system;
- Since the volume of the gas system is fixed (illustrated by curve ① in Figure 2-1), the gas remaining in the system will expand to fill the entire system volume, leading to a reduction of pressure (curve ②) and reduction of temperature (curve ③);
- The average molecular weight and the composition of the discharged gas will remain constant (curves ④, bar charts ⑤ and ⑥).

**Figure 2-1: Vapor phase characteristics for the depressurization of a gas system**

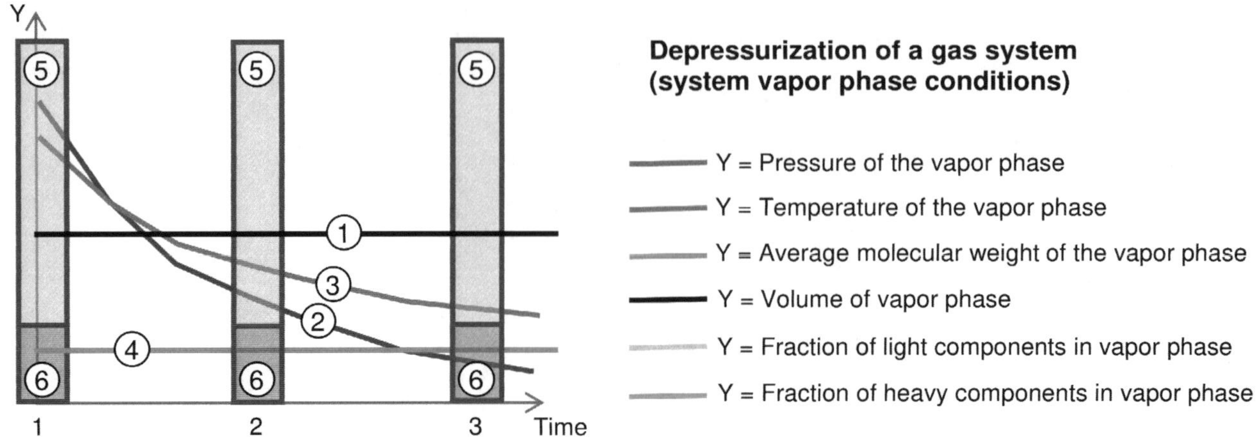


### *2.2 Multi-component flashing liquid system*

With a multi-component flashing liquid system, the gas fills only a fraction of the system volume (the vapor phase), the rest being filled with the boiling liquid phase. As gas is withdrawn from the vapor phase, the gas remaining in the system expands leading to a decrease of system pressure. The presence of the boiling liquid phase however introduces additional phenomena:

- As pressure drops, a fraction of the liquid is vaporized to reach new boiling conditions under the lower system pressure;
- The volume of the gas is not fixed. Since a fraction of the liquid inventory is vaporized, the liquid mass inventory decreases. The liquid being incompressible, a reduction of mass inventory leads in a decrease of the liquid phase volume and an expansion of the vapor phase volume (illustrated by curve ① in Figure 2-2).
- The composition of the vapor phase becomes proportionately leaner in light components and richer in heavy components as the pressure decreases over time (bar charts ⑤ and ⑥).
- Consequently, the average molecular weight of the vapor phase increases over time (curve ④).
- Temperature of the liquid and the vapor phases decrease gradually over time due to the expansion of the gas in the vapor phase and the vaporization of the liquid (curve ③). Note that the temperature decreases slower than for the gas case (Figure 2-1) because of the presence of the liquid.

**Figure 2-2: Vapor phase characteristics for the depressurization of a multi-component flashing liquid system**

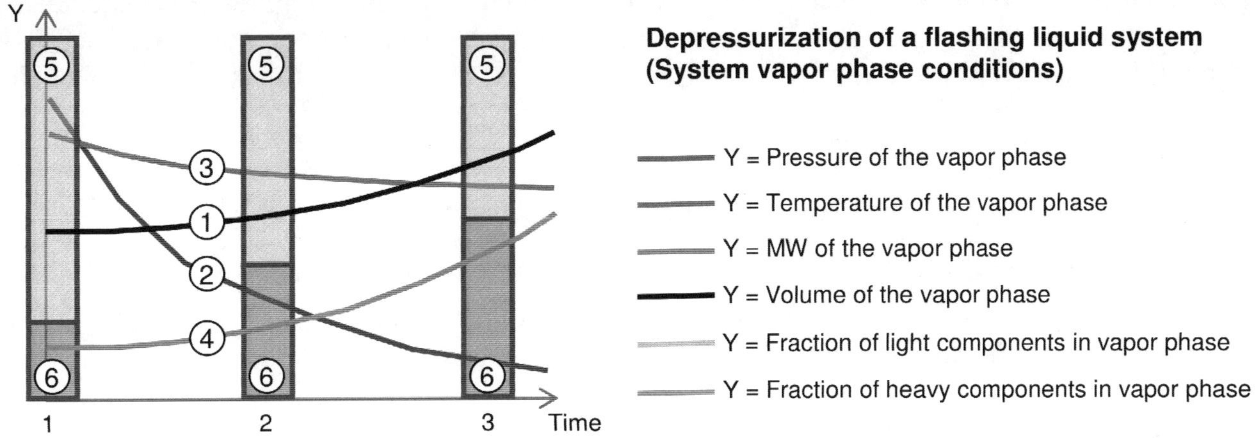


### *2.3 Important difference*

One important difference between these two systems is the increase in density of the vapor phase over time. For the gas system, it is driven only by the drop of temperature over time. For the multi-component flashing liquid system, it is driven by both the drop of temperature over time and the increase of the average molecular weight over time.

For dispersion, the density of the vapor phase is of primary importance since a light gas discharge would disperse like a buoyant cloud (raising in atmosphere) but a heavy discharge would disperse like a heavy cloud (slumping on the ground). Thus, for a multi-component flashing liquid system, both the change in the temperature and the average molecular weight of the vapor over time need to be considered.

The rest of the paper will relate to the multi-component flashing liquid system only. Examples will present toxic cloud dispersions because they emphasize the different behaviors of the cloud. For flammable dispersions, those differences would be more subtle in our example.

## 3. Dispersion modeling of a discharged gas

To model the dispersion of the cloud following a gas discharge, two approaches are generally used:

1. Dispersion modeling based on a single constant discharge rate and
2. Dispersion modeling based on multiple fixed discharge rates (referred as "time-varying modeling")

In case #1, the single fixed discharge rate is used to represent the entire duration of the discharge. This is a simplification that does not take into account the change of pressure over time.

In case #2, the effects of a pressure decrease over time can be accounted by using a time-varying release model which divides the discharge into multiple segments and uses a separate fixed discharge rate for each segment. Thus the inputs into the dispersion model "varies with time".

### 3.1 *Dispersion modeling based on a single fixed discharge rate*

This approach is often used since it presents the advantage of being simple and adequate for many situations: The modeling requires very few input data and outputs are relatively straightforward and easy to interpret.

Two types of fixed rate can be chosen:

1. Case A: A single fixed rate equal to the initial discharge rate
2. Case B: A single fixed rate equal to an intermediate discharge rate

This approach presents a few limitations: the difficulty to justify the choice of an adequate fixed rate and therefore, the difficulty to defend the reasonable conservatism of modeling results.

#### 3.1.1 Case A: single fixed rate equal to the initial discharge rate

The evaluation of the initial discharge rate is based on the initial process conditions in the multi-component flashing liquid system. If the pressure in the system drops slowly, the discharge is nearly constant and therefore using the initial discharge rate would provide accurate results.

In the case where the pressure drops quickly, this will often lead to conservative dispersion results since the initial discharge rate is higher than the average discharge rate over the depressurization duration.

The graph in Figure 3-1 illustrates toxic results (100 ppm $H_2S$ concentration) for a 10 minute depressurization (discharge duration of 10 minutes) of a flashing liquid system containing a light crude oil extracted from a well. In the initial conditions, the vapor phase in equilibrium with the crude oil has an average molecular weight of 26 g/mol (main components ranging from Methane to n-Decane, plus a small amount of $H_2S$, $N_2$ and $CO_2$).

**Figure 3-1: Side-view of a 10min-depressurization – Cloud Footprint (F1.5) at initial conditions**

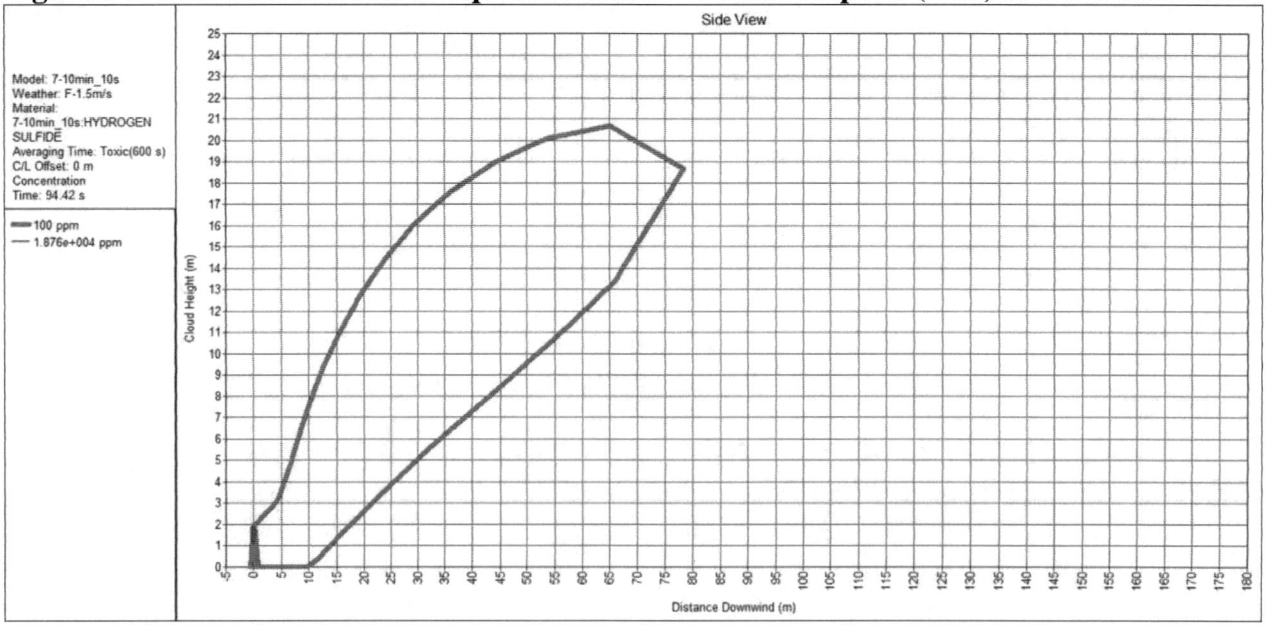

In the initial discharge conditions, the gas is lighter than air hence the dispersing cloud ascends in the atmosphere as illustrated by the side view plot. Based on this modeling, one would conclude that there is no risk of being exposed to a 100 ppm $H_2S$-cloud at ground level beyond a 10m radius from the source.

### 3.1.2 Case B: single fixed rate equal to an intermediate discharge rate

Choosing which intermediate rate to use is a challenge: too high intermediate rate may lead to conservative dispersion results and too low may lead to optimistic dispersion results.

Using the same example as for case A) with an intermediate rate corresponding to discharge conditions at 3.5 minutes would lead to results illustrated in Figure 3-2.

**Figure 3-2: Side-view of a 10min-depressurization – Cloud Footprint (F1.5) at 3.5min conditions**

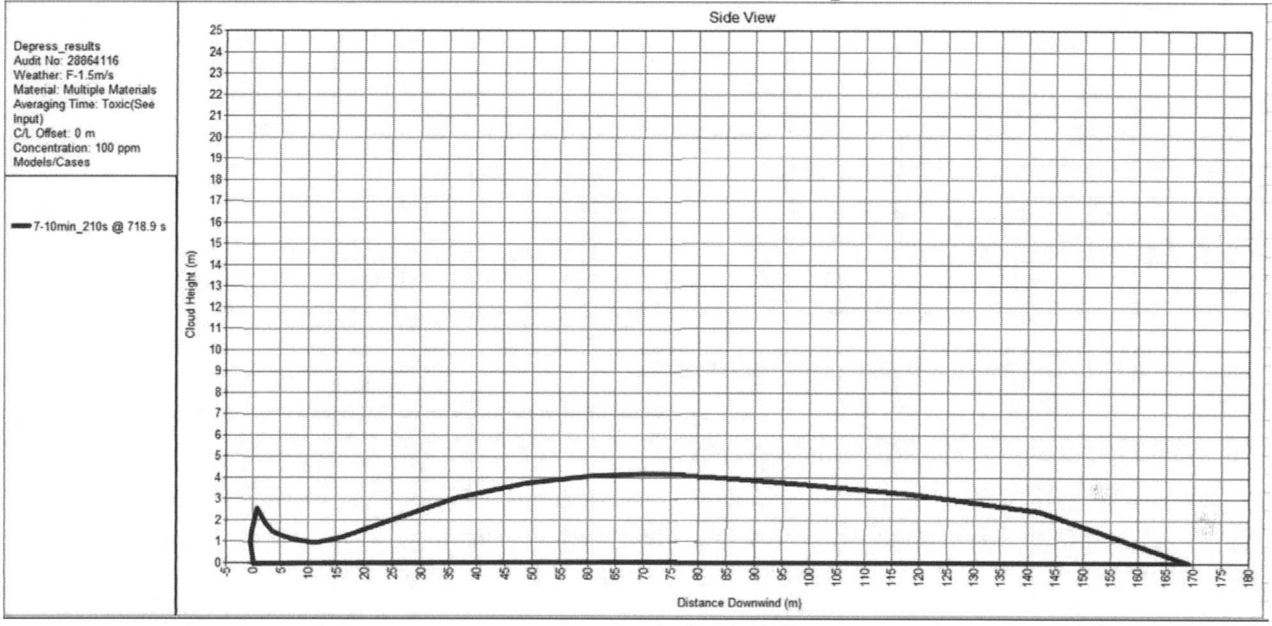


In this intermediate discharge condition, the discharged gas is heavier than air (molecular weight of 33 g/mol and cold discharge temperature) hence the cloud descends and disperses at ground level. Based on this modeling, one would conclude that there is a toxic risk at ground level up to 170m from the source. Consequences would obviously be much worse that for case A).

Therefore, considering that the dispersing cloud transitions from a buoyant vapor mixture early in the depressurization event to a dense vapor mixture later in the depressurization event, what would be the correct fixed discharge to use?

If the location of concern is somewhere above the ground, case A) might provide better results than case B). But if the location of concern is at ground level, case B) would provide better results.

Since toxic risk is dependent not only on concentration but also on exposure duration, accurate results should also account for exposure duration. Assuming that someone can be exposed to the toxic cloud during the whole duration of the depressurization might therefore lead to over-conservative results.

### *3.2 More rigorous modeling based on time-varying discharge rates*

For a multi-component flashing liquid system, both the change in the temperature and the average molecular weight of the vapor over time need to be considered. This is accomplished by modeling the time-varying discharge as multiple discrete steps. Each step will be represented by a different discharge rate, gas temperature and gas composition. Figure 3-3 illustrates an example of a time-varying discharge model using six discrete steps.

**Figure 3-3: TV dispersion by six steady-states dispersions – 10s, 2min, 3.5min, 5min, 7min, 9 min**

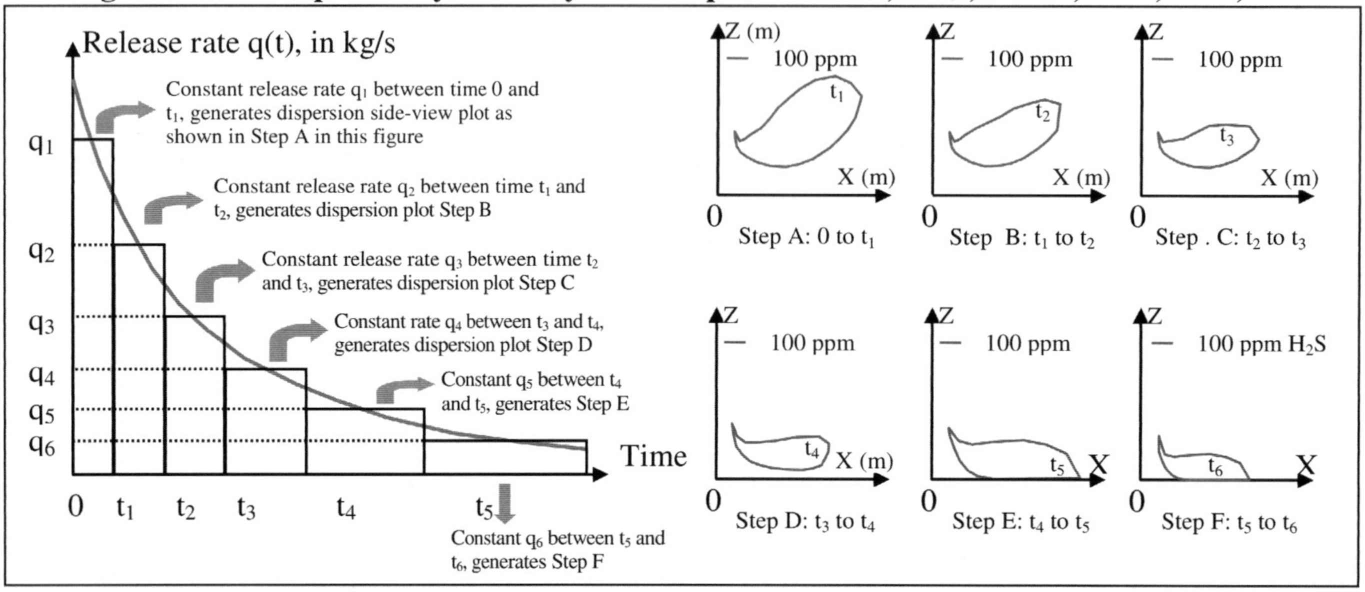


For each step, the dispersion of the cloud is modeled using as inputs, the fixed discharge rate, composition and temperature values from the corresponding quasi-steady state discharge. Dispersion results from each step can be illustrated by the side-view concentration plots, as shown in Figure 3-3 (Step A, Step B, Step C, Step D, Step E and Step F).

To facilitate the analysis, results for all steps can be superimposed on a single figure as presented in Figure 3-4.

**Figure 3-4: TV dispersion summary with six discrete steps**

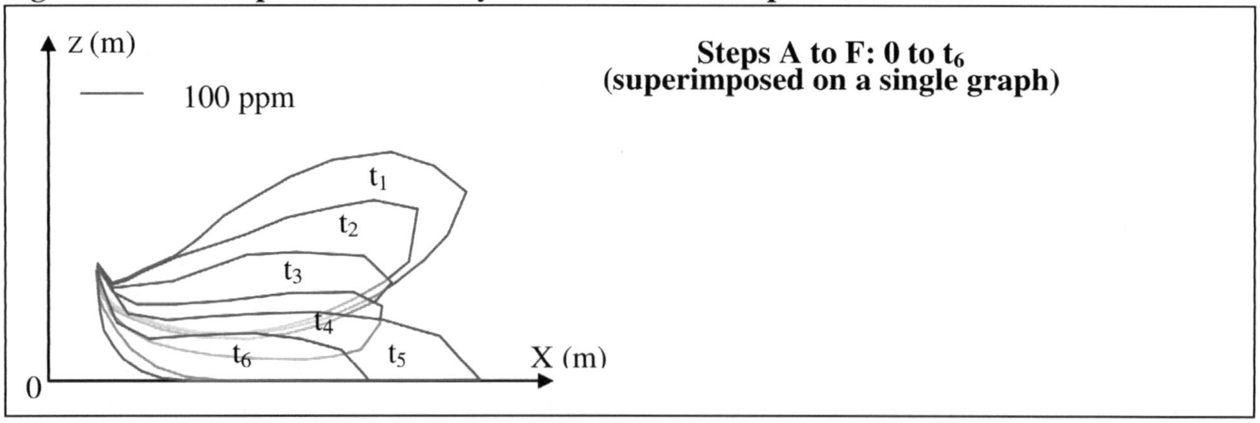

This figure reflects the dynamic nature of the dispersing cloud which in this example starts as a buoyant cloud and transitions to a slumping cloud. The use of a single discharge dispersion modeling could only reflect one of these steps. Depending on the location of interest, attempting to rely on a single fixed discharge release could result in misleading conclusions.

## 4. ERM recommended approach to the modeling of the multi-component flashing liquid depressurization and cloud dispersion

To model properly this complex release and dispersion case, one shall:

- Model the time-varying discharge using multiple steps. Each step will be represented by a different discharge rate, gas temperature and gas composition;
- Model the dispersion individually for each step.

Commercial dispersion programs like PHAST [1] are currently unable to model the depressurization of a vessel filled with a multi-component flashing liquid. Therefore, the multi-component time-varying discharge shall be performed with a chemical process simulation software like HYSYS [2], PRO/II [3], ASPEN PLUS [4], or through the use of a specific modeling tool developed internally for this purpose. Once modelled, the depressurization shall provide the following outputs:

- Gas pressure inside the vessel over time
- Gas discharge rate over time
- Gas temperature inside the vessel over time
- Gas composition over time

These outputs can then be used to generate multiple fixed-discharge-rate dispersion cases that will be modelled with a commercial dispersion modeling program like PHAST.

In the example chosen for this paper, the multi-component time-varying discharge is performed with HYSYS specific depressurization module. Then based on HYSYS outputs, six PHAST dispersion cases are defined to represent six steps along the time-varying discharge. These steps correspond to 10s, 2min, 3.5min, 5min, 7min and 9min. The 100 ppm $H_2S$ side-view iso-concentrations for each of the six steady-states' discharge modeling are presented on a single graph in Figure 4-1.

**Figure 4-1: Side-view of a 10min-depressurization – Cloud Footprint (F1.5) at 10s, 2min, 3.5min, 5min, 7min**

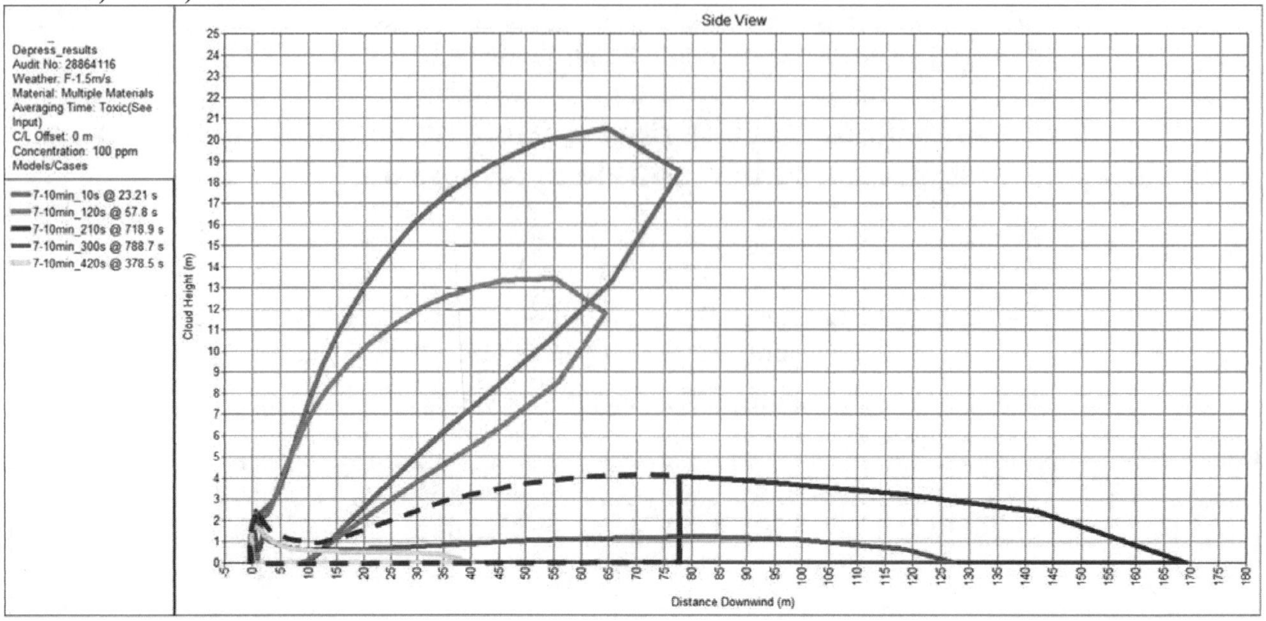


Based on these results, it can be concluded that a receptor located at 100 meters from the leak is exposed to the 100 ppm toxic effect of $H_2S$ only after the first 2 minutes of the depressurization and is no longer exposed after 7 minutes. Hence the exposure duration would be approximately 5 minutes. This result differs radically from the results based on the dispersion modeling performed with a single fixed rate as shown in Figure 3-1 (toxic exposure would be evaluated as 0 minute) and Figure 3-2 (toxic exposure would be evaluated as 10 minutes).

## 5. Conclusion

When modeling the dispersion following the depressurization of a multi-component flashing liquid system, the use of a single fixed discharge rate might provide results that are either too conservative or too optimistic.

A multi-steps/multiple dispersion source modeling approach as presented in this paper would more rigorously represent the phenomenon involved in the depressurization of a multi-component flashing liquid system. Results would therefore be more representative of the consequences. However, this modeling approach presents the following limitations:

- This approach is relatively complex and requires the use of a specific tool like a chemical process simulation software, to model the depressurization discharge.
- The depressurization module built in the chemical process simulation software do not provide the final velocity and the final temperature of the gas after expansion to atmospheric conditions. These parameters would have to be evaluated separately for each dispersion case before the modeling with a dispersion software like PHAST.

A future paper should investigate whether an intermediately complex approach could be defined, using solely a dispersion modeling software like PHAST.

## 6. References


[1] PHAST (Process industry Hazard Analysis Software Tool), an integrated consequence modeling software package developed by DNV GL to analyze situations which present hazards to life, property and the environment, and to quantify their severity (http://www.dnv.com/services/software/products/phast_safeti/phast/).

[2] HYSYS, a steady-state process simulator (process simulation) for process design and operational analysis for process engineers in the chemical, petroleum, natural gas, solids processing, and polymer industries, initially developed by Hyprotech and now developed and released by Aspen Technology (http://www.aspentech.com/hysys/).

[3] SimSci PRO/II, a steady-state process simulator (process simulation) for process design and operational analysis for process engineers in the chemical, petroleum, natural gas, solids processing, and polymer industries, developed initially by SimSci and now released by Invensys (http://software.invensys.com/products/simsci/design/pro-ii/).

[4] ASPEN Plus, a steady-state process simulator (process simulation) for process design and operational analysis for process engineers in the chemical and specialty chemical organizations and related industries to design and improve process plant. Aspen Plus is developed and released by Aspen Technology. (https://www.aspentech.com/products/aspen-plus.aspx)

# 349966 Quantitative Analysis of Environmental and Societal Risk for Onshore FUEL Pipelines

**Abstract:**

Combustible materials transportation by pipeline has been established as a safe and efficient mean. Due to the low accidental frequency of incidents associated, it is socially accepted as favorable. However, like any other industrial practice it can be dangerous and represents risks to the society, environment and infrastructure. Since production sites are often away from consumption centers, increased demand in Colombia and the world has led to increased transportation needs, forcing the growth of pipeline networks.

Carrying out a-priori risk analysis could provide engineering tools to support decision making, related to the compatibility of activities within a territory (Land-use Planning). This allows taking preventive measures with respect to potential accidents. Hence, implementing scenario building enables to explore and clarify chain of possible events, including actions and consequences. Also, it is important to define event probabilities; this is done using Event Tree Analysis (ETA). Subsequent phase consists of consequence analysis, where simulation is performed in order to determine effects intensity and then vulnerability values. Once all this is accomplished, risk assessment is completed.

So, this work aims to develop a methodology for the assessment of societal and environmental risk associated with: loss of containment in onshore fuel liquid pipelines and consequences of critical events, as radiation, overpressure and contamination effects. As risk values vary along pipelines due to differences in environmental and societal conditions, analysis is performed for segments of a pipeline, including analysis of causes and types of events.

For societal risk assessment the methodology includes some guidelines to develop a segmentation process. Risk changes along the pipeline due to differences in environmental and societal conditions in order to define manageable pipeline sections for analysis. It is also necessary to identify the probable events and their probabilities; in this case MIMAH methodology and historic analysis is used. Then consequence analysis is performed for each segment, determining the intensity of overpressure and radiation effects caused by the previously defined events. This determination is done by means of simulations performed on specialized software and then using *probit* equations the vulnerability is estimated. With vulnerability and probability values known, risk can be estimated. As a result, risk values for people are determined and expressed as individual risk contours and FN curves for societal risk.

In addition to the societal impacts, assessment of environmental contamination is also done. It is known that hydrocarbon releases are a real hazard to the environment as they produce contamination to soil and groundwater. In Colombian topography, mountains and slopes are mostly present. This generates scenarios out of the right of way making analysis of onshore hydrocarbon spillages even more relevant in this territory. Therefore, environmental effects on soil are determined considering mass balance of infiltrated and evaporated material in horizontal and inclined surfaces. Consequently, effective volume remaining on soil surface is determined to establish amount of material available for migration pathways. Environmental damage is then estimated taking into account depth of infiltrated material, associated to soil excavation and off-site disposal, soil clean-up and groundwater clean-up costs. As a result, environmental risk values are presented as risk indexes in terms of remediation action costs, defining an environmental local monetary damage. Contamination is analyzed in order to determine average soil volumes which need to be treated, considering intervention, mitigation and remediation times.

This methodology is targeting to support decision-making in Colombia in early stages of analysis, with respect to land-use planning, from a societal and environmental point of view. A total risk index is calculated, in order to be used as a comparative tool for ranking different projects or pipeline sections. This total risk index is calculated turning both societal and environmental risk in equivalent terms, meaning monetary terms. It is represented as an FM curve (frequency vs. millions of dollars, euros, etc.). Simultaneous assessment of both human and environmental risk is remarkable, as it integrates and supports decision making.

This methodology is applied on a case study to illustrate how it can support decision making (related to Land-use planning and to alternatives choice).Taking into account international societal risk criteria, it will be determined whether these values can be accepted or should be reduced. Therefore, coupling risk assessment of environmental contamination and damage to exposed population allows quantifying risk and thus, supporting decision making within the territory.

**Small Scale Experimental Study of Vaporization Rate of Liquid Nitrogen Released on Ice**

Nirupama Gopalaswami[1]
M. Sam Mannan[1]
[1]Mary Kay O'Connor Process Safety Center
Artie McFerrin Department of Chemical Engineering Department
Texas A&M University System
College Station, Texas 77843-3122, USA
nirupama89@tamu.edu
mannan@tamu.edu

Luc Vechot[2]
Tomasz Olewski[2]
[2]Mary Kay O'Connor Process Safety Center - Qatar
Texas A&M University at Qatar, PO Box 23874, Education City, Doha, Qatar

Luc.vechot@qatar.tamu.edu
Tomasz.olewski@qatar.tamu.edu

Prepared for Presentation at
American Institute of Chemical Engineers
2014 Spring Meeting
10th Global Congress on Process Safety
New Orleans, LA
March 30 – April 2, 2014

UNPUBLISHED

# Small scale experimental study of vaporization rates of liquid nitrogen released on ice

Nirupama Gopalaswami[1], Luc Vechot[2], Tomasz Olewski[2] and M. Sam Mannan[1]
[1]Mary Kay O'Connor Process Safety Center
Artie McFerrin Department of Chemical Engineering Department
Texas A&M University System
College Station, Texas 77843-3122, USA
[2]Mary Kay O'Connor Process Safety Center - Qatar
Texas A&M University at Qatar, PO Box 23874, Education City, Doha, Qatar
nirupama89@tamu.edu
mannan@tamu.edu
Luc.vechot@qatar.tamu.edu
Tomasz.olewski@qatar.tamu.edu

**Keywords:** Source term, liquid nitrogen, conduction, ice, vaporization

## Abstract

One of the LNG accident scenarios is the collision of an LNG carrier on an iceberg while transporting on the sea. A collision can result in damages to the vessel and lead to the leakage of contents on ice or an ice-water mixture. When cryogenic liquid comes in contact with ice, it undergoes rapid vaporization and boiling due to the difference in temperature between the ice and cryogenic liquid. This process is different from the heat transfer between water and cryogenic liquid as ice is a solid and thus heat transport to the pool occurs through the conduction. In this paper, the heat transfer ~~phenomena between ice and cryogenic liquid has~~phenomena between ice and cryogenic liquid have been studied and the boil-off rates have been quantified. A small scale experiment was conducted to investigate the heat transfer phenomenon of cryogenics released on ice. The amount of cryogenic liquid spilled was varied to determine its effect on vaporization rates and the temperature of ice was monitored. Additionally the vaporization rates were determined by direct measurement of the mass loss during the experiment. As expected, the vaporization rates were found to be a function of ice temperature and were found to be independent of time for long duration releases. The one dimensional conduction model was validated against experimental results. The predicted temperatures and heat flux were found to be in good agreement with the experimental data.

## 1. Introduction

One of the commonly considered LNG accident scenarios is the collision of an LNG ship with an iceberg while transporting on the sea. Detailed consequences analysis considering various scenarios and hypotheses through calculations are required for LNG carriers. It is also important to take into account the fatigue effect of vibrations in risk analysis when LNG ship traverses through ice. Consequence analysis in such cases depends on features of the ice and the structural strength of the LNG carrier. Two of the common types of collision that can occur are the side collision between vessel and iceberg and bow collision [1]. A collision can result in damage of the vessel and allow leakage of contents on ice or ice-water mixture. Such a scenario is common in areas where ice is predominantly present throughout the year like the Arctic Circle. Upon spillage of contents, one of the most important factors that affect cryogenic liquid release on ice is the heat flux. This process is different from the heat transfer between water and cryogenic liquid as ice is a solid and conduction phenomena is more dominating than convection due to boiling.

Currently, very few experiments have been performed by releasing cryogenic liquids on ice. In 1971, Nakanishi observed that when cryogenic liquids like Condensed Pipeline Gas (CPG, with composition $CH_4$ 92.7%, $N_2$ 7.3%; boiling point about -111.6K) and Liquid Nitrogen ($LN_2$, boiling point -77 K) are poured on an ice layer maintained at -150 °C, there is no vapor film formation under the cryogen's layer and both liquids are boiling in nucleate regime [2]. When the experiment was repeated by pouring onto an ice layer maintained at -5 °C, a vapor film was observed to occur for liquid nitrogen alone, but not for CPG. It was concluded when CPG was spilled on ice, it stays in film boiling for a brief period. This might have been due to the fact that the temperature difference between ice and CPG would have been small to sustain film boiling for a long period. This phenomenon was not addressed in detail in the research work.

In another study, small scale experiments were performed with liquid nitrogen and LNG (Burgess, Murphy and Zabetakis, 1970a). The primary objective of this study was to measure the heat flux and the vaporization rates. The experiments were performed at laboratory scale, (spill area - 0.18 $m^2$) where cryogenic liquid was poured onto a flat tray (30.5 x 61 cm) of ice. The ice was made 4 cm thick and covered the entire surface of the tray such that the cryogenic liquid contacted only ice. LNG and LN2 were released on ice. An array of thermocouples was placed below the ice to measure the time dependent temperature and associated heat transfer. However, wrinkling of the ice-cryogen interface disturbed the position of thermocouples and a spatial variation of heat transfer was not obtained in the test. The amount of cryogenic liquid poured was varied and the vaporization rate was reported for the experiment. The boil-off rates of nitrogen on ice varied from 0.08-0.12 kg $m^{-2}s^{-1}$. The small scale tests showed a lot of variation in the boil-off rates reported for nitrogen. The vaporization rates were reported for discrete values of time.

A heat transfer model for growing ice thickness involving conduction and convection phenomena for release of cryogenic liquids on shallow water was undertaken by Vesovic [3].However, the model was not validated with any experimental data. In all these experiments the source term, the heat flux from the substrate (ice) to the pool was calculated by measuring the temperature difference between the substrate and cryogenic liquid pool with time. This was possible by measuring the temperature difference between the substrate and cryogenic pool and using analytical expressions to determine the heat flux. The conclusions of small scale experiments had limited effects due to the small confined space of the boiling cell. Most of the tests were also qualitative in nature and currently there are very few quantitative results for cryogenic liquids released on ice.

An experimental investigation is undertaken here to improve the understanding of cryogenic liquid releases on ice and to provide a benchmark qualitative and quantitative results. The small scale experimental study is performed to verify the heat transfer mode of cryogenic liquids released on ice and independently validate the one dimensional conduction for cryogenic liquid released on ice.

## 2. Experimental Setup and Procedure

The experimental setup designed for small scale spill of liquid nitrogen on ice is shown in Figure 1. The experiments were performed in a wind tunnel present at the Fire Station 2 of the Ras Laffan Industrial City, Qatar. The wind tunnel was 12 m long, 2 m wide and 0.855 m tall offering a convenient space for safe dispersal of vapor. The setup consists of a metallic cylinder of inner diameter 58.5 cm and height 87.5 cm made of carbon steel. A stainless steel chimney with dimensions of 35 x 35 x 80 cm was placed inside the metal cylinder to protect the cylinder from fractures resulting from a cryogenic liquid spill. The stainless steel chimney was square in cross-section of 0.1225 $m^2$ and devoid of top and bottom walls, which offered a well-defined area for the spill on water. The material was robust in cryogenic liquid temperature and offered additional protection for the external cylinder. The stainless steel chimney was lowered into the metallic cylinder using a steel support in such a way that it was half filled with water at any point of time during the experiment. Such a setup made possible for the liquid nitrogen to interact only with water and stainless steel walls rather than the cylinder, which was susceptible to fractures at cryogenic liquid temperatures. An ice slab of dimensions 25 x 25 x 10 cm was placed inside the stainless steel chimney on top of water. The thermocouple to measure the temperature of the ice was embedded into the ice slab. Additional ice cubes were added around the ice slab to create an ice-water mixture and to limit contact of $LN_2$ with water.

The temperature of water, liquid nitrogen and nitrogen vapor was monitored by distributed N-type thermocouples. The thermocouples were mounted on two polycarbonate boards and each included sixteen thermocouples distributed vertically. One board was mounted near the wall of stainless steel box and another was held in the center of the box. The dimensions of the thermocouples relative to water level are given in Table 1. The polycarbonate board was suspended into the metallic cylinder in such a way that 11 thermocouples measured the temperature below water, three measured liquid nitrogen temperature and two measured liquid nitrogen vapor temperature. Each of these thermocouples was calibrated for measuring liquid nitrogen and ice temperature and the temperature was measured every second. The thermocouples were distributed with smaller distances (<0.5 cm) near the water-liquid nitrogen surface. The position of each thermocouple and cryogenic discharge hose nozzle is shown in

Table 1. The cryogenic hose was placed in the middle of the setup to ensure smooth discharge of liquid nitrogen on ice. The width of the tunnel was considered X axis and length of the wind tunnel was considered as the Y axis (and the positive value indicates downwind direction). The entire setup was placed on a balance (maximum load – 300 kg and sensitivity – 100 g) to monitor the change of mass when nitrogen vaporizes on water. The sensors' output was recorded every second (1 Hz) by a Data Acquisition system (DAQ), and the recorded data were sent to a computer for processing.

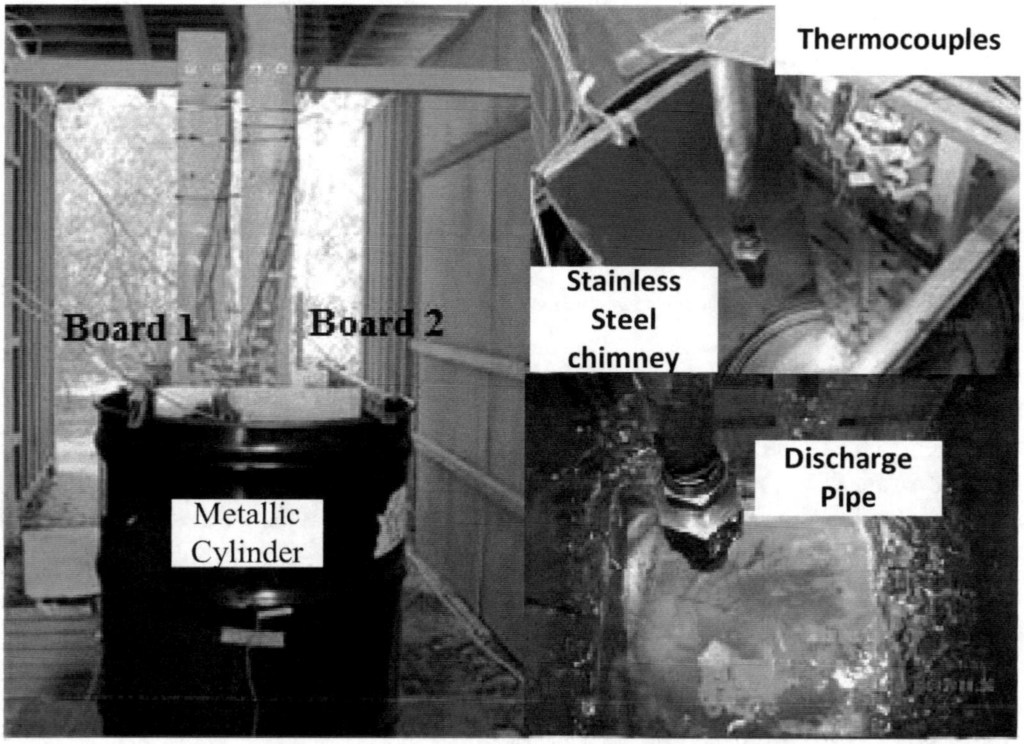

Figure 1: Top and lateral views of experiment setup

Table 1: Thermocouple positions

| Thermocouple number | Position with respect to cylinder's corner (cm) | | | Thermocouple number | Position with respect to cylinder's corner (cm) | | |
|---|---|---|---|---|---|---|---|
| | X | Y | Z* | | X | Y | Z* |
| TC-201 | 14 | 26.6 | -10.25 | TC-204 | 26 | 26 | -9.85 |
| TC-202 | 14 | 26.6 | -8.15 | TC-205 | 26 | 26 | -7.7 |
| TC-203 | 14 | 26.6 | -6.25 | TC-216 | 26 | 26 | -5.67 |
| TC-206 | 14 | 26.6 | -4.25 | TC-218 | 26 | 26 | -3.65 |
| TC-207 | 14 | 26.6 | -3.25 | TC-219 | 26 | 26 | -2.65 |
| TC-208 | 14 | 26.6 | -2.35 | TC-222 | 26 | 26 | -2.25 |
| TC-209 | 14 | 26.6 | -1.25 | TC-224 | 26 | 26 | -0.75 |
| TC-210 | 14 | 26.6 | -0.8 | TC-225 | 26 | 26 | -0.25 |
| TC-211 | 14 | 26.6 | -0.25 | TC-226 | 26 | 26 | 0.35 |
| TC-212 | 14 | 26.6 | 0.25 | TC-227 | 26 | 26 | 0.75 |
| TC-213 | 14 | 26.6 | 0.65 | TC-228 | 26 | 26 | 1.25 |
| TC-214 | 14 | 26.6 | 1.65 | TC-229 | 26 | 26 | 2.35 |
| TC-215 | 14 | 26.6 | 2.65 | TC-230 | 26 | 26 | 3.25 |
| TC-217 | 14 | 26.6 | 4.75 | TC-231 | 26 | 26 | 5.35 |
| TC-220 | 14 | 26.6 | 7.75 | TC-232 | 26 | 26 | 8.25 |
| TC-221 | 14 | 26.6 | 22.25 | TC-233 | 26 | 26 | 23.45 |
| Discharge Pipe | 14 | 14.5 | 13.75 | | | | |

*Distance from top ice surface; positive value means above the ice surface, negative one below the ice surface (inside ice).

The liquid nitrogen was discharged continuously from a tank of 180 liter capacity through a cryogenic liquid hose of length 15 m and internal diameter 0.0127 m, until desired amount of $LN_2$ was reached. The mass of the box was recorded every 1 second and used to measure the $LN_2$ vaporization rate. The time of opening the cryogenic liquid valve and the time when liquid nitrogen contacted with the water was noted. The filling rate of $LN_2$ was then calculated from the time and mass data (where filling rate referred to discharge rate ~~substracted~~subtracted by vaporization rate during the spill). The time to complete vaporization of liquid nitrogen was also noted. The discharges were repeated several times and the spilled amount of liquid nitrogen varied. The temperature of ice decreased after each experimental run due to release of liquid nitrogen on it. A summary of the six experimental runs with initial temperature of water bulk, initial temperature of ice and the amount of liquid nitrogen is given in Table 2.

Table 2: Summary of experiments

| Run Number | Total Amount Spilled, (kg) | Release rate, ($kg.s^{-1}$) | Initial temperature of ice, (°C) | Water Temperature below ice, (°C) |
|---|---|---|---|---|
| Run 1 | 4.3 | 0.02 | -11 | 8 |
| Run 2 | 10.5 | 0.08 | -5 | 13 |
| Run 3 | 5.3 | 0.05 | -66 | 10 |
| Run 4 | 2.6 | 0.02 | -51 | 9 |
| Run 5 | 8.0 | 0.03 | -44 | 8.3 |
| Run 6 | 13.6 | 0.01 | -119 | 7.4 |

## 3. Theory

When a cryogenic liquid contacts the ice surface, a large temperature difference exists between ice and liquid nitrogen which results in rapid vaporization of liquid nitrogen. The heat transfer from ice to liquid nitrogen is predominantly controlled by conduction phenomenon. In this study, one dimensional semi-infinite unsteady heat conduction equation with ideal contact of liquid nitrogen and ice is used to describe the heat transfer process [4]. The heat flux $q$ from ice to liquid nitrogen is given by

$$q = \left(\frac{k}{\sqrt{(\pi\alpha t)}}\right)(T_i - T_b)\exp\left(-\frac{z^2}{4\alpha t}\right) \qquad (1)$$

The heat flux is also proportional to inverted square root of time and is given by

$$q = \left(\frac{k}{\sqrt{(\pi\alpha)}}\right)(T_i - T_b)\exp\left(-\frac{z^2}{4\alpha t}\right) * t^{-0.5} \qquad (2)$$

The temperature $T$ inside the ice at the $z$ depth is given by

$$T = T_i - (T_i - T_b).erfc\left(\frac{z}{2\sqrt{\alpha t}}\right) \qquad (3)$$

~~$q = \left(\frac{k}{\sqrt{(\pi\alpha t)}}\right)(T_i - T_b)\exp\left(-\frac{z^2}{4\alpha t}\right)$ (1)~~

~~The temperature $T$ inside the ice at the $z$ depth is given by~~

~~$T = T_i - (T_i - T_b).erfc\left(\frac{z}{2\sqrt{\alpha t}}\right)$ (2)~~

Where the $T_i$ represents the initial temperature of ice, $T_b$ is the temperature of liquid nitrogen, $k$ is the thermal conductivity and $\alpha$ is the thermal diffusivity of ice. A perfect thermal contact between liquid nitrogen and ice is assumed in calculating the temperature variation of ice.

## 4. Results and discussion

The results of small scale experiments are discussed in subsequent sections. Section 4.1 discusses the vaporization flux determined from mass loss data obtained during the experiment. Section 4.2 and 4.3 discusses the validation of model with experimental data.

### 4.1. Vaporization Mass Flux

The mass flux of vapor (vaporization rate) was obtained from direct measurement of mass loss data obtained during the experiment (Figure 2). An averaging time of 60 seconds was used to compute the vaporization rates of the liquid nitrogen on ice. The selection of averaging time was based on a sensitivity analysis shown in Figure 3. Four averaging times were chosen to calculate the mass vaporization flux of LN2 vapor. An averaging time of 10 and 20 seconds resulted in large fluctuations in vaporization flux due to low resolution in mass measurement (maximum capacity of balance was 300 kg and 100 g resolution). An averaging time of 30 seconds resulted in fairly smooth curve but still with periodic fluctuations. An averaging time of 60 seconds was found to give low variations and thus sufficiently describing the vaporization mass flux without much loss in the actual physics. The spill area included the constant cross sectional area of the stainless steel chimney and a variable area based on the height of the liquid nitrogen interacting with the stainless steel. The chimney cross sectional area of 0.1225 $m^2$ was used to determine the vaporization rates.

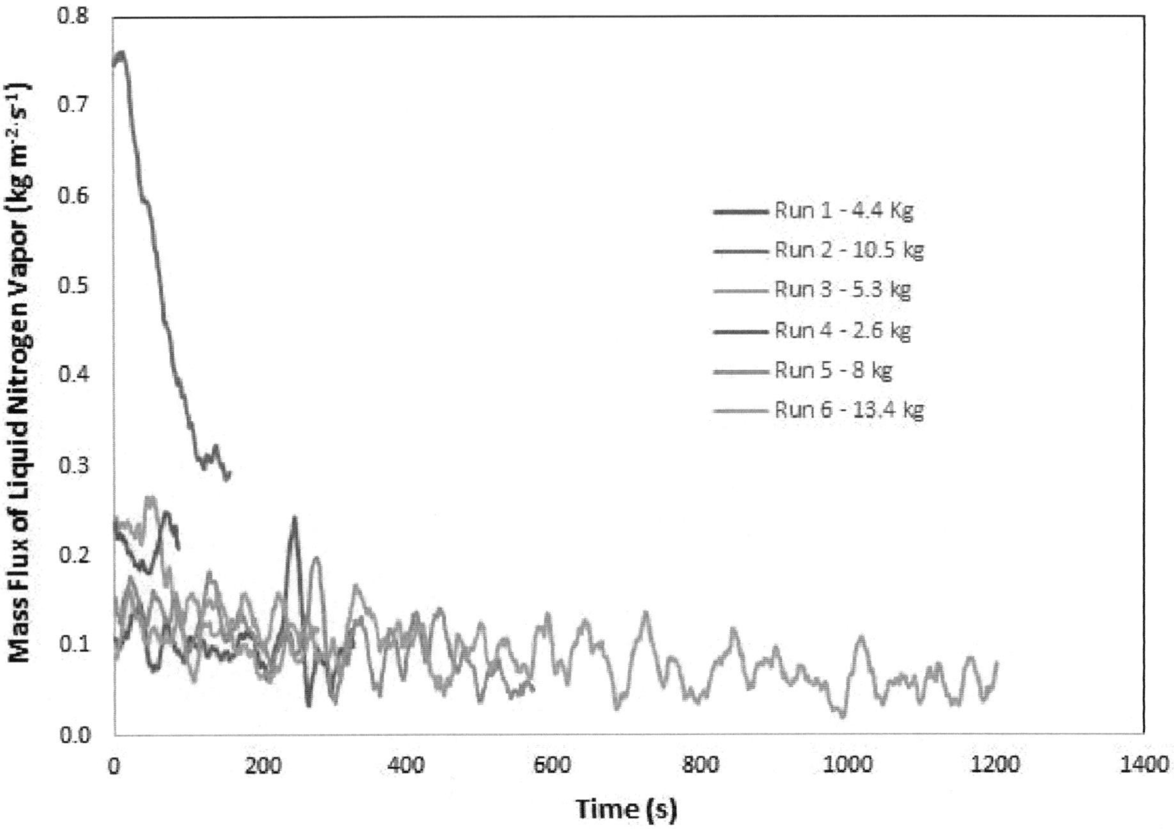


Figure 2: Vaporization Mass Flux

Atypical for conduction, a decreasing trend of vaporization rate was observed when the ice temperature was relatively high (-5 °C at run 2). However, the mass flux became comparatively constant, when the ice temperature dropped down below -11°C, (run 1 and runs 3 to 6). However, small peak in vaporization rate was observed during run 1 (red line in Figure 2). This is with the agreement to the result obtained at the Bureau of Mines [5], where a decreasing trend was observed at the initial stage of the spill on ice.

Initially during the start of experimental runs, the ice block was transparent in nature. With the release of liquid nitrogen in every test, the thickness and opacity of ice increased. Small ice balls were formed during each experiment on the surface of ice, which might be attributed to freezing of condensed water from air and their aggregation. The ice balls and frost on the walls of the equipment are shown in Figure 3. It has been observed that the growing ice thickness in the pool created resistance to heat transfer and thus controlled the heat flux supplied to the cryogenic pool. An average heat flux of 36.6 $kW/m^2$ was measured in the tests, which represents an average vaporization flux of 0.18 $kgm^{-2}s^{-1}$. The values of vaporization rates were compared with experiments conducted by Bureau of Mines (1970). The average heat flux obtained in their experiment was around 25.85 $kW/m^2$ with corresponding vaporization flux determined to be 0.13 kg $m^{-2}s^{-1}$. The variability can be attributed to the lack of water and ice mixture that was not present in the previous experiments. A combination of ice and water mixture results in heat fluxes that are intermediate between heat flux obtained due to convection and conduction alone.

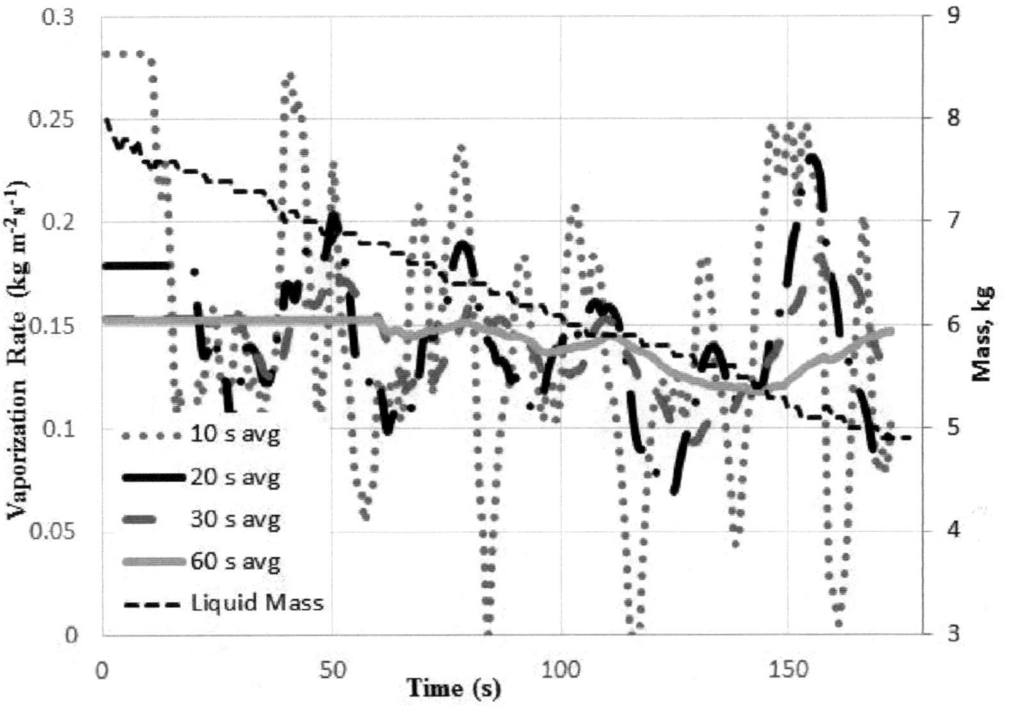


Figure 3: Vaporization flux for different averaging times

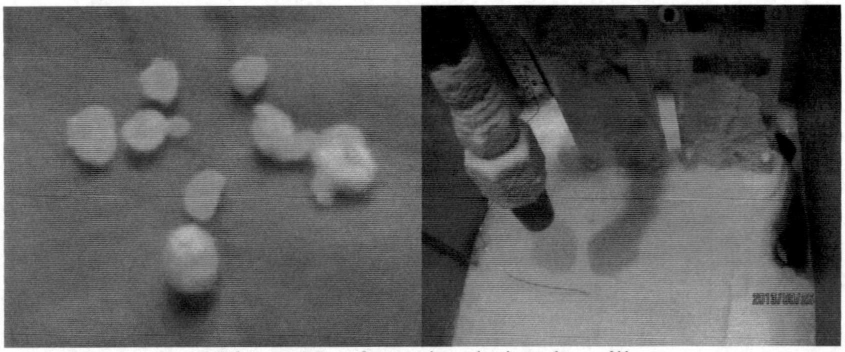

Figure 4 Ice formation during the spill

### 4.2. Temperature profile of ice

The variation of ice temperature during the spill was analytically calculated from equation (2) and compared with experimental values at seven different depths, and the result obtained for experimental run 5 is shown in Figure 5. The solid lines represent the temperature measured during the experiment and the dashed lines represent the predicted temperatures. A good agreement was observed between the experimental and predicted temperature. It is to be noted that the initial temperature profile of ice in this run (run 5) was an effect of previous spill and this explains lower temperature near to the ice top surface and higher values for thermocouples placed farther from the ice surface. Also, the thermocouples placed near the top surface are cooled earlier than the thermocouples below, due to the release of liquid nitrogen. A deviation in predicted temperatures is observed after 750 seconds, which might be an effect of heat transfer from warmer water below ice. This might also be due to the variation in thermal properties of ice with temperature, which is not accounted for in the model.

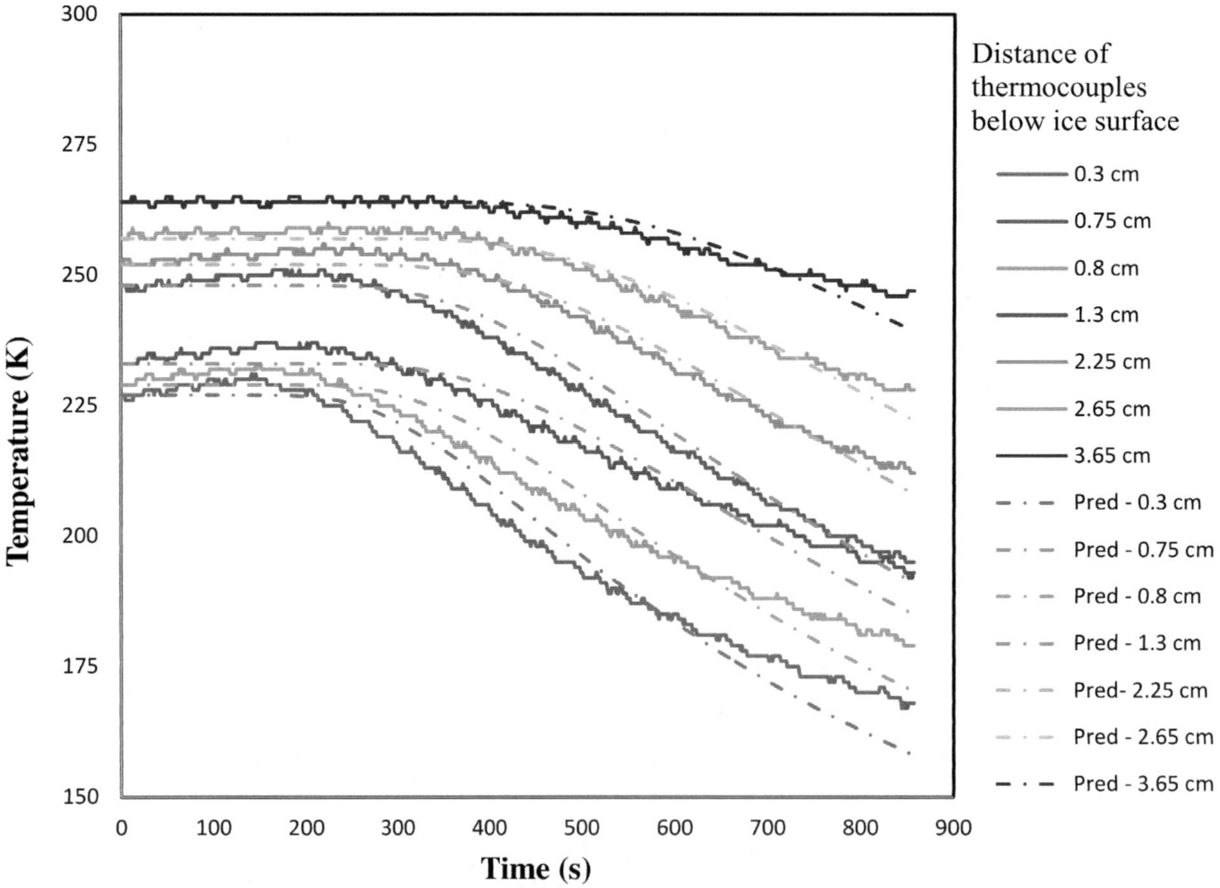


Figure 5: Temperature profiles of thermocouple below water surface

### 4.3. Comparison of experimental and predicted transient heat flux

A comparison of the predicted heat flux and experiment heat flux as the function of time is given for run 5 is given in Figure 7. The trend for one experimental run is shown here since all the experiments except run 2 followed similar trends in data. Figure 7(a) represents the comparison of predicted heat flux obtained from equation (1) and experimental heat flux with respect to time. The heat flux obtained from experiment was found to be varying above and below the predicted heat flux. However, the variations of experiment were comparatively low. Figure 7 (b) represents the comparison of predicted heat flux obtained from equation (2) and experimental heat flux with respect to inverted square root of time. The results seem to have good agreement except for the run 2. However, run 2 was a case where the liquid nitrogen was spilled on a mixture of ice and water, thus it cannot be well modeled by the conduction and perhaps is closer to the convection as for the spill on water.

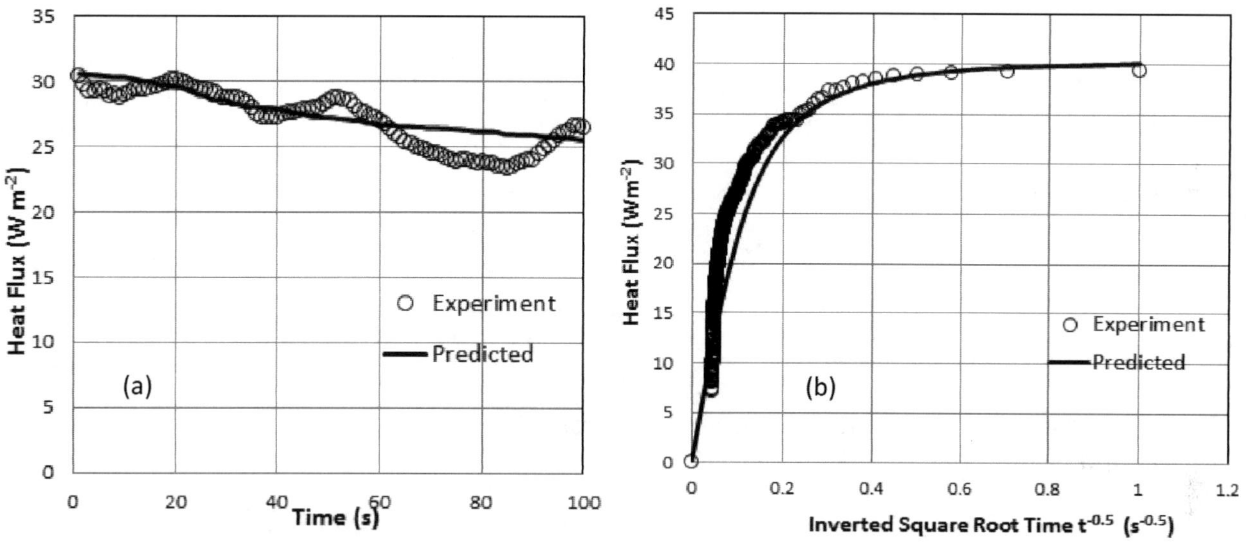


Figure 6 Comparison of Experimental and Predicted Heat Flux for Run 5
(a) Time (b) Inverted Square Root of Time for Run 5

## 5. Conclusion

An experimental investigation of small scale releases of liquid nitrogen on ice was performed to improve the understanding of the source term modelling of cryogenic liquid spilled on ice. The vaporization rates obtained were independent of the amount of liquid nitrogen spilled, and as expected, were a function of the ice temperature. A very good prediction agreement of the temperature inside the ice obtained by 1D ideal conduction model was observed. The performance of the theoretical model in predicting heat flux for the release on ice was found to be fair, but not for ice-water mixture. The vaporization rates were found to be constant for long duration releases of liquid nitrogen on ice.

## 6. Acknowledgements

The authors would like to thank BP Global Gas SPU for their financial support and guidance. They would also like to acknowledge the support of Qatar Petroleum (QP) for offering facilities at Ras Laffan Industrial City. The authors personally thank QP's staff who worked with LNG research team of Texas A&M University at Qatar for the experiment.

## 7. References

[1] K. A. Dahl, "Analysis of accidental iceberg impacts with LNG tank carriers," Norwegian University of Science and Technology, 2012.

[2] E. Nakanishi and R. C. Reid, "Liquefied Natural Gas-Water Reactions," *Chem. Eng. Prog.*, vol. 67, no. 12, pp. 36–41, 1971.

[3] V. Vesovic, "The influence of ice formation on vaporization of LNG on water surfaces," *J. Hazard. Mater.*, vol. 140, no. 3, pp. 518–526, Feb. 2007.

[4] H. S. Carslaw and J. C. Jaeger, *Conduction of Heat in Solids*, 2nd ed. Oxford University Press, USA, 1986, p. 520.

[5] D. Burgess, J. Biordi, and J. Murphy, "Hazards of spillage of LNG into water," Pittsburg, Pennslyvania, 1972.

# 348750 Consequences Analysis Associated with the Failure of the Safety Interlock System of Methanator (I-351) of the Ammonia Plant of Fertilizantes Nitrogenados De Venezuela, C.E.C. (FertiNitro)

**Abstract:**

Consequences Analysis associated with the failure of the safety interlock system of methanator (I-351) of the Ammonia Plant of Fertilizantes Nitrogenados de Venezuela, C.E.C. (FertiNitro).

Fertilizantes Nitrogenados de Venezuela, C.E.C. (FertiNitro) is a petrochemical company, dedicated to the production and sale of anhydrous ammonia and granular urea high quality, located within the Oil and Petrochemical Complex General José Antonio Anzoategui, in the town of José, north of the Anzoategui state, adjacent to National Highway José Antonio Anzoategui (troncal 9), between Barcelona and Puerto Píritu Population. 30% of the production of these fertilizers are for the domestic market and 70% is exported to the international commercial market.

Throughout the value chain of Fertinitro, ammonia and urea are produced by transforming raw materials such as natural gas (which comes from the gas extraction plant of PDVSA) and atmospheric air. Fertinitro is formed by two identical trains, which in turn are cut by two anhydrous ammonia plants whose maximum production capacity is 1,800 tons per day each, and two granular urea plants, with maximum production capacity of 2,200 tons per day each. To achieve optimum performance of such plants, they are fed by the Industrial Services Unit and then the final product is sent to the storage unit and dispatch for further marketing.

Ammonia plant is divided into several sections (Desulfurization, Reformation, conversion, carbon dioxide removal, methanation, Synthesis Gas Compression, Synthesis and Ammonia Refrigeration). In the methanation section, the reactor is also known as methanator reactor (12/22-R-301), which introduces the process gas coming from the sections of Conversion and Carbon Dioxide Removal respectively (rich in hydrogen and Nitrogen with small traces of carbon monoxide and carbon dioxide).

Considering that the presence of carbon monoxide (CO) and carbon dioxide ($CO_2$) into the reactor 12/22-R-501 (located upstream of methanation section) produce adverse effects design intention inside thereof, the catalyst content in the methanator promotes an exothermic reaction converting the traces of CO and $CO_2$ to methane ($CH_4$). The resultant process gas methanation section contains primarily hydrogen ($H_2$), Nitrogen ($N_2$) with trace amounts of methane ($CH_4$) and argon (Ar), the latter two being inert in the ammonia synthesis section.

The inlet temperature to the methanation reactor is controlled by the Temperature Indicator Controller 12/22-TIC-3353 that, in split-range, the first controls the "by pass" associated with the exchanger gas / gas 12/22-E-311 A / B generating the opening of the temperatura control valve 12/22-TV-3353[a], and if is necessary additional heating, the 12/22-TIC-3353 sends a signal for the opening of the valve 12/22-TV-3353B and therefore increases the amount of the converted gas that passes through the heater 12/22-E-211 of the methanation reactor.

On the other hand, the temperatures in the catalyst bed is measured at four different levels in order to monitor the temperature profile. If there is a pass of CO and / or CO2 in the methanator due to poor operation of the CO conversion section and / or CO2 removal section, the temperature rise through the methanator may be so high that exceed the design temperature of the methanator. To avoid such an incident, it has been installed in the catalyst bed a high temperature switch to trigger the safety interlock system of the methanator (I-351). If the I-351 is triggered, the flow of process gas that is directed to the methanator is cut, being automatically vented downstream of the CO2 absorber (belonging to the CO2 removal section) through 12/22-PV- 3350 and downstream of the CO converter of low temperature through 12/22-PV-2076 12/22-PV-2076 a and B.

The I-351 activates the manual air vent on the top of methanator, which also can be operated safely from the control room. This action allows the gas to be vented through reverse flow of the methanator to protect the vessel and the catalyst from overheating during the release. When the I-351 is activated, the 12/22-HY-3050 decrease the pressure at 2 kg/cm2g per minute, opening the 12/22-HV-3050.

In that sense, due to hazardous materials that are handled in this equipment, plus the operating conditions, it is considered a high-risk equipment, that is why there is a need to realise an consequences analysis associated with the failure of the safety interlock system of methanator (I-351) of the Ammonia Plant of Fertilizantes Nitrogenados de Venezuela, C.E.C. (FertiNitro).

# 349822 Building Siting Evaluation: A New Software Tool for the Determination of Blast Loads from Potential Vapour Cloud Explosions (VCEs)

**Abstract:**

A central element of building siting evaluation—especially where vapour cloud explosion (VCE) hazards are credible—is the calculation of the blast parameters/loads to buildings of interest. This determination is typically approached in two ways. Using simplified models that utilize blast curves (notably the TNO Multi-Energy and Baker Strehlow Tang models), or advanced methodologies using computational fluid dynamics (CFD) technology. Simplified models are more widely used as they offer a good level of characterization potential and are more pragmatic to use. The latter (i.e. CFD) can be resource and time intensive and is typically reserved for instances where the simplified models are limited (e.g. in the determination of near field blast loads).

A rigorous application of the simplified approaches would typically involve dispersion modelling (under specific conditions e.g. weather and wind speed) for the identified loss of containment scenario (or scenarios) to evaluate the size of the resulting flammable cloud and the potential for (and degree of) interaction with any identified regions of congestion and/or confinement. However, such a task is not trivial and to reduce the calculation burden this analytical step is sometimes bypassed and a simplifying assumption that the entire region is full of flammable gas is made. Whilst such an assumption ensures a conservative, cautious approach to the analysis, it does not reflect reality and can result in outcomes that are excessive to requirements – the upshot of which can be sub-optimal resource allocation. Consequently, there is a need for an approach that allows for a more realistic analysis of VCE potential.

This paper describes an advanced implementation of the two most widely used vapour cloud explosion models mentioned above, namely the TNO Multi Energy (ME) Method and the Baker-Strehlow-Tang (BST) model, into Phast, the well-known general purpose consequence analysis software package developed by DNV GL Software. A key aspect of this new addition is the capability to examine the interaction of a dispersing dynamic flammable cloud profile with congested/confined regions in all three dimensions (length, width and height) and in multiple directions. This is in direct response to the aforementioned challenge and the intention is to provide a robust platform by which a more detailed analysis of the VCE hazards can be evaluated without recourse to simplification.

The implementation of both models is extensive and includes all the key parameters that have a bearing on the magnitude of a vapour cloud explosion. A case study of the tool as applied to a hypothetical occupied building siting challenge is presented and the value of using the tool discussed.

# Modeling of time-varying dispersion from ground-level liquid pools or vapor area sources

**Henk W.M. Witlox**
**DNV GL Software**
**3 Cathedral Street, London SE1 9DE, UK**
**henk.witlox@dnvgl.com**

**Mike Harper**
**DNV Software**
**Exchange Street, Stockport, SK0 3ET**
**michael.harper@dnvgl.com**

**Maria Fernandez**
**DNV Software**
**3 Cathedral Street, London SE1 9DE, UK**
**michael.harper@dnvgl.com**



Prepared for Presentation at
American Institute of Chemical Engineers
2014 Spring Meeting
10th Global Congress on Process Safety
New Orleans, LA
March 30 – April 2, 2014

UNPUBLISHED

# Modeling of time-varying dispersion from ground-level liquid pools or vapor area sources

**Henk W.M. Witlox**
**DNV GL Software**
**3 Cathedral Street, London SE1 9DE, UK**
**henk.witlox@dnvgl.com**

**Mike Harper**
**DNV GL Software**

**Maria Fernandez**
**DNV GL Software**

**Keywords:** dispersion, time-varying, pool, along-wind diffusion

## Abstract

The current paper discusses a new enhanced dispersion formulation accounting for time-varying effects resulting from a time-varying ground-level liquid pool or vapor area source. This new formulation is implemented in the Unified Dispersion Model (UDM), and is planned to be included in a future version of Phast. The new model includes a considerably improved model for linking between the pool and the cloud. It furthermore includes along-wind-diffusion effects (due to ambient turbulence) at the upwind and downwind edges of the cloud. This results in reduced concentrations while the cloud travels in the downwind direction which is particularly relevant for toxic releases. The new formulation presumes a number of 'observers' to be released at successive times from the upwind edge of the pool. The UDM first carries out 'pseudo steady-state' calculations for each observer, where the 'release data' correspond to the time at which the observer is released, and where the observer accumulates vapor while travelling over the evaporating pool. Subsequently effects of along-wind diffusion (AWD) are included by means of Gaussian integration over the downwind distance, whereby the downwind dispersion coefficient is chosen to be consistent with the current finite-duration correction (FDC) formulation adopted in the UDM.

First the above new UDM formulation has been verified against the time-varying HGSYSTEM model HEGADAS-T. The UDM provides more accurate results in the near-field because of more accurate observer calculations while observers are moving over the pool. Subsequently the UDM FDC and new UDM models have been validated against experimental data including finite-duration carbon-dioxide vapor ground-level area releases (Kit Fox experiments), and accurate results were obtained using both formulations.

# 1. Introduction

Many accidents involve releases of hazardous chemicals into the atmosphere. Commonly used atmospheric dispersion models are limited to continuous or instantaneous releases only. They do not accurately account for time-varying effects following time-varying releases i.e. resulting from a pressure drop in a vessel or pipe, and/or from a time-varying pool (following rainout or from direct spill). These effects include along-wind-diffusion (due to ambient turbulence) resulting in reduced concentrations.

The consequence modelling package Phast includes the Unified Dispersion Model UDM (Witlox and Holt [1], Witlox and Harper [2]) for modelling two-phase outdoor releases including potential rainout. The UDM in the current Phast (versions 6.7 and 7.1) only accounts for along-wind-diffusion (AWD) for instantaneous releases or finite-duration continuous releases that do not rain out.

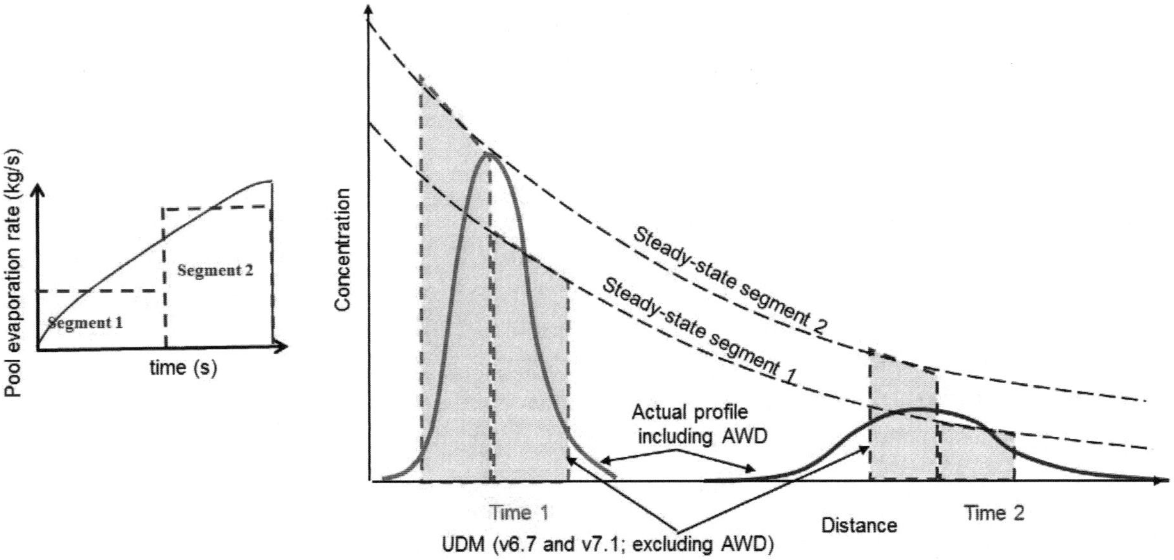


**Figure 1: UDM time-varying dispersion – old segment method excluding AWD**

For dispersion directly from an evaporating pool (Figure 1), the evaporated mass from the pool is divided into equal-mass segments. Subsequently the UDM model carries out steady-state dispersion calculations for each segment, and determines for successive times the concentration as a function of distance from these segment data as shown in Figure 1 by the dashed curves. Thus the current UDM, henceforth referred to as the old UDM or old model, does not apply along-wind-diffusion at the upwind and downwind edges of the cloud, or between segments, which may lead to significant over-prediction of concentration and under-prediction of duration in the far-field (see Figure 1 at time 2). This is particularly important for toxic releases, where dispersion calculations are required to be carried out to small concentration levels such as ERPG levels. It is less important for flammable releases with calculations to relatively high concentration levels only, such as LFL or 0.5LFL.

The Shell consequence modelling package HGSYSTEM 3.0 (Post [3]) includes the time-dependent dispersion model HEGADAS-T (Witlox [4]) for modelling the time-dependent dispersion of a heavy-gas cloud moving with the ambient wind. It can be used to model the dispersion downwind of a time-dependent ground-level source (unpressurised release). The transient behaviour of the cloud is approximated by a quasi-steady-state description in which so-called "observers" are released upwind of the source at a series of times. These observers travel with the cloud. For each observer, the observed concentration is set from steady-state HEGADAS-S calculations using the observed source/transition data. Thus by calculating the position of each observer at a given time t, the concentration c is set for a number of downwind distances. Subsequently the actual concentration is set from Gaussian integration with respect to the downwind distance x of the above observer concentrations as illustrated in Figure 2. This involves a downwind dispersion coefficient $\sigma_x$, which allows along-wind diffusion to be taken into account.

A new enhanced dispersion formulation has been developed to account for time-varying effects resulting from a time-varying release. It generalises the above HEGADAS-T formulation to account for along-wind-diffusion both for ground-level unpressurised releases (e.g. evaporating pools) and for elevated two-phase pressurized releases including potential rainout. This method has been implemented in a new version of the UDM model, which is to be included in a future version of Phast. The current paper considers the case of an unpressurised release from a ground-level liquid or vapor area source only. Previous separate papers (Witlox and Harper [5], [6]) have dealt with the cases of an elevated pressurized release without rainout and an elevated pressurized release with rainout.

## 2. New UDM along-wind diffusion formulation

In the case of a liquid pool, first pool evaporation calculations are carried out to determine the time-varying pool spreading/evaporation data [radius (m), evaporation rate (kg/s), temperature (K)]; see Fernandez et al. [7] for details.

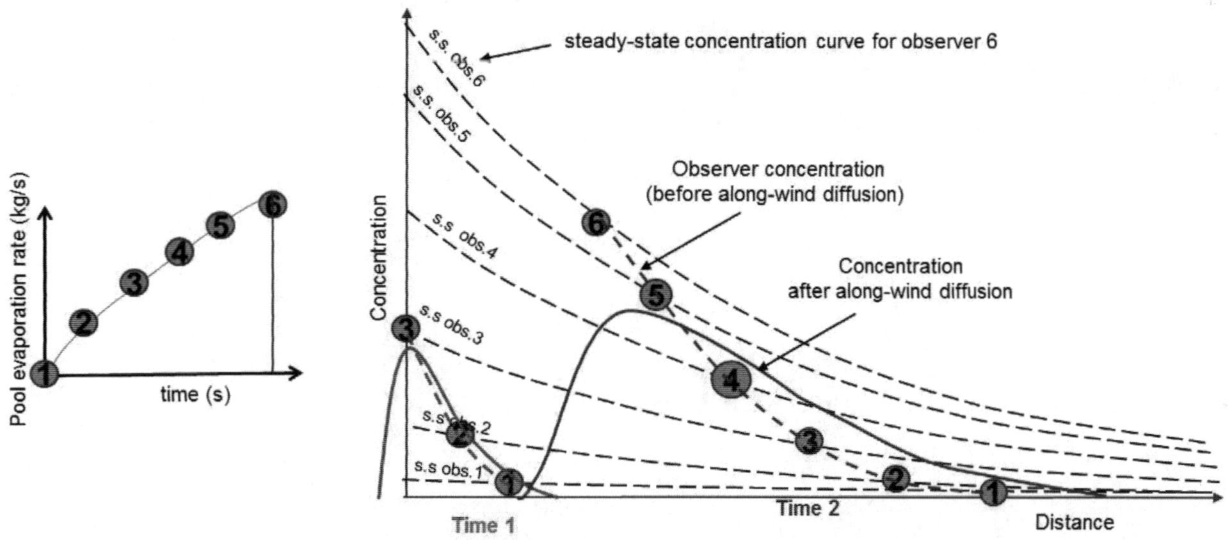


**Figure 2: UDM time-varying dispersion –new observer method including AWD**

Subsequently 'observers' are released from the upwind edge of the pool at intervals which can be chosen to correspond to equal-mass increments as for the old UDM method (see Figure 2 for the case of 6 release observers). These observers are assumed to move along the cloud centre-line (located at the ground) at the UDM cloud centroid speed. This is illustrated by Figure 3, where the first observer (1) is released at the start of pool evaporation, and subsequent observers (2,3) start from the upwind edge of the spreading pool.

For each observer steady-steady calculations based on the observed source data are carried out to evaluate the observer concentration prior to including effects of along-wind diffusion; see Figure 2. Primary variables and associated differential equations are as follows (see [1], [2] for details):

- Mass of moist air added to the cloud (kg/s) – air entrainment law describing turbulent mixing of air with the cloud, accounting for heavy-gas entrainment and/or passive entrainment
- Excess downwind cloud momentum (kg m/s$^2$) – conservation of cloud momentum
- Position of downwind edge of cloud (m) – relation between cloud position and speed
- Heat conduction from substrate (J/s) – heat transfer equation between substrate and cloud
- Water evaporated from water substrate (kg/s) – water-vapor transfer equation between water substrate and cloud
- Cross-wind dispersion coefficient (m) - empirical spreading law in case of heavy-gas dispersion and based on formula for ambient crosswind dispersion coefficient in case of passive dispersion

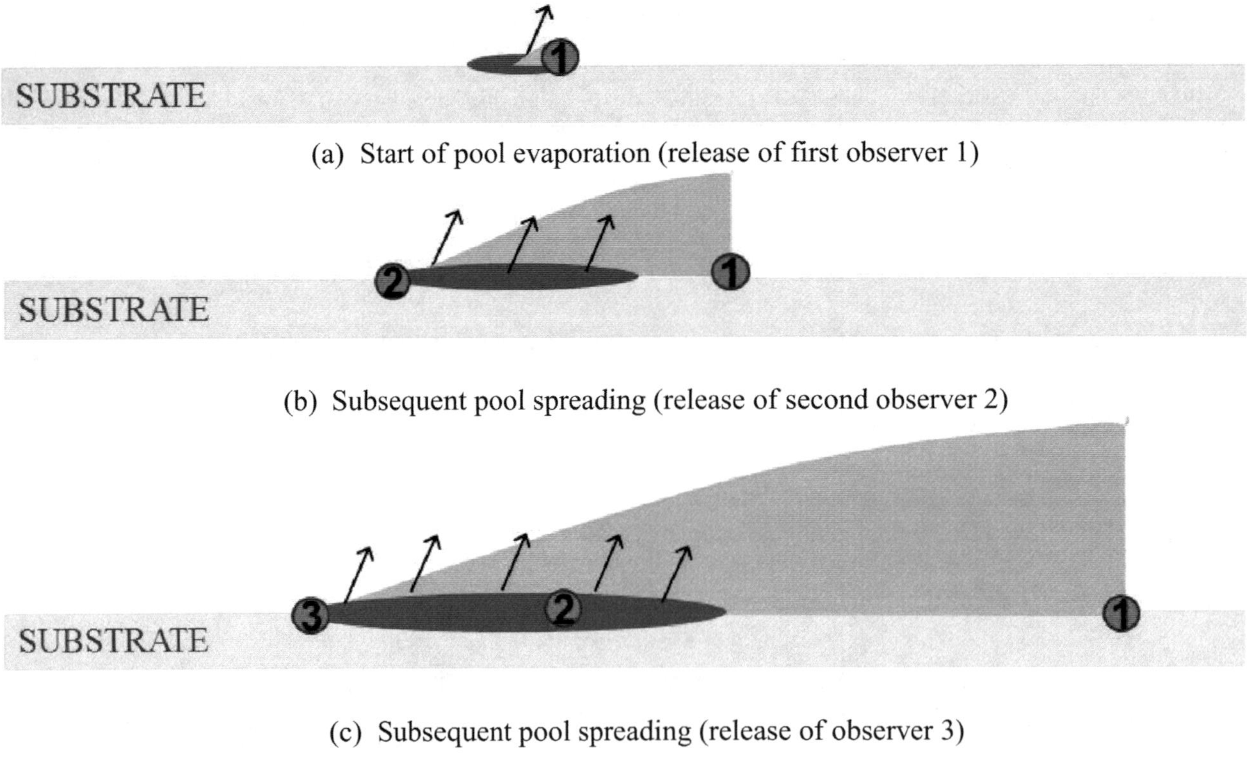

(a) Start of pool evaporation (release of first observer 1)

(b) Subsequent pool spreading (release of second observer 2)

(c) Subsequent pool spreading (release of observer 3)

**Figure 3: New UDM dispersion phases for time-varying dispersion from pool**

While the observer moves above the pool, the above observer dispersion equation (momentum conservation, cloud crosswind gravity spreading, heat and water-vapor transfer from the substrate) are modified to account for the pool vapor added back to the cloud. Two equations are solved for two additional primary variables: the mass of component present in the cloud (to account for mass added from the pool) and the component enthalpy (to account for component enthalpy added from the pool).

Additional equations are solved to derive the so-called 'secondary' variables from the 'primary' variables, which includes nonlinear equations for cloud thermodynamics (isenthalpic equation to evaluate cloud temperature) and cloud geometry. Furthermore an empirical concentration profile at each calculated downwind distance x is adopted to evaluate the observer concentration C as function of crosswind distance y, and vertical height z.

Inclusion of along-wind diffusion

The actual plume concentration c(x,y,z,t) including effects of along-wind-diffusion is a function of time t, distance x downwind of the release location, crosswind distance y, and vertical height z; this function is calculated by means of Gaussian integration of the observer concentration $C(\xi,y,z,t)$,

$$c(x,y,z,t) = \int_0^\infty \frac{C(\xi,y,z,t)}{(2\pi)^{1/2}\,\sigma_x(\xi)} \exp\left\{-\frac{(x-\xi)^2}{2\sigma_x^2(\xi)}\right\} d\xi \qquad (1)$$

In the above equation $\xi$ is the downwind distance from the release point at time t of an observer travelling with the cloud in the downwind direction. At this position the observer observes the concentration $C(\xi,y,z,t)$. In Equation ( 1 ) along-wind diffusion is taken into account by assuming that the concentration $C(\xi,y,z,t)$ spreads out around $\xi$ according to a Gaussian distribution with a downwind dispersion coefficient $\sigma_x = \sigma_x(\xi)$. Figure 2 depicts the pre-AWD observer concentration C and the post-AWD concentration c at a short time after the release (time 1; limited AWD effects), and at a larger time after the release (time 2; larger AWD effects).

The downwind dispersion coefficient $\sigma_x$ is evaluated as function of downwind distance $\xi$ according to the formulation by Ermak [8] [9]. This assumption is consistent with the current finite-duration-correction (FDC) formulation adopted in the UDM [2] to apply AWD effects for finite-duration releases; see Witlox and Harper [6] for further details.

Averaging time effect because of time-varying release rate

In addition to the averaging time effect of wind meander, the user can optionally apply additional time-averaging of time-dependent concentrations resulting from time-varying pool evaporation,

$$[c(x,y,z,t)]_{avg} = \frac{1}{t_{av}} \int_{t-t_{av}/2}^{t+t_{av}/2} c(x,y,z,\tau)\,d\tau \qquad (2)$$

## 3. UDM verification against HGSYSTEM for dispersion from pool

The case of an instantaneous spill of 10000 kg of butane liquid is considered with both cases of a bund (radius 10m) and no bund (pool on water; dispersion over land). Ambient data applied are stability class D, wind speed at 10m height of 5m/s, temperature 300K and humidity 70%. Furthermore a maximum pool duration of 1 minute is applied, and thus no pool evaporation is assumed to occur after 1 minute.

Figure 4 shows the values predicted by the UDM of butane pool radius and evaporation rate for both cases with and without a bund. It is seen that the immediate butane rainout within the bund results in a virtually steady-state pool of 10 meter radius, 12.3 kg/s evaporation rate and a temperature of 271.8K.

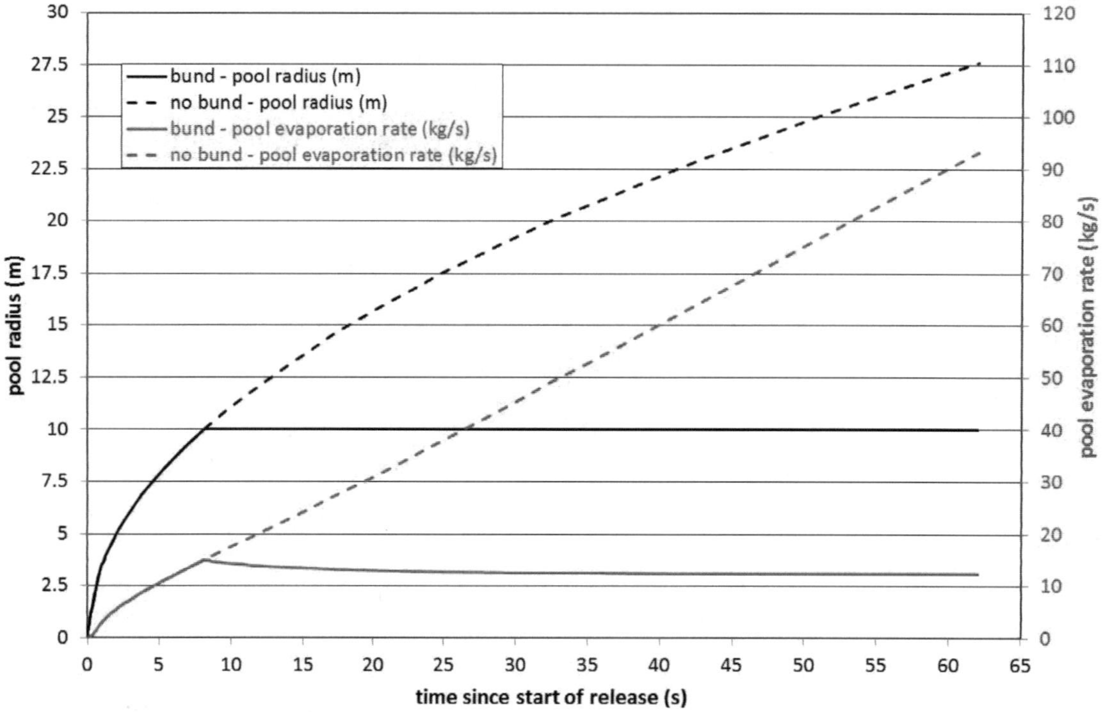


**Figure 4: UDM pool predictions (instantaneous 10000kg spill of butane on water)**

Figure 5 includes results of the following calculations for the case of a bunded pool:

- HGSYSTEM 3.0 calculations (the rectangular HEGADAS source was chosen to be square with the same area as the UDM circular pool)
    - steady-state HEGADAS-S calculations (ground-level steady-state heavy-gas dispersion from area source) based on the steady-state pool data mentioned above.
    - FDC finite-duration correction (using Ermak's formula) to above HEGADAS-S results (60 seconds duration) using the post-processor program POSTHS
    - HEGADAS-T calculations (ground-level time-varying heavy-gas dispersion from area source). In line with the new UDM model, no gravity-shape correction was

applied and the along-wind-diffusion coefficient was chosen to be based on Ermak's formula (non-default HEGADAS-T options)

- UDM calculations
    - Steady-state UDM calculations based on the above steady-state pool data
    - FDC correction to the above steady-state UDM results (60 seconds duration)
    - UDM AWD calculations, with 30 observers released from the upwind edge of the pool

Figure 5a includes the steady-state HEGADAS-S and UDM results of the centre-line ground-level concentrations in the near-field for the case of a steady-state pool. HEGADAS-S assumes a uniform concentration (top-hat profile) above the area source (about 15% mole fraction), which is envisaged to lead to an over-prediction near the source. On the other hand, the UDM model more rigorously solves the dispersion equations across the pool allowing for a variation of the concentration across the pool. The very rapid decrease predicted by the UDM of the concentration near the upwind edge of the pool is a numerical artefact and may not occur in reality. Further downwind from the pool a closer match is observed between the UDM and HEGADAS-S data.

Figure 5b includes results of all the above calculations. It includes results of the time-varying dispersion calculations by UDM AWD and HEGADAS-T at output times 300s, 600s and 1200s. It is seen that the maximum values of these curves correctly touch the maximum centre-line concentration (maximum value over all times) predicted by UDM FDC and HEGADAS-S FDC, respectively. This confirms that the AWD integration is carried out correctly by both programs and is consistent with the FDC formulation. Also adequate agreement between the UDM and HEGADAS FDC results is observed. It is seen that initially the HEGADAS-T cloud is heavier (larger concentrations) and therefore moves more slowly than the UDM AWD cloud, but this effect reduces in the far-field when the predictions become closer.

Likewise, Figure 6 includes UDM AWD and HEGADAS-T results at times 300s, 600s and 1200s for the case without the bund. Here the UDM calculated pool data were specified as input to HEGADAS-T with a step size of 2 seconds. The observed differences are confirmed to be very similar to that seen for the case with the bund (compare Figures 5b and 6).

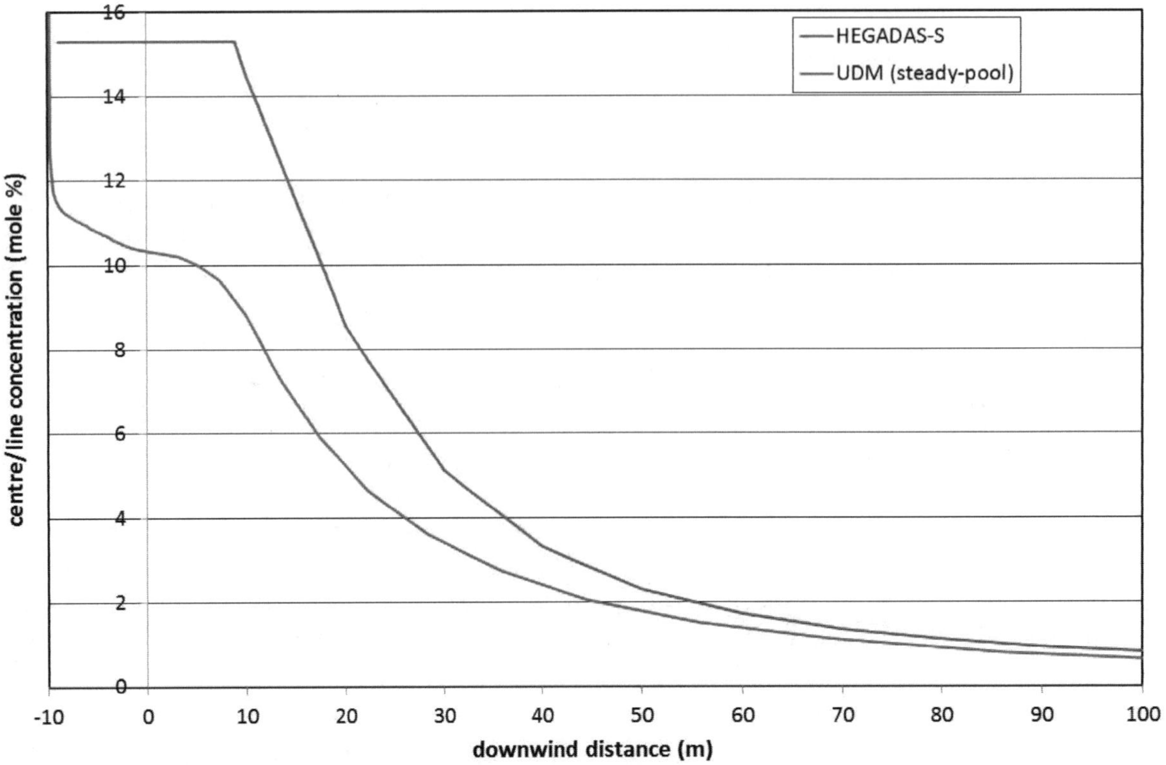


(a) Near-field centre-line concentrations (above pool and downwind of the pool)

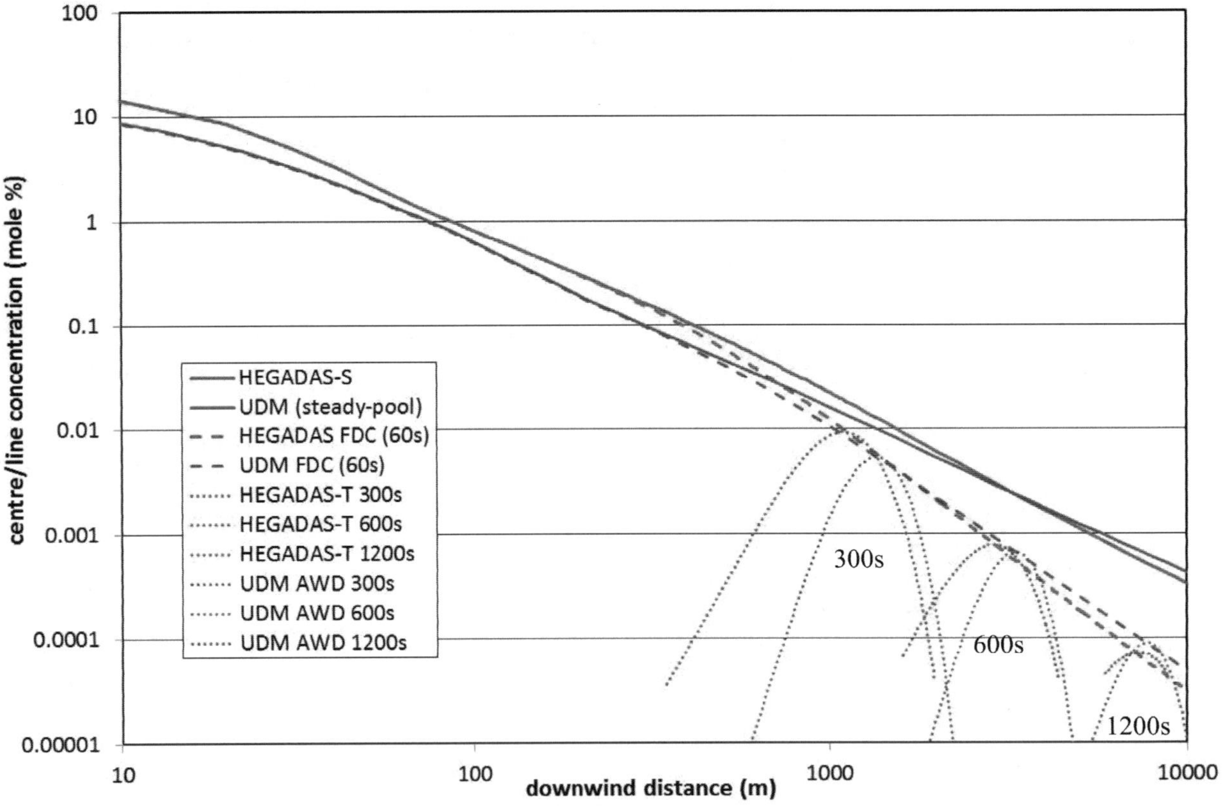


(b) Far-field concentrations

**Figure 5: UDM verification against HEGADAS for spill of butane with bund (radius 10m)**

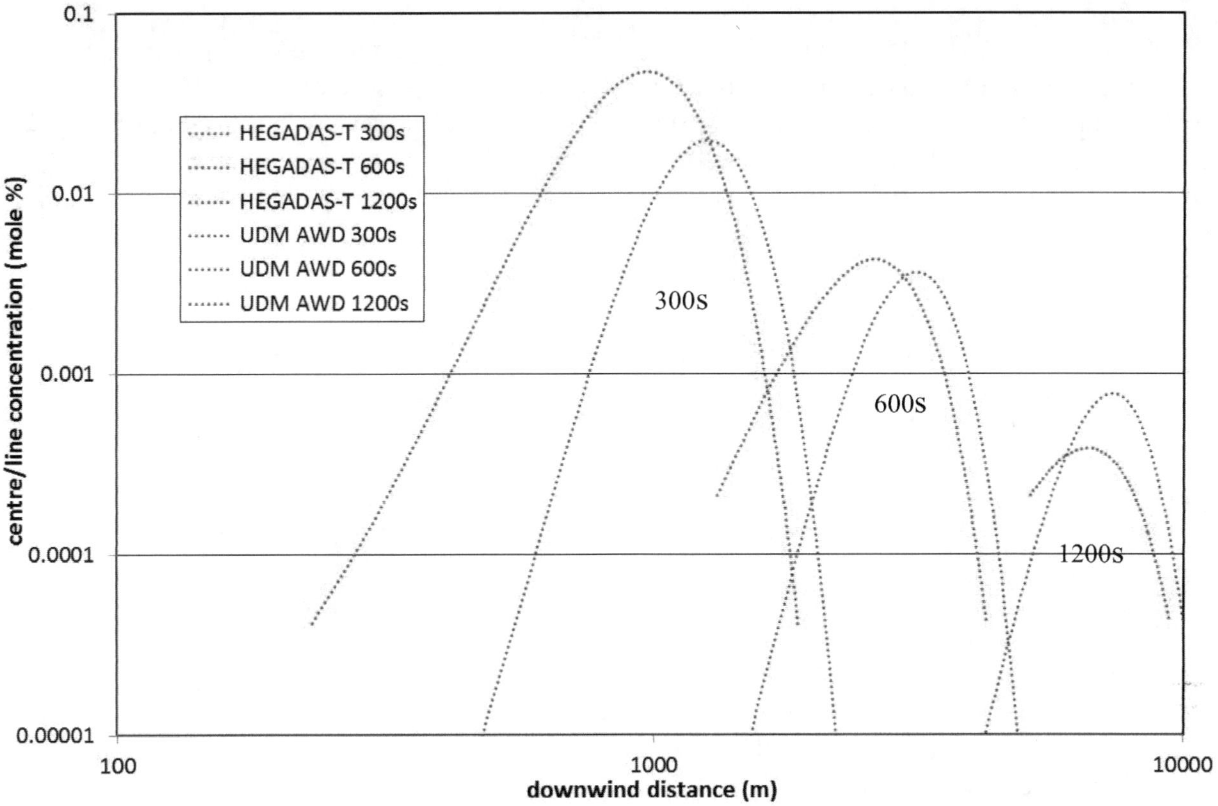


**Figure 6: UDM verification against HEGADAS for spill of butane without bund**

## 4. UDM validation against Kit Fox experiments ($CO_2$ area source)

In these experiments the $CO_2$ was released as a dense vapor from a 1.5m x 1.5m ground-level area source for continuous and 20-second finite-duration releases, during both neutral and stable conditions. In this paper results are reported of experiments with a uniform surface roughness (URA; roughness estimated between 0.01 and 0.02 m; adopted value 0.01m); see Table 1.

**Table 1. List of Kit Fox experiments for UDM validation**

| Kit Fox Series (URA) | duration (s) | flow rate (kg/s) | release velocity (m/s) | Experiments grouped by stability class | | |
|---|---|---|---|---|---|---|
| | | | | D | E | F |
| continuous | $\geq$ 120 | 1.5-2.1 | 0.4-0.6 | 604, 805 | 702,808,605,703,705 | 606,811,709, 609,712 |
| puff | 20 | 1.1-1.8 | 0.3-0.5 | 801,601,803,802, 804,602,603,806 | 807,809,706,810,812 | 704,708,710,607,711, 608,714 |

UDM input data have been obtained from the MDA database given by Hanna and Chang [10]. The ground-level area source is modeled as a circular source with diameter 1.69 m and corresponding source area A equal to that of the actual 1.5m x1.5m square source. The release velocity (m/s) is obtained by dividing the release rate (kg/s) by the source area A ($m^2$) and the $CO_2$ vapour density ($kg/m^3$).

In a previous paper (Witlox et al. [11]) results were reported for the above experiments (as well as other experiments), where the velocity was imposed as a horizontal instead of a vertical velocity, and where the plume was modeled as a horizontal jet. In the current paper the above new logic is applied where observers are released upstream of the source and where the ground-level cloud is modeled physically more correctly above the source and in the near-field. Below results are presented using the 'finite-duration-correction' model (UDM FDC) as well as results using the new time-dependent UDM AWD formulation.

Figure 7 compares steady-state, UDM FDC and UDM AWD results for ground-level centre-line concentrations (maximum value over all times) against experimental data for experiment KF0706 (URA puff with 20 second duration, stability class E, wind-speed 2.66m/s). Both approaches of a horizontal jet (Figure 7a) and pool source (Figure 7b) are considered, as well as the effect of additional time-averaging [(see Equation (2)], using an averaging time of 20s as adopted for the experimental values. The following conclusions can be drawn:

- In the near-field the AWD and (un-averaged) FDC results are close to the steady-state results, and increasingly deviate from the steady-state results with increasing downwind distance because of effects of along-wind diffusion.
- In the far-field the UDM AWD results show less effects of along-wind diffusion than the FDC results. The reason for this discrepancy is being investigated.
- Since the release duration of 20s is identical to the averaging time, the effect of time-averaging should be limited in the near-field (no AWD effects), more pronounced in the intermediate distance (more time-variation over the averaging period of 20s because of AWD effects), and less in the far-field (relatively less variation over the averaging period of 20s because of increasing cloud duration due to increasing AWD effects). This is confirmed by the AWD results, but not in the near-field for the FDC results. Thus the analytical FDC results seem to be less correct in the near-field.
- At the experimental data points the results between horizontal jet and pool source are very close.

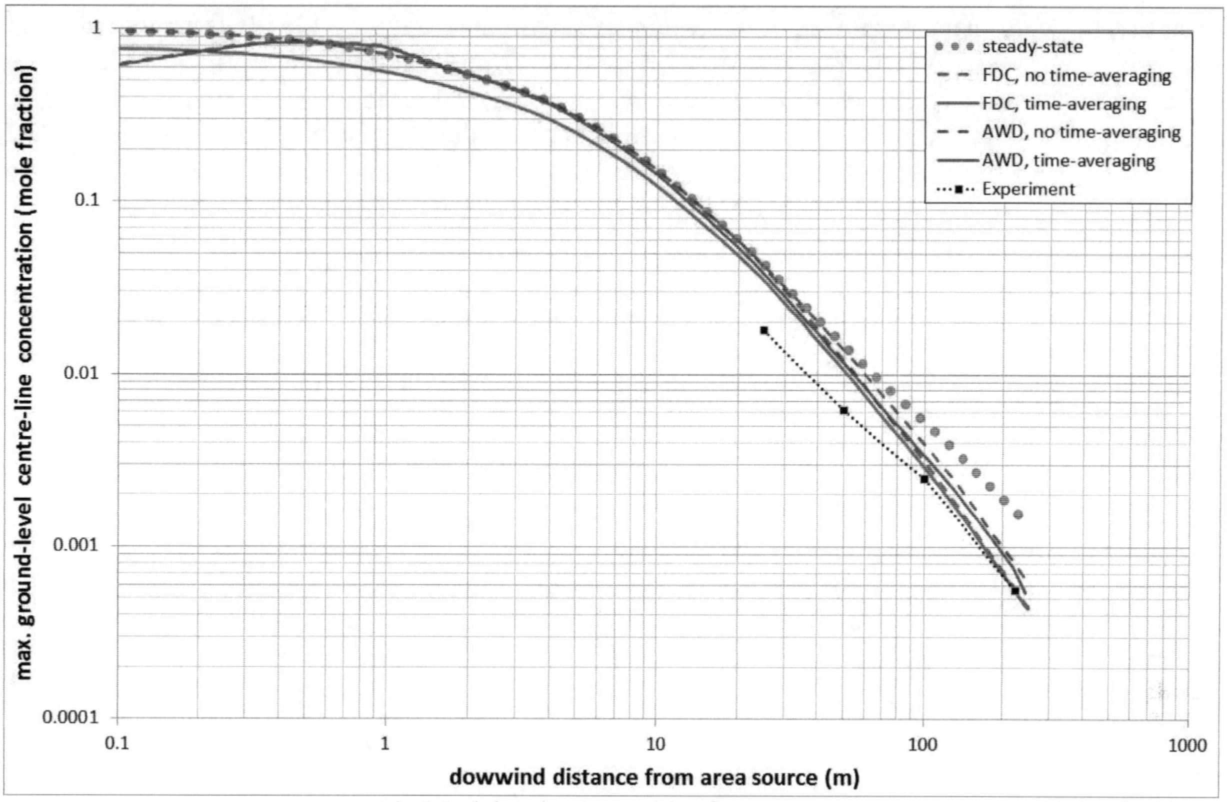


(a) Model as horizontal jet from area source

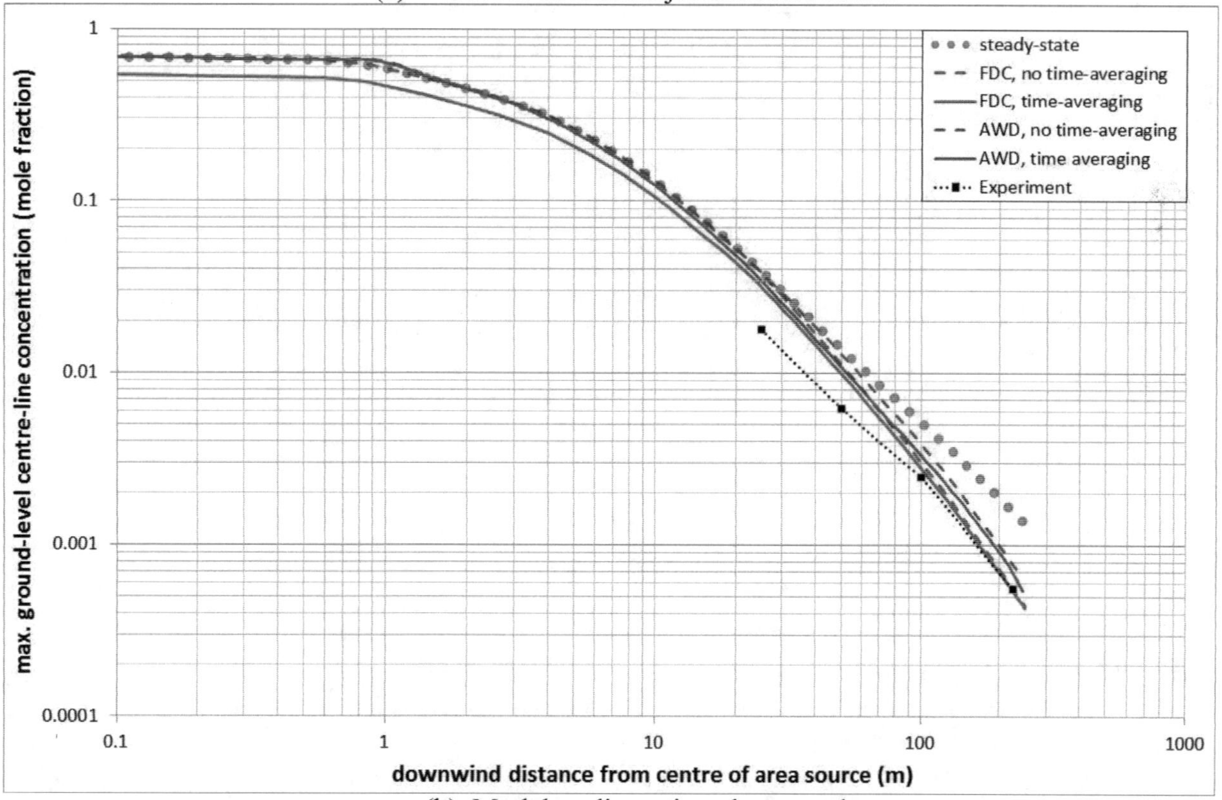


(b) Model as dispersion above pool

**Figure 7. UDM predicted maximum concentrations for Kit Fox experiment KF0706 (20s release)**

For a given experimental dataset, it is common practice [Hanna et al. [10]] to calculate the geometric mean bias MG (averaged ratio of observed to predicted concentrations; MG<1 over-prediction and MG>1 under-prediction) and the geometric variance VG (variation from mean; minimum value = 1). Ideally, MG and VG would both equal 1.0. Geometric mean bias (MG) values of 0.5 and 2.0 represent a factor of 2 in over-predicting and under-predicting the mean, respectively. Likewise, a geometric variance (VG) of about 1.6 indicates scatter from observed data to predicted data by a factor of 2.

A graphical presentation of the overall MG and VG validation results for arc-wise concentration is shown in Figure 8 for both FDC and AWD models presuming a pool source including additional time averaging. The overall results can be summarized as follows:

- URA continuous (black markers): excellent prediction for both concentration and cloud widths, and very close consistent agreement between FDC and AWD results
- URA puff (red markers): The FDC method (solid markers) results in concentration over-prediction for neutral stability class D, accurate results for stability class E and in under-prediction for stability class F. The AWD method (hollow markers) results in increased over-prediction for D, E and improved results for F (slight under-prediction).

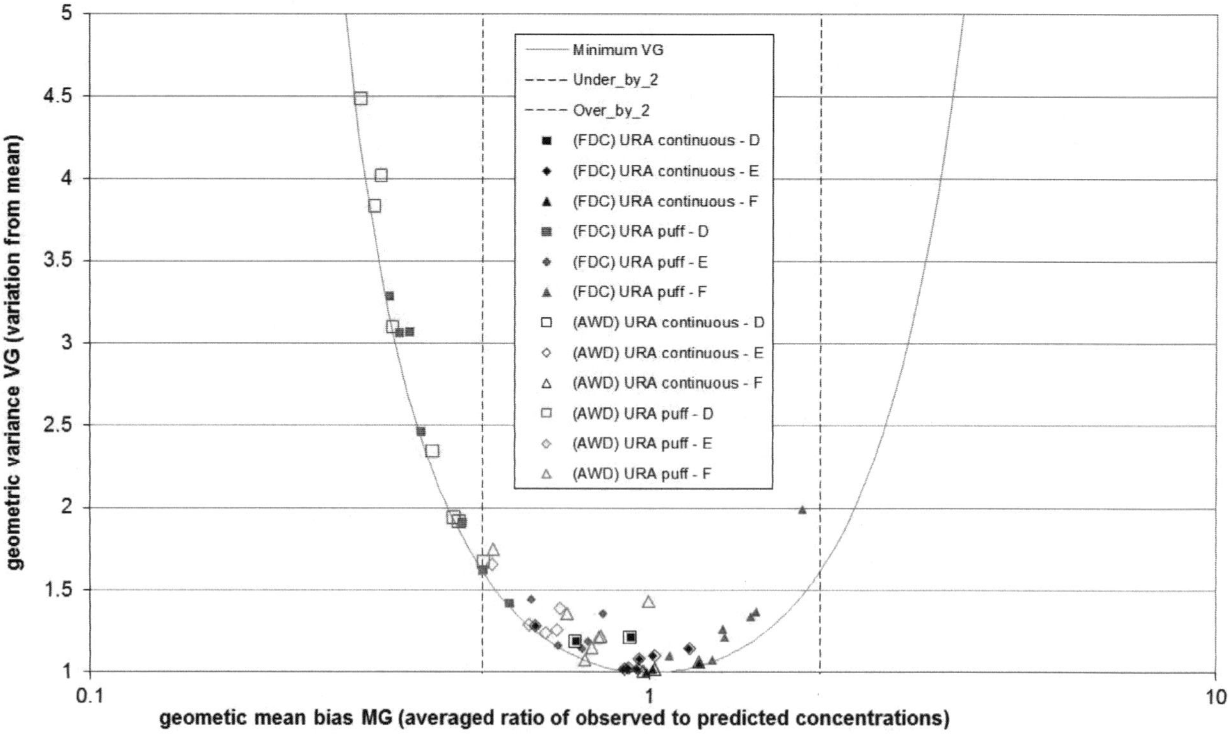


**Figure 8. UDM validation statistics for Kit Fox URA experiments**
(case of pool source including additional time-averaging; FDC and AWD results)

## 5. Conclusions and future work

The following main conclusions can be drawn:

- The Phast dispersion model UDM has been extended to improve modeling for time-varying ground-level releases. The new formulation presumes a number of 'observers' to be released at successive times from the upwind edge of the pool. The UDM first carries out pseudo steady-state calculations for each observer, accounting for vapor accumulation over the pool. The effects of along-wind diffusion are included by means of Gaussian integration over the downwind distance, and additional time-averaging can be carried out subsequently.
- In the old model (Figure 1) the equations were discontinuously modified at the original point of rainout to account for the vapor added back from the entire pool, and no along-wind-diffusion was applied. In the new model the cloud adapts continuously to the presence of an evaporating pool.
- The new method is based on a generalization of the HGSYSTEM model HEGADAS-T of dispersion from area sources.
- The new UDM AWD model has been verified against HEGADAS-T for the case of dispersion from a ground-level area source and good agreement was obtained
- The new UDM FDC and UDM AWD models have been validated against the Kit Fox experiments ($CO_2$ area source). Accurate results were obtained using both the FDC and AWD formulations, with most accurate results by FDC for stability classes D, E and by AWD for F.

The current further work is being planned:

- Extension of UDM instantaneous model to account for along-wind-diffusion after rainout.
- The current new model includes effects of along-wind-dispersion resulting from ambient turbulence corresponding to passive dispersion. For heavy-gas continuous or time-varying releases, it only takes into account gravity spreading in the cross-wind direction and not gravity spreading in the downwind direction. Thus the model should be extended to include effects of along-wind gravity spreading, which are particularly relevant in the near-field in the case of small wind speeds
- Validation against additional experimental data
- Implementation of new UDM AWD model into Phast
- Further refinement of empirical formulas for along-wind-diffusion coefficient $\sigma_x$

## References

[1] Witlox, H.W.M., and Holt, A., "A unified model for jet, heavy and passive dispersion including droplet rainout and re-evaporation", *International Conference and Workshop on Modelling the Consequences of Accidental Releases of Hazardous Materials*, CCPS, San Francisco, California, September 28 – October 1, pp. 315-344, 1999

[2] Witlox, H.W.M and Harper, M.., "UDM theory manual", Phast 6.7 technical documentation, DNV Software, 2011

[3] Post, L. (editor), Section 7C in "HGSYSTEM 3.0 Technical Reference Manual", Report TNER.94.059, Thornton Research Centre, Shell Research, Chester, England, 1994

[4] Witlox, H.W.M., "The HEGADAS model for ground-level heavy-gas dispersion, II. Time-dependent model", Atmospheric Environment, Vol. 28, No. 18, pp. 2933-2946, 1994

[5] Witlox, H.W.M, and Harper, M., "Modelling of time-varying dispersion for releases including potential rainout", Process Safety Progress, In Press, 2013; also published at Loss Prevention Symposium, 9th Global Congress on Process Safety, San Antonio, Texas, April 28 – May 1, 2013

[6] Witlox, H.W.M., Harper, M., Natarajan, S., and Williamson, A., "Modelling of time-varying dispersion for elevated pressurised releases without rainout", 2013 International Symposium, Mary Kay O'Connor Process Safety Centre, Texas A&M University, College Station, Texas, 22-24 October 2013

[7] Fernandez, M.I., Harper, M., Mahgerefteh, H., and Witlox, H.W.M., "An integral model for pool spreading, vaporisation and dissolution of hydrocarbon mixtures", pp. 466-472,Symposium Series No. 158, Hazards XXIII, Southport, England, 13-15 November 2012, ICHEME

[8] Ermak, D.L., "Unpublished notes on downwind spreading formulation in finite-duration release version of SLAB", Lawrence Livermore National Laboratory, California, 1986

[9] Ermak, D.L., "User's Manual for SLAB: an atmospheric dispersion model for denser-than-air releases, Lawrence Livermore National Laboratory, Livermore, California, February 1989

[10] Hanna, S.R., Chang, J.C., New MDA database version of Kit Fox, Provided as part of official CD together with overall Kit Fox report by Hanna, Chang and Briggs, 1999

[11] Witlox, H.W.M. Harper, M., Oke,A. and Stene, J., "Validation of discharge and atmospheric dispersion for unpressurised and pressurised carbon dioxide releases", Journal of Process Safety and Environmental Protection 92 , pp. 3-16, 2014

## Acknowledgment

Financial support of the work reported in this paper was provided by DNV Software, RIVM (Dutch Government), and TOTAL. The contents of this paper including any opinions and/or conclusions expressed, are those of the authors alone and do not necessarily reflect the policy of these organizations.

# 349889 Nanotechnology on Removing Arsenic Using Modified Carbon NANO- Tubes (MCNTs)

**Abstract:**

This study explains the percentage removal of arsenic from water by using modified multi- walled carbon nanotubes (MCNTs) as adsorbents. Apart from the raw carbon nanotubes, various metals has been impregnated on the large surface area of the CNT namely iron, titanium, silver, gold, copper and analyzed further to compare the percentile removal of each. The study investigates further the effect of the presence of carboxylic functional group (-COOH) functionalized on the surface of the carbon nanotubes on the removal of arsenic. To study the influence of the arsenic removal from water, the independent variables considered in this part of the paper are pH, running time and agitation speed. The concentration of the arsenic solution used in throughout the experiment was 1000 µg/L i.e. 1 ppm. The presence of Nano materials in the CNTs was characterized by using field emission scanning electron microscopy (FESEM) and transmission electron microscopy (TEM) in order to measure the lengths and the diameters of the CNT. Results of the study showed that iron and titanium impregnated on the surface of the CNT has the highest removal percentage compared to pure CNT and the functionalized CNT with carboxylic group. Almost 70% of the arsenic was removed from the water that contained 1 ppm arsenic contamination. The highest removal was at pH 3, contact time of 24 hours and agitation speed of 200 RPM. Part of the study is also dedicated to investigate the effect of dosage in the removal of arsenic from water.

# Possible Two Physical Hazard Scenarios for Polystyrene Foams, Based on Life Cycle Stages

**Toyoaki Nakarai, Graduate School of Environment and Informaion Sciences, Yokohama National University, Yokohama, Japan**
nakarai@ynu.ac.jp

**Atsumi Miyake, Graduate School of Environment and Information Sciences, Yokohama National University, Yokohama, Japan, atsumi@ynu.ac.jp, Satoru Yoshino, College of Industrial Technology, Nihon University, Narashino, Japan, yoshino.satoru@nihon-u.ac.jp**

Prepared for Presentation at
American Institute of Chemical Engineers
2014 Spring Meeting
10th Global Congress on Process Safety
New Orleans, LA
March 30 – April 2, 2014

UNPUBLISHED

# Possible Two Physical Hazard Scenarios for Polystyrene Foams, Based on Life Cycle Stages

**Toyoaki Nakarai, Graduate School of Environment and Informaion Sciences, Yokohama National University, Yokohama, Japan**
nakarai@ynu.ac.jp

**Atsumi Miyake, Graduate School of Environment and Information Sciences, Yokohama National University, Yokohama, Japan, Satoru Yoshino, College of Industrial Technology, Nihon University, Narashino, Japan**

**Keywords:** Polystyrene, Foam, Life cycle, Physical hazard, Scenario, HBCD

## Abstract

Polystyrene foams are good thermal insulators and therefore widely used in building insulation materials, such as roofs, and/or walls, down to foundation. However these foams have significant physical hazards (fire, etc.), because flammable blowing agent has come to be used, and foamed polystyrene itself is combustible. And so, HBCD (Hexabromocyclododecane) is generally used as flame retardant in polystyrene insulation foams, while currently a global ban on HBCD is being considered under the framework of Stockholm Convention on POPs (Persistent Organic Pollutants), due to persistence, toxicity and eco-toxicity.

Under these circumstances, attempts to find environmental-compatible flame retardants are being performed by, such as ENFIRO, a European Commission-funded project of offering a prototypical case study on substitution options for specific brominated flame retardants for plastics, based on flammability, environmental and toxicological risks.

Regarding the toxic potency in fire hazard caused by plastic burning, a model, based on the calculation of a FED (Fractional effective Exposure Dose) of mixture of smoke gases, had been previously developed by Building and Fire Research Laboratory, NIST, where some most toxic chemicals in $LC_{50}$, played important role.

On the other hand, for chemical process, the incident scenarios are related to the immediate impact of unplanned releases of material and energy, where the effects of fire, explosion or acute toxicity are generally targeted. In case of flammable blowing gas release from polystyrene foam to the atmosphere, for example, in a warehouse of low air exchange rate, it is possible to cause fire by static electricity as the ignition source.

This paper describes the possible two physical scenarios for polystyrene foams, based on life cycle stages: 1) Production/import, 2) Transportation, 3) Industrial use, 4) Private use, 5) Waste & disposal. One is acutely toxic gas release scenario by incomplete combustion case (nearly 400°C), etc., and the other is flammable blowing gas release scenario directly from the foam itself (nearly 100°C). The evidence was shown mainly by using gas chromatography.

The acutely toxic gas concentration from newly fabricated EPS, were compared with different flame retardants (HBCD, FP-2200) heated by electric furnace at 400°C, and analyzed by gas chromatography. The detected CO gas concentration was higher for EPS foams with HBCD than foams of tested phosphorous retardant.

# 1. Introduction

This paper does not describe ecological/environmental hazards related to HBCD flame retardant under Stockholm Convention on POPs, but possible two main physical hazards for insulation foams made by polystyrene.

## *1.1 Physical Hazards and Risk Assessment Types*

The three major types of risk assessment follow essentially the same logic but differ in emphasis. R. Kolluru presents an overview of physical (safety), health, and ecological/environmental risk assessments including 'major process steps', 'typical endpoints', and 'typical applications' in his "Risk Assessment and Management Handbook" [1]. All the assessments start with hazard identification or problem definition. A hazard is a chemical, biological, or physical agent (including electromagnetic radiation) or set of conditions that presents a source of risk but not risk per se.

In physical (safety) hazard analysis, the end points are well defined: fatalities, injuries, and economic losses. The impact is immediate and transparent, and the cause-effect relationship is generally clear. In contrast, considerable uncertainty pervades health risk analyses and also ecological/environmental risk analyses.

## *1.2 ENFIRO Project*

ENFIRO, the EU-funded collaborative research project, offer a prototypical case study on substitution options for specific brominated flame retardants (BFRs) [2]. It started out based on the assumption that not enough environmental and health hazard data were known for the alternatives to the brominated fire retardants such as HBCD. In order to make the evaluation fully comprehensive, it was decided to compare fire performance as well as a life cycle assessment of halogen free flame retardants (FRs) versus BFRs. 17 halogen-free flame retardants as alternatives for three BFRs were selected.

Three FR/product combinations (e.g. metal-based FRs, phosphorous-based and nanoclay-based FRs in printed circuit boards, paints and foams) are studied for: environmental and toxicological risks, viability of industrial implementation, fire safety, application of the FR into products (electronic equipment, specific plastics, paints and textiles). And overall it was found that alternative materials have much lower tendency to bioaccumulate in fatty tissue than the studied brominated flame retardants.

In above EU project, alternative FRs for plastic foams are studied, however, this paper describes about the physical hazards of polystyrene foams from the stand point of hazardous gas release.

### *1.3 Toxic Data by NIST Fire Hazard Analysis*

It has long been realized that most fire victims die from smoke inhalations rather than from burns. However, there has also been long-term controversy over how to characterize the potential for harm from fire-generated smoke and an apparatus from which to obtain accurate data.

A comprehensive methodology has been developed for obtaining and using smoke toxicity data in fire hazard analysis. The methodology comprises: focusing of smoke inhalation deaths: bench scale $LC_{50}$ measurement method, using a minimal number of test animals: interpreting gas measurement data using the N-Gas Model, etc. A small fraction of the chamber atmosphere is removed for chemical analysis of CO, $CO_2$, $O_2$, HCN, HCl, HBr, and NOx. The Model has been developed, to enable the use of these data to obtain approximate $LC_{50}$ values, based on the calculation of FED of mixture of these gases. For the $LC_{50}$ measurement, animals are exposed to the smoke for 30 minutes, where the lethality is noted both during this time and for 14 days afterwards, including the delayed effect of those animals who received a mortal insult [3, 4].

The toxic data by the study of NIST indicate that a few toxic gases play important role for toxic fire hazards.

## 2. Polystyrene Insulation Foams

### *2.1 Production Flow*

EPS and XPS, two foam types for polystyrene, are both rigid, thermoplastic materials. Starting from the polymerization of styrene, additives such as flame retardants, blowing agent are incorporated during or at the end of foam production.

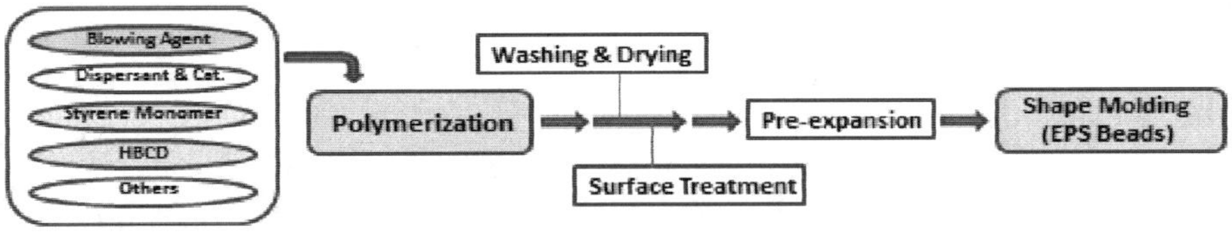


**Figure 1. Expanded polystyrene (EPS) manufacturing process**

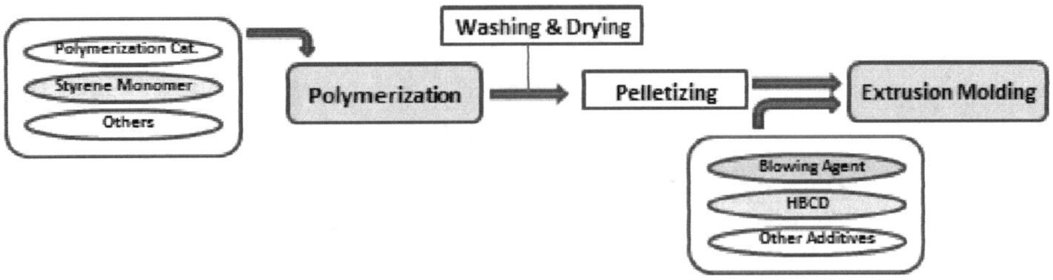


**Figure 2. Extruded polystyrene (XPS) manufacturing process**

2.1.1 EPS production

Foamable polystyrene beads are generally produced by two basic methods; 1) suspension polymerization of styrene into spherical beads containing a blowing agent, 2) incorporation of a blowing agent during the extrusion process of bulk foaming. Since extruded EPS is normally

processed directly to packaging applications, not to insulation materials [5], then Figure 1 showed suspension polymerization process.
The term of suspension polymerization means a process in which water-insoluble monomers are dispersed as liquid droplets with suspension stabilizer, and vigorous stirring to produce polymer particles as a dispersed solid phase. The polymerization takes place within the styrene monomer droplets, where oil-soluble initiators are used as starting substance. And the blowing agent and flame retardant, HBCD, and some additives are applied in this polymerization step.
EPS particles based on above suspension polymerization, are converted into final foam through pre-expansion, which is achieved due to small amounts of gas contained in the polystyrene bead. The gas expands when heat of steam is set on, and the large EPS blocks beads can be fabricated.

#### 2.1.2 XPS production

To produce XPS foam, polymerized styrene is converted into pellets by extrusion machine. These pellets are, then, fed into extruder along with special additives and a blowing agent, shown as Figure 2. Within the extruder the mixture is combined and melted under controlled conditions of high temperature and pressure into a viscous plastic fluid. The hot, thin liquid is then forced in a continuous process through a die. As it emerges from the die, it expands to foam, is shaped, cooled, and then trimmed to dimension.

### *2.2 Flame Retardant*

Since plastic foam itself is combustible, then flame retardant is essentially necessary for building insulation material. Under the framework of the Stockholm Convention on Persistent Organic Pollutants (POPs), a global ban on HBCD (hexabromocyclododecane), a flame retardant used mainly for polystyrene foams, is currently being considered.
EPS board is reported to contain typically 0.5% HBCD by weight in the final product [6] and the maximum concentration of HBCD in EPS beads is assumed to be 0.7% [7]. In XPS, HBCD is used in up to 3% loading to meet technical and flammability foam requirements [8].
Based on the standards such as JIS A9511, classification of insulation board, combustibility, and thermal conductivity are required. Flame retardants are used to clear the requirements of, for example, "the flame disappears within three seconds, there is no remainder dust, and it does not burn exceeding the combustion limit instruction line". However, this kind of standard does not mention physical hazards based on blowing agents and their requirements for use [9].

## 3. The Possible Physical Hazards of Polystyrene Insulation Foams

### *3.1 Physical Hazards and ISO 14040 LCA Frameworks* [10, 11]

LCA (Life Cycle Assessment) provides a framework, approach, and methods for identifying and evaluating environmental burdens associated with life cycle of chemicals from cradle to grave. Since the 1970's, there have been several efforts to develop LCA methodologies, mainly in the USA and Europe. As a result of these efforts, consensus has been achieved on an overall LCA framework and a well-defined inventory methodology has been outlined. For example, LCA procedures are a part of the ISO 14000 environmental management standards as ISO14040 and ISO14044. ISO14040: 2006 describes the principles and framework for life cycle assessment including 'the definition of the goal and scope of the LCA', 'the life cycle inventory analysis phase', 'the life cycle impact assessment phase', and 'the life cycle interpretation phase'. However, the LCA standard methodology noted above is 'not a framework to evaluate the effects of physical hazards, but rather to evaluate the environmental effects of chemicals'.

In this manuscript a chemical's life cycle is described in totally five life stages: in fixed facilities, four life stages (raw material collection/storage/import/production, industrial use/distribution, commercial product private use, waste & disposal), and additional one in transport stage (waterways, roads, railways and pipelines).

### *3.2 The Outline of Life Cycle Physical Hazards of Polystyrene Insulation Foams*

Figure 3 shows the outline of life cycle physical hazards of polystyrene insulation foams (EPS, XPS), divided into 5 stages: Production/import, Industrial use, Private use, Waste disposal and Transportation. The main physical hazards are 1) flammable gas release caused by abnormal storage temperature (below 100°C), and 2) carbon monoxide (CO) release, heated at higher temperature (nearly 400°C), as found in normal case of hydrocarbon heat decomposition.

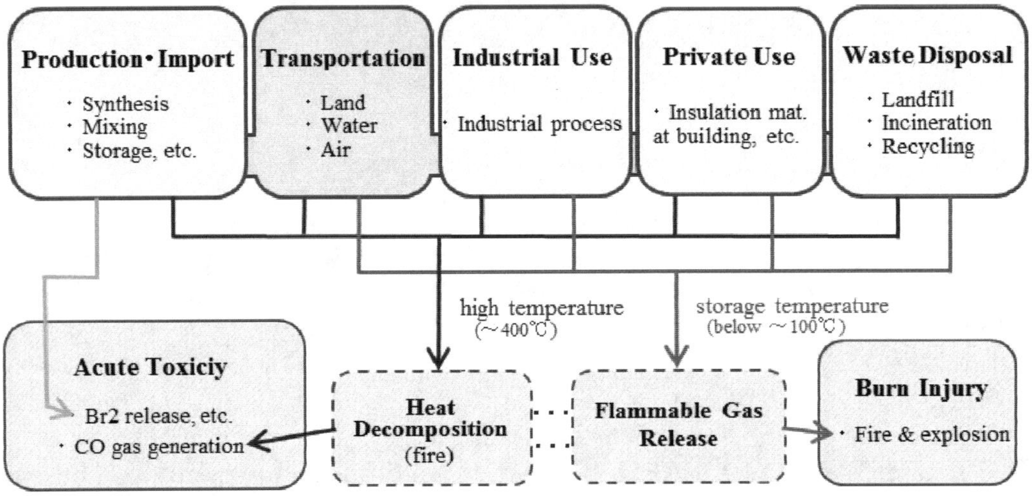


**Figure 3: The outline of life cycle physical hazards for polystyrene insulation foams**

Raw materials necessary for fabrication of XPS products, such as polystyrene, flame retardant, additives like dyestuffs and blowing agents constitute "Production /import stage".

During the "Industrial use stage", the raw materials above are fed to an extruder continuously under high pressure as shown in Figure 2. After foamable gel is formed, it is then cooled just before the orifice, with foam structure created by evaporation of blowing agent. The introduction of chemically stable blowing agent into the molten polymer, at normal pressure, reduces the foam product density, by forming a myriad of closed cells. The foam is then trimmed to the desired shape: the foam boards are packed into shrink-wrapped bundles on pallets, and then stored for curing over a certain period, until shipment to the next stage [12].

At "Private use stage", XPS boards are mainly applied as insulation materials for office buildings, residential houses, refrigerated warehouses, Tatami floor mats, and as road construction material, where light foam product is necessary to solve soft soil problems.

As for EPS products, blowing agents are generally introduced in the stage of polymerization, and the gases are contained in the polymer beads. At the "Industrial use stage", by heating these polymer beads, most of EPS foam products are fabricated in shape molds as Figure 1, due to the expansion of polymerized beads through pre-expansion process.

In each stage, physical hazards were identified based on accident/incident information of polymer foams. Hereinafter, we define physical hazards as those caused by mainly high energy events, such as fire, explosion and/or acute toxicity, leading to death by heat, blast overpressure, and acutely toxic chemicals.
Thus, flammable blowing agent, from polymer insulation foam, will become the cause of physical hazards with toxic hazardous gas, because it may catch fire in a certain gas concentration, with the existence of ignition source.

# 4. Experiments

## *4.1 FID-Gas Chromatography Analysis*

The following experiments of commercially available XPS foams by gas chromatography (flame ionization detector) showed evidence of possible blowing gas release from the polystyrene insulation foam board.

### 4.1.1 Methodology

Polystyrene foam from Company A, Company B and also Company C, was cut by hot-wire to the dimensions of W8, L8, D8 mm, and then 5 samples were placed in 16 ml test tubes with a septum cap (made of silicone) for gas chromatography analysis. Test tubes were heated at 90˚C for 1 hour in an oil bath. Released gas in the test tubes was injected into a gas chromatography (GC-2010, SHIMAZU), equipped with ULBON capillary column HR-1, 0.25mm in diameter and 30m in length. Column temperature was 35˚C, and helium carrier gas was used for FID detection, together with hydrogen gas. Figure 4 shows the retention time comparison between XPS released gas and standard gas, which was made by Sumitomo Seika Chemicals, Co., Ltd., with n-pentane (Kanto Chemical, Co., Ltd.) added.

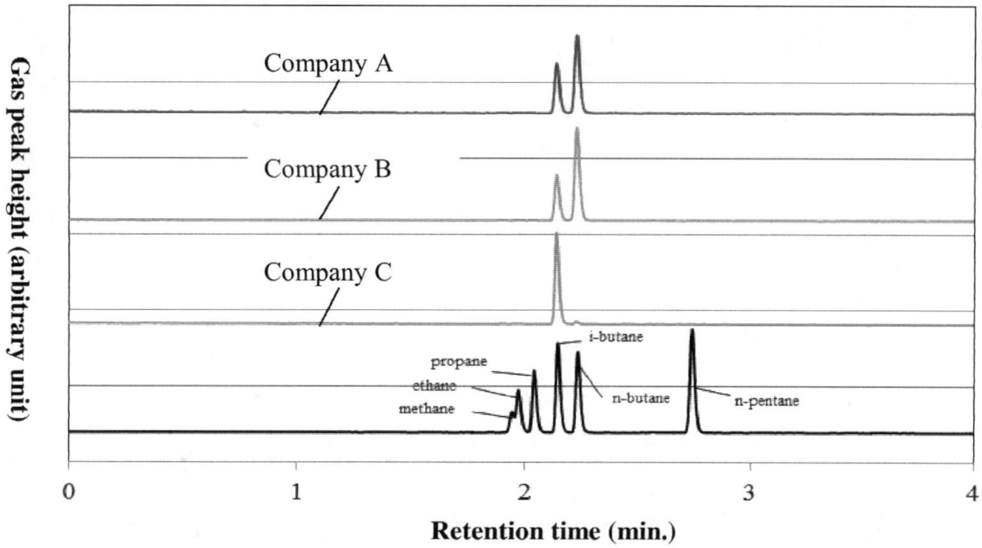


**Figure 4: Gas chromatograph comparison between XPS released gas and standard gas**

4.1.2 Results

Qualitative analysis by FID-GC of foams, heated at 90°C for 1 hour, detected two main kinds of gases, as shown in Figure 4. Comparison with commercially available standard gas, containing methane, propane, iso-butane, and n-butane, with pentane added, found retention times for released gas peaks to be the same as those of iso-butane and n-butane in foams of Company A and Company B, however only iso-butane for Company C.

### *4.2 TCD-Gas Chromatography Analysis (1) -commercially available foams-*

The following experiments by TCD-GC showed evidence of carbon monoxide generation from the polystyrene foam board, heated at 400°C. The Thermal Conductivity Detector (TCD) is a universal detector and can detect air, hydrogen, carbon monoxide, carbon dioxide, nitrogen, inorganic gases and many other compounds. For most organic molecules, the sensitivity of the TCD is lower compared to the Flame Ionization Detector.

4.2.1 Methodology

Commercially available XPS foams from Company B and Company C, were cut by hot-wire and 100mg each was prepared, together with commercially available EPS bead from Company E. Then, each by each, they are placed in 20ml test tube (outer diameter;11mm $\phi$ , length;75mm) in Titanium sphere type vessel (63.5mm $\phi$ ) for gas chromatography analysis. The metal vessel, which contained glass test tube in the neck part (inner diameter;11.5mm $\phi$ ), was placed in ceramic tube type electric furnace at 400°C, and was kept for 1 hour.
Gas release from test tube in sphere vessel due to heat decomposition, was caught by micro syringe with silicone cap, and then injected into a gas chromatography (GC-2012, SHIMAZU), equipped with SHINCARBON ST column made by Shinwa Chemical Industries Ltd. Column temperature was 40°C for 3minutes at first stage, and then heated up to 200°C in 10°C/min., and then kept additional 3 minutes at 200°C, with helium carrier gas.

4.2.2 Results

Figure 5 shows the gas chromatograph comparison of decomposition gases from two kinds of XPS foams (Company B,C; 100mg each), one EPS bead (Company E; 100mg) and CO standard gas, made by Sumitomo Seika Chemicals. X axis is retention time in minutes and Y axis is gas peak height in arbitrary unit.
Detected released gases from two foams and EPS bead, in 400°C for 1hour, showed the similar chromatographs.

### *4.3 TCD-Gas Chromatography Analysis (2) -commercially nonavailable foams-*

Three kinds of EPS foams (flame retardants of HBCD, FP-2200, and no retardants) were heated in 400°C, and thereafter released decomposition gas, containing CO, was compared by TCD-GC analysis.

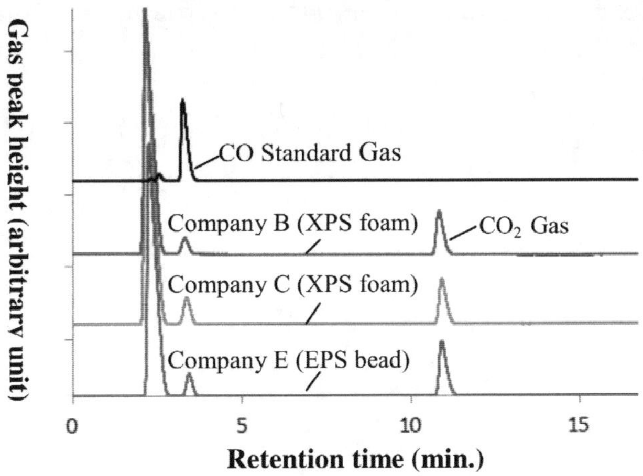


**Figure 5: Gas chromatograph comparison of released gas from XPS foams, EPS bead and CO standard gas**

4.3.1 Methodology

Suspension polymerized beads and 1.0wt% flame retardant was placed in glass bottle, and pre-expanded (100˚C, 30minutes). And then EPS foams were fabricated by using metal mold, heating in 100˚C for 20minutes. Here, HBCD was from Wako Pure Chemicals Industries, Ltd., and FP-2200 (Phosphorus/intumescent) from ADEKA Corporation.

As mentioned before, released gas of heat decomposition by test tube in sphere vessel, was injected by micro syringe into TCD-GC (GC-2012, SHIMAZU) for analysis. In each test, 140mg of EPS foam materials were used.

4.3.2 Results

Three kinds of EPS foams (flame retardants of HBCD, PF-2200 and no retardants) were shape molded, and then decomposed by heating in 400˚C, for 1 hour. Table 1 shows CO gas concentration, by TCD-GC analysis: Concentration of CO gas from EPS (HBCD, 1.0wt%) was 3.9%, while EPS foam without flame retardant was 2.8%, and EPS (PF-2200, 1.0wt%), phosphorous flame retardant, was 2.2%. The density of fabricated EPS foams was 0.04g/cm$^3$.

**Table 1. Released CO gas concentration analyzed by TCD-GC**

| Polystyrene foams | | Fire Retardant | CO gas conc. |
|---|---|---|---|
| EPS | Fabricated foams of Company D beads | HBCD 1.0wt% | 3.9% |
| | | PF-2200 1.0wt% | 2.2% |
| | | - | 2.8% |

### *4.4 Analysis in Polystyrene Foams*

The amount of HBCD flame retardant in polystyrene foams was measured by energy dispersive X-ray fluorescence spectrometer, EDX1800B, Skyray Instrument Inc.

Suspension polymerized beads, with no halogen based retardant, were mixed and expanded with certain amount of HBCD at 100°C for 30minutes in 100ml glass bottle, and calibration curve was prepared.

Based on above data, HBCD content in polystyrene foams and beads were measured. EPS molded piece from Company C, contained 0.6wt%. HBCD in beads from Company D and E were not detectable. However XPS molded foams contained retardant a little higher: Company A article was 6.1wt%, Company B was 2.4wt%, and Company C was 2.5wt% as shown in Table 2.

**Table 2. X-ray fluorescence analysis (XRF) of HBCD in polystyrene foams**

| Polystyrene foams | | | HBCD analysis by XRF |
|---|---|---|---|
| XPS | Company A | foam | 6.1wt% |
| | Company B | foam | 2.4wt% |
| | Company C | foam | 2.5wt% |
| EPS | Company C | foam | 0.6wt% |
| | Company D, E | bead | ND |

## 5. Result and Discussion

### *5.1 HSE Information Sheet and Experimental Results*

HSE information sheet titled "Fire and explosion risks from pentane in expandable polystyrene (EPS)" suggests that it is necessary to calculate air exchange rate to reduce risk at the storage site [13].

In Japan, following a warehouse fire and explosion accident (1982), a notification for physical hazard for EPS beads was sent to local prefectural governments by the Fire and Disaster Management Agency (FDMA) [14]. The accident cause was attributed to beads in flexible containers with no explosion-proof structure for cooling equipment switching, in a low temperature (2-3°C) warehouse [15]. HSE's suggestion will also be useful in reducing physical risks from the flammable gas used in XPS.

### *5.2 Fire Retardant*

HBCD, hexabromocyclododecane, is used as fire retardant for XPS insulation products, because it prevents fire through direct contact between ignition source and polystyrene foam. It is not yet known whether HBCD has capabilities as a flammable gas barrier or not.

Hence, additional countermeasures are necessary for physical risk (fire, explosion), such as sufficient air ventilation for flammable gas just after wire cutting of large quantities of foam, and/or excluding ignition sources, etc., in indoor storage areas or underground construction sites.

HBCD content in EPS was found to be lower than those of commercially available XPS, made by Company B or C. This was below 3.0wt% as the literature shows [6, 7 and 8]. However XPS foam from Company A was found to contain HBCD more than 3.0wt%.

### *5.3 Gas Chromatography Experiments*

In FID-GC analysis, sample heating temperature of 90°C was selected, supposing much more gas release at abnormal storage temperature. In Figure 4, n-pentane was not found in three kinds of

commercially purchasable XPS foams, because no gas peaks were detected between 2.5minutes and 3.0minutes retention time, as n-pentane in standard gas.
We understand that only two gas peaks were found for foams from Company A and Company B and one gas peak from Company C, because the concentration of these butane gases was high.

In TCD-GC, the detected gases from decomposed XPS at 400°C, for 1 hour, were $N_2$, CO, and $CO_2$. In Figure 5, three commercially available polystyrene materials showed the similar gas chromatographs.
The CO gas concentration from newly molded EPS (HBCD, 1.0wt%) was higher than that of EPS with 1.0wt% of phosphorous flame retardant (FP-2200). Intumescent Char seems to decrease CO gas concentration in the latter EPS foam, compared with bromine flame retardant.

Heating studies for EPS foams were performed at the temperature between 200°C and 400°C, and it was reported that CO gas increased much more at 400°C [16]. So, 400°C was selected for foam or bead decomposition temperature.

## 6. Conclusion

Possible two kinds of physical hazard scenarios for polystyrene foams (EPS, XPS) were shown in this paper for their life cycle: 1)Production/import, 2)Transportation, 3)Industrial use, 4)Private use, 5)Waste disposal. One scenario is acutely toxic gas release by incomplete combustion case (nearly 400°C), etc., and the other is flammable blowing gas release from the foam itself (nearly 100°C). The evidence was shown mainly by experiments with two kinds of gas chromatography (FID-GC, TCD-GC).

Qualitative analysis by FID-GC of foams heated at 90°C for 1 hour detected two main kinds of gases. Comparison with commercially available standard gas, containing methane, propane, iso-butane, and n-butane, with pentane added, found retention times for released gas peaks to be the same as those of iso-butane and n-butane for Company A and Company B, however only iso-butane for Company C foam. However, pentane was not found by FID-GC analysis for commercially available 3 kinds of XPS foams.

In TCD-GC, the detected gases from decomposed XPS at 400°C, for 1 hour, were $N_2$, CO, and $CO_2$, where $O_2$ was not found. The three commercially available polystyrene foams showed the similar gas chromatographs as Figure 5.

The CO gas concentration from newly fabricated EPS foams, with different flame retardant (HBCD, FP-2200) was compared, by gas chromatography analysis. The CO gas concentration was higher for EPS foams with 1.0wt% HBCD than EPS foams of tested phosphorous retardant.

HBCD content in EPS foam was found to be lower than those of commercially available XPS made by Company B or C, which was below 3.0wt% as the literature shows. However XPS foam from Company A was found to contain HBCD more than 3.0wt%.

## 7. References


[1] R. Kolluru, S. Bartell, R. Pitblado and S. Stricoff, "Risk Assessment and Management Handbook for Environmental, Health, and Safety Professionals", New York: McGraw-Hill, Inc., 1996

[2] http://www.enfiro.eu/ (accessed on February 27, 2014)

[3] V. Babrauskas, B.C. Levin, R.G. Gann, et al., "Toxic Potency Measurement for Fire Hazard Analysis", NIST Special Publication 827, 1991

[4] R.G. Gann, V. Brabrauskas, E. Brawn, B.C. Levin, et al., "Toxic Data for Fire Hazard Analysis", Proceedings of 12th Joint Meeting of the UJNR Panel on Fire Research and Safety, Oct. 27-Nov.2, 1992

[5] J. Scheirs and D. Priddy, "Modern Styrene Polymers: Polystyrene and Styrenic Copolymers ", John Wiley&Sons, Ltd., 2003

[6] G. Morose, "An Overview of Alternatives to Tetrabromobisphenol A (TBBPA) and Hexabromcyclododecane(HBCD)", Lowell Center for Sustainable Production, University of Massachusetts Lowel, 2006

[7] European Chemical Bureau, European Commission. Joint Research Centre. EU Risk Assessment Report, Hexabromocyclododecane, CAS No: 25637-99-4, EINECS No: 247-148-4, 2008

[8] J. Troitzsch, "The relevance of hexabromocyclododecane for polystyrene EPS/XPS foams to meet fire safety requirements as construction products in Europe. Expert opinion for the European HBCD Industry Working Group", 2008.

[9] Y.Ohara, K.Tanaka, T.Hayashi, H.Tomita and S. Motani, "The Development of a Non-Fluorocarbon Based Extruded Polystyrene Foam which Contains a Halogen-Free Blowing Agent", Bull. Chem. Soc. Jpn, 77, 599-605, 2004

[10] T. Nakarai, S. Yoshino, I. Nishimaru, M. Kutsumi, A. Miyake, "Accident/Incident Scenarios based on Life Stages Across the Complete Chemical/Product Lifecycle", 7th Global Congress on Process Safety (GCPS2011), March 2011

[11] M. Finkbeiner, A. Inaba, R.B.H. Tan, K. Christiansen, and H.J. Kueppel, "The New International Standards for Life Cycle Assessment: ISO14040 and ISO14044", Int. J.LCA 11(2) 80-85p, 2006

[12] Fire and Disaster Management Agency, Fire Safety Countermeasures for Expandable Polystyrene Beads, etc., Notification, Dec. 14, 1982

[13] Health and Safety Executive 2001, Fire and Explosion Risks from Pentane in Expandable Polystyrene (EPS), HSE Information Sheet, http://www.hse.gov.uk/pubns/ppis1.pdf

[14] Fire and Disaster Management Agency, Fire Safety Countermeasures for Expandable Polystyrene Beads, etc., Notification, Dec. 14, 1982

[15] H. Koseki, Warehouse Fire at Yokkaichi City, Shoubou Shuuhou, No.36, 25-29p, 1982

[16] S. Yoshino, T. Nakarai, I. Nishimaru, A. Miyake, "Hazard analysis for plastics contained HBCD based on physical risk scenarios", Society of Environmental Sciences, Japan, Annual Meeting, Sep. 14, 2012

## Acknowledgements

Part of this study was promoted by the "Effective Cooperation of Science and Technology Policies" based on special coordination funds for promoting science and technology in 2007-2009, by the Ministry of Education, Culture, Sports, Science and Technology, and was also financially supported partly by the Environment Research and Technology Development Fund (c-1003) of the Ministry of the Environment, Japan, for 2010-2012.

# 349296 Problems Encountered in the Development of a Process Safety Climate Tool

**Abstract:**

Following on from the successful launch of the Health and Safety Laboratory's (HSL) Safety Climate Tool in 2010, Human factors specialists at HSL have been investigating the feasibility of developing of a process specific safety climate tool. There is an appetite in industry for a safety culture tool aimed specifically at major hazards and market research identified a gap for a tool that would enable the perceptions of workforce on process safety to be collected within an organisation. HSL have identified a number of key topics from the literature and market research to be included in a tool, these will be described. The paper will also discuss the challenges HSL have faced when developing a process specific tool with a valid/ reliable question set, and how this relates to the HSL's SCT.

# 348745 Upper Explosible Limits for Combustible Dusts

**Abstract:**

Upper Explosible Limits for Combustible Dusts

by Richard W. Prugh

Principal Process Safety Engineer

Chilworth Technology, Inc., a DEKRA Company

113 Campus Drive, Princeton, New Jersey 08540

609-799-4449; richard.prugh@dekra.com

Abstract

Although Upper Flammable/Explosible Limits are sharply defined for flammable liquids and gases, combustible dusts do not appear to have well-defined Upper Explosible Limits. The causes for this behavior include the typically-wide particle-size distribution of dusts and difficulty in obtaining a uniform dispersion of dust in the test equipment. During an explosibility test, the larger particles may act as inert materials, and the smaller particles represent a low concentration in the surrounding mixture of air and inert particles, and they burn readily. If the composition and heat of combustion of the combustible dust are known, thermodynamic calculations can yield the concentration at which the combustion temperature is insufficient for propagation of flame through a cloud of small particles. Examples are given for a typical carbohydrate dust and a metal dust.

# Agglomeration Effect on Combustion and Explosion Properties of Nanoparticles

**Jiaqi Zhang**
**Mary Kay O'Connor Process Safety Center,**
**Artie McFerrin Department of Chemical Engineering, Texas A&M University**
**College Station, Texas 77843-3122, USA**
**jiaqizhang@tamu.edu**

**Yi Liu, Hao Chen, and M. Sam Mannan**
**Mary Kay O'Connor Process Safety Center,**
**Artie McFerrin Department of Chemical Engineering, Texas A&M University**
**College Station, Texas 77843-3122, USA**
**mannan@tamu.edu**

Prepared for Presentation at
American Institute of Chemical Engineers
2014 Spring Meeting
10th Global Congress on Process Safety
New Orleans, LA
March 30 – April 2, 2014

UNPUBLISHED

# Agglomeration Effect on Combustion and Explosion Properties of Nanoparticles

**Jiaqi Zhang**
**Mary Kay O'Connor Process Safety Center,**
**Artie McFerrin Department of Chemical Engineering, Texas A&M University**
**College Station, Texas 77843-3122**
**jiaqizhang@tamu.edu**

**Yi Liu, Hao Chen, M. Sam Mannan**
**Mary Kay O'Connor Process Safety Center,**
**Artie McFerrin Department of Chemical Engineering, Texas A&M University**

**Keywords:** Dust explosion, Agglomeration, Nanoparticles

## Abstract

Nanoparticles have been demonstrated to possess significantly high ignition sensitivities compared with micro-particles, *e.g.*, much lower minimum ignition energy (MIE) and feasible spontaneous ignition phenomena, and have been predicted to have a high deflagration index of dust explosion ($K_{St}$) and a low minimum explosive concentration (MEC). However, these predictions are inconsistent with experimental observations, which were attributed to the agglomerations of nanoparticles. Multi-scale agglomerations, including soft and hard agglomerations, have been observed in nanofiber systems, which may serve as proper platforms to validate this assumption. This study tested the MECs of carbon nanofibers (CNFs), and found that MECs decreased after the milling process. This finding calls for attention to a widely used processing method in nanomaterial industries. Comparing to MECs of micro-sized graphite particles, this study supports the hypothesis that it is the size of agglomerates rather than that of single nanoparticles can be used as an index to estimate CNFs' MEC. Theoretical analyses were performed for carbonaceous particles with different sizes at different surface temperature to explain how specific surface area affects particles' combustion and the limiting step in combustion process was also discussed.

## 1. Introduction

Particle size plays a significant role in affecting dusts' combustion and explosion behaviors. Minimum ignition energy (MIE) was found to reduce exponentially as particle size decreases; the "plateau effect" of particle size on MEC exists, which means that the minimum explosive concentration (MEC) declines with reduced particle size until reaching a limiting stage where particles are too fine. Thus, the explosion severity can be significantly increased by reducing particle size [1]. This trend was summarized from much experimental data of micro-size particles [1].

However, when it comes to nanoparticles, the situation becomes much more complicated. On one side, some experimental data and observations indicate that nanoparticles possess significantly high ignition sensitivities; most metal nanoparticles' MIE is lower than 1 mJ while corresponding micro-size particles' MIE are ~20 mJ [2]. Furthermore, an explosibility research of several different types of carbon nanotubes (CNTs) reached a conclusion that the explosion severity is indeed higher for shorter CNTs [3]. On the other hand, much research states that nanoparticles' explosibility is limited by limiting dispersibility and high coagulation rates [4]. For example, an explosibility research of aluminum nanoparticles found that explosion of 200 nm particles' cloud was stronger than that of 100 nm particles [5]. One possible explanation is the larger agglomerates of 100 nm particles [6]. Another possible reason is the different oxidation degree [5].

Since it is difficult to exclude the influences of other factors in the aluminum nanoparticle research, this study intended to choose samples with identical properties, except for the different sizes, to study agglomerations' effect on combustion and explosion behaviors. There are both milled and unmilled carbon nanofibers (CNFs). Considering the probability that different combustion and explosion behaviors are dominated by different properties of combustible dusts, this study focuses on agglomerations' effect on MECs.

In this study, MECs were identified according to the ASTM standard E 1515 [7]. A scanning electron microscope (SEM) was used to measure the size gaps between different samples. Theoretical analyses were also applied to explore the mechanisms.

## 2. Samples and Experiments

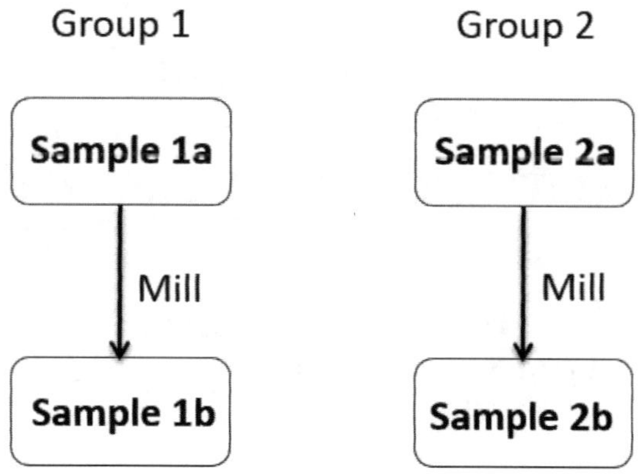


**Figure 1. Samples**

Two groups of CNFs were used in this study, as shown in Figure 1. There are two samples in each group. Samples 1a and 1b are in group 1; samples 2a and 2b are in group 2. Sample 2a was obtained by annealing sample 1a at 3000 °C. They are CNFs with 100 nm diameter and 10-30 μm length. Samples 1b and 2b were obtained by milling samples 1a and 2a with improved bulk densities up to 12 pounds per cubic feet (pcf). Consequently, samples in the same group are of quite similar properties, except for the particle sizes.

### 2.1 MEC tests experiments

MEC tests were conducted in a customized 36 L dust explosion vessel, as shown in Figure 2. It consists of a vaccum system, an air reservoir, a fast acting valve, a dust container, a rebound nozzle, an ignition system, and pressure transducers. Compared to the standard 20 L sphere, besides the difference of volume, this equipment's dust container is also located at a different location – the bottom of the vessel. Therefore, the dispersion process may be slightly different. However, after calibrated several essential parameters, it generates very similar data as the 20 L sphere does [8].

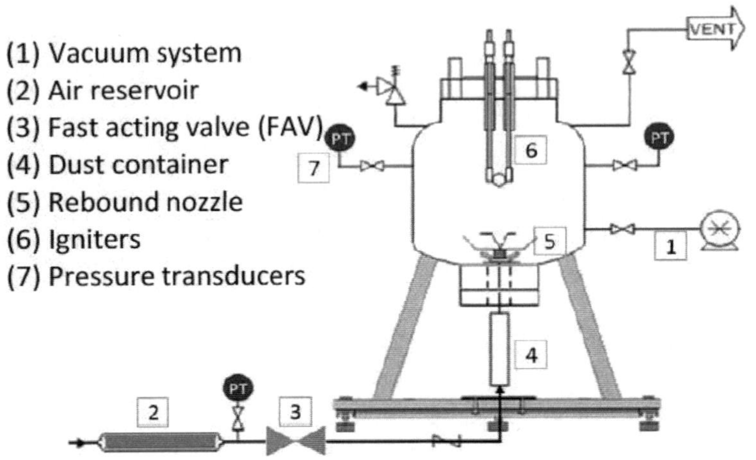


**Figure 2. Schematic diagram of 36 L dust explosion vessel [8]**

Following the standard ASTM E 1515 [7], this study identified the MECs by increasing the dust concentration until the pressure ratio (PR) ≥ 2.0:

$$PR = \left(P_{ex,a} - \Delta P_{ignitor}\right) / P_{ignition} \qquad \text{[Eq.1]}$$

Where $P_{ex,a}$ is the maximum explosion pressure; $P_{ignition}$ is the absolute pressure when the ignitor is activated; and $\Delta P_{ignitor}$ is the pressure rise in the chamber due to the occupied energy of the ignitor in air at atmospheric pressure.

### 2.2 SEM characterization

A JOEL JSM-7500F field emission scanning electron microscope (FE-SEM) was used to characterize the nanoparticles. Dry sampling was adopted to study the differences between the agglomeration sizes of the original and milled nanoparticles. Wet sampling was also performed to observe the size and length of individual particles

## 3. Results and discussion

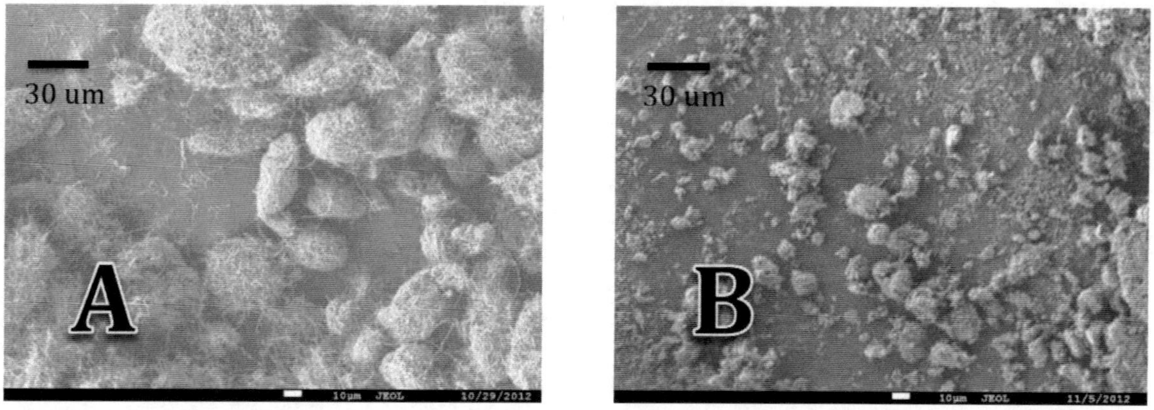


**Figure 3. SEM images for A) unmilled sample (sample 1a), magnification: 400X; B) milled sample (sample 1b) magnification: 400X**

Figure 3 shows that CNFs agglomerated together and formed micro-sized spheres, rather than single fibers. It also indicates that the milling process effectively reduced the size of agglomerated spheres from approximately 30 µm to less than 10µm.

Sample 1a and 1b were dispersed in a 90% isopropyl alcohol solution. The resulting samples are shown in Figure 4. From Figure 4, it can be seen that the milling process also reduced the length of a considerable portion of fibers from about 10µm to 5µm or less, although long fibers still existed.

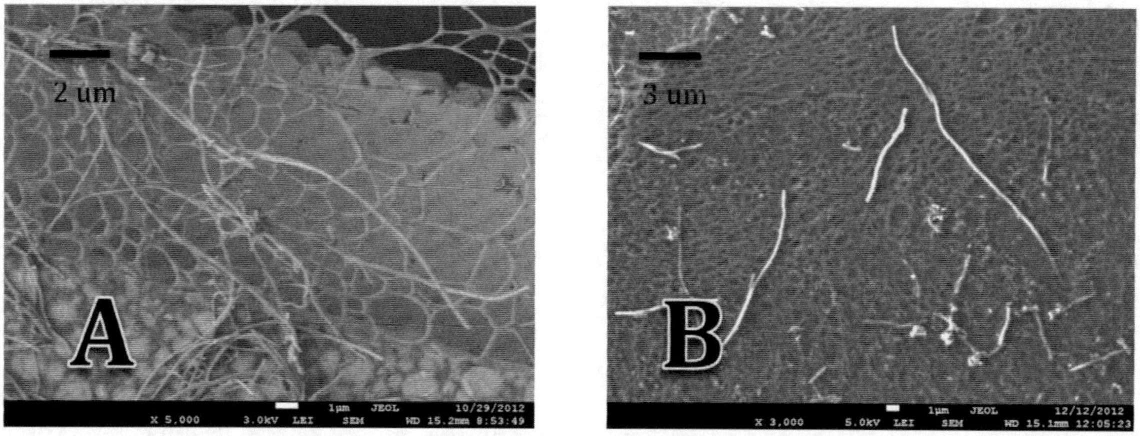


**Figure 4. SEM pictures for A) unmilled sample (dissolved sample 1a), magnification 5000X; B) milled sample (dissolved sample 1b), magnification 3000X**

MECs of sample 1-4 were also measured and corresponding results are shown in Table 1. It is easy to find that milled samples have much lower MECs, which means higher risks of explosion.

The differences between MECs show that the milling process increases the explosion risks of CNFs. Given that the milling method is widely used in customizing CNFs, and no specific attention is given to those customized samples, this finding calls for attention to this "blind spot".

**Table 1. MECs of Samples**

| | Sample | Description | MEC ($g/m^3$) |
|---|---|---|---|
| **Group 1** | 1a | Unmilled | >300* |
| | 1b | Milled | 105 |
| **Group 2** | 2a | Unmilled | >300* |
| | 2b | Milled | 210 |

* Due to the samples' low density and dust container's limited volume, MECs of samples 1a and 2a were not able to be obtained. The achieved maximum concentration was 300 $g/m^3$. Therefore, their MECs were identified as larger than 300 $g/m^3$.

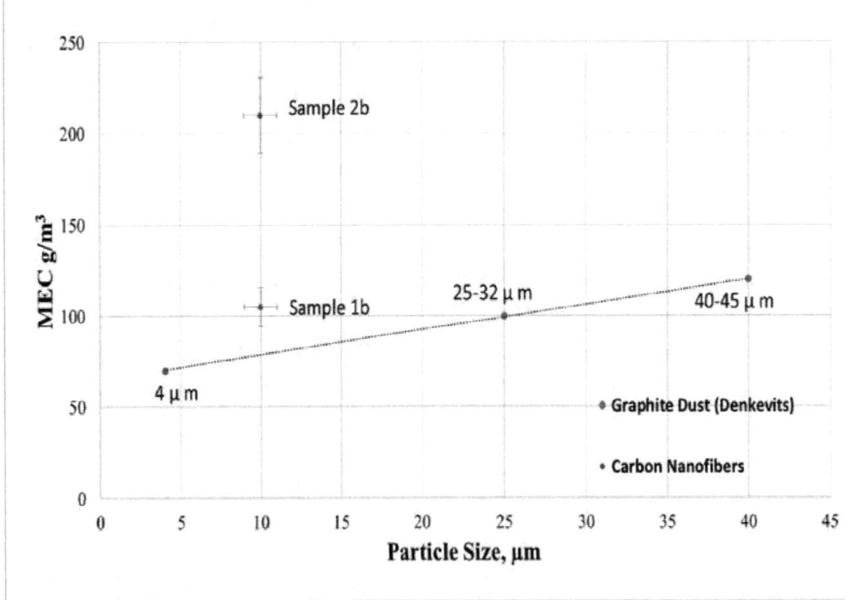


**Figure 5. MEC of graphite dust and carbon nanofibers vs. dust particle size**

Lower explosive concentration of some graphite dust (4 μm – 45 μm) are between 70 $g/m^3$ and 120 $g/m^3$ as shown in Figure 5 [9]. The MECs also exhibit a decreasing trend when particle size goes down. While MECs of carbon nanofibers are even higher than these graphite dust, it supports the hypothesis that the size of agglomerates rather than that of single nanoparticles can be used as an index to estimate CNFs' MEC. Figure 5 also shows that MECs of CNFs of similar agglomerates sizes are higher than their corresponding graphite dust. It might be caused by several other factors. Further study can be done to find them out.

## 4. Theoretical Analyses

Theoretical analyses were also performed to illustrate how the particle sizes affect carbon particles' combustion behavior at its first stage. In order to simplify the analyses, only single agglomeration's combustion process was considered in this research.

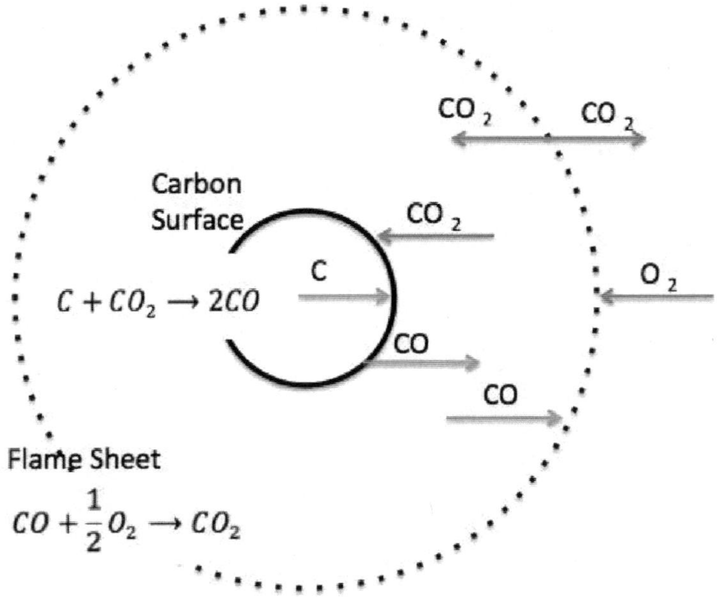


**Figure 6. Mass flow at the carbon surface and the flame sheet reproduced based on [10]**

In the classic two-film model, proposed by Burke and Schumann [11], developed by Held [12], and supported by experiments involving large particles [13], final combustion takes place in an outer layer, but the formed carbon dioxide attacks the carbon surface according to the reaction:

$$C + CO_2 \rightarrow 2CO \qquad \text{[Eq.2]}$$

The CO generated at the surfaces diffuses outward and is consumed at a flame sheet where it meets an inward-diffusing flow of $O_2$ and reacts stoichiometric proportions:

$$CO + \frac{1}{2}O_2 \rightarrow CO_2 \qquad \text{[Eq.3]}$$

Figure 5 shows the mass flow at the carbon surface and the flame sheet for this model.

Diffusion of oxygen and reaction on the particle surface are two main parts in the combustion process. Either of them can be the limiting step. The following section will compare the combustion time of single particle to identify the limiting step for the combustion of particles we are interested in. The following section illustrates the calculation example for a 30 μm particle with 1500 K surface temperature. For diffusion, air comes from unburnt zone, therefore, a temperature of 300 K and a pressure of 101.3 kPa is used in calculation. For reaction, since it happens on the hot particle surface, a temperature of 1500 K and a pressure of 516.7 kPa is used.

The collision integral of air, $\Omega_D$ is a function of temperature and intermolecular potential field for one molecule of A (Oxygen) and one molecule of B (Nitrogen). It can be approximated by the following expression [14]:

$$\Omega_D = \frac{a}{\left(T^*\right)^b} + \frac{c}{\exp\left(dT^*\right)} + \frac{e}{\exp(fT^*)} + \frac{g}{\exp\left(hT^*\right)} \qquad \text{[Eq.4]}$$

The parameters in this equation are listed in the following table:

**Table 2. Parameters for collision integral [13]**

| Parameter | Value | Parameter | Value |
|---|---|---|---|
| □* | κT/ε$_{AB}$ | **e** | 1.03587 |
| **a** | 1.06036 | **f** | 1.52996 |
| **b** | 0.15610 | **g** | 1.76474 |
| **c** | 0.19300 | **h** | 3.89411 |
| **d** | 0.47635 | **ε$_{AB}$ /κ, K** (for air) | 97 |

While assuming the temperature of air coming to the particle surface is 300 K:

$$T^* = \frac{T}{\varepsilon_{AB}/K} = \frac{300K}{97K} = 3.09 \qquad \text{[Eq.5]}$$

$$\Omega_D = \frac{1.06036}{(3.09)^{0.15610}} + \frac{0.19300}{\exp(0.47635\times 3.09)} + \frac{1.03587}{\exp(1.52996\times 3.09)} + \frac{1.76474}{\exp(3.89411\times 3.09)} = 0.9424$$

[Eq.6]

Then the diffusivity of air can be determined by using the Wilke and Lee Method [15]:

$$D_{AB} = \frac{\left[3.03 - \left(\frac{0.98}{M_{AB}^{1/2}}\right)\right]\left(10^{-3}\right)T^{3/2}}{PM_{AB}^{1/2}\sigma_{AB}^2\Omega_D} \qquad \text{[Eq.7]}$$

$D_{AB}$ : Diffusion coefficient, cm²/s

$M_A, M_B$ : Molecular weights of A and B, respectively

T: Temperature, K

P: Pressure, bar

$\sigma_{AB}$ : "Collision diameter," a Lennard-Jones parameter, angstrom

$\Omega_D$ : Diffusion collision integral, dimensionless

Then

$$M_{AB} = 2\times\left(\frac{1}{M_A} + \frac{1}{M_B}\right)^{-1} = 29.867 \qquad \text{[Eq.8]}$$

$$\sigma_{AB} = 3.620 \text{ Å} \qquad \text{[Eq.9]}$$

$$D_{AB} = \frac{\left[3.03 - \left(\frac{0.98}{29.867^{0.5}}\right)\right](10^{-3})(300^{3/2})}{1 \times 29.867^{0.5} \times 3.620^2 \times 0.9424} = 0.198 cm^2/s \qquad \text{[Eq.10]}$$

If the reaction is controlled by a diffusion step, then the combustion time, $\tau_b$, can be calculated [6]

$$\tau_b = \frac{\rho_s d_p^2}{8\rho_g D_{O_2}(Y_{O_{2,\infty}}) \Big/ \left(MW_{O_2} \big/ 2MW_C\right)} \qquad \text{[Eq.11]}$$

$\rho_s, \rho_g$ : Density of solid particle and air, kg/m$^3$

$d_p$ : Diameter of particle

$Y_{O_{2,\infty}}$ : Oxygen mass fraction in air, 0.233

$MW_{O_2}, MW_C$: Molecular weight of oxygen and carbon

For 30 μm agglomerates

$$\tau_b = \frac{1600 kg/m^3 \times (30\mu m)^2}{8 \times 0.198 cm^2/s \times 1.204 kg/m^3 \times 0.233 \Big/ \left(32 \big/ 2 \times 12\right)} = 43.126 ms \qquad \text{[Eq.12]}$$

If the reaction is controlled by a surface reaction step, then the combustion time can be calculated [6]

$$\tau_b = \frac{\rho_s d_p}{2k_1 \left(P_{tot} MW_{gas} \big/ RT_s\right)\left(MW_C \big/ MW_{O_2}\right) Y_{O_2,s}} \qquad \text{[Eq.13]}$$

$$k_1 = 3.007 \times 10^5 \times \exp\left(-\frac{17966}{T_s}\right) \qquad \text{[Eq.14]}$$

$k_1$ : Kinetic constant of combustion, m/s

$T_s$ : Temperature at the particle surface

For 30 μm agglomerates with the surface temperature of 1500 K,

$$k_1 = 3.007 \times 10^5 \times \exp\left(-\frac{17966}{1500}\right) = 1.890 m/s \qquad \text{[Eq.15]}$$

$$\tau_b = \frac{1600kg/m^3 \times 30\mu m}{2\times 1.890m/s \times \left( \frac{516.7kPa \times 29.867g/mol}{8.314J/(mol\cdot K)\times 1500K} \right)(12/32)\times 0.233}$$

$$=156.6ms$$

[Eq.16]

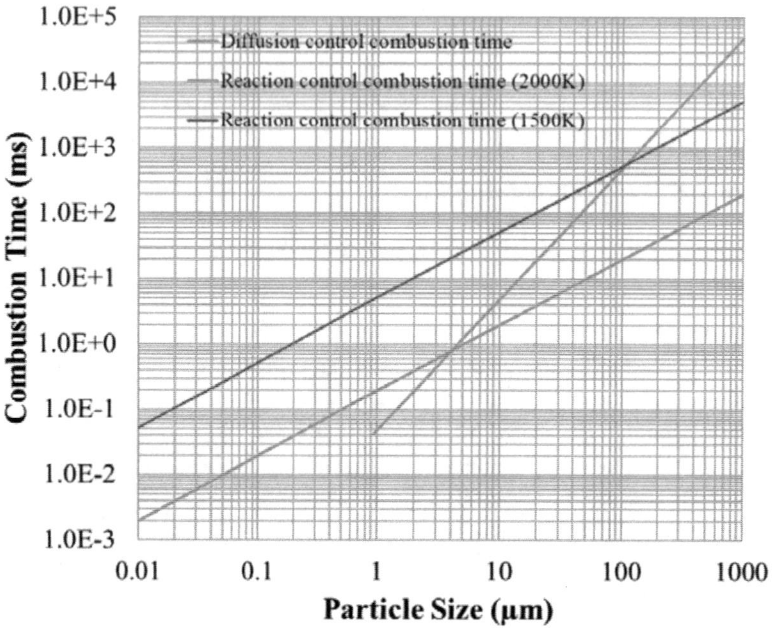


**Figure 7. Combustion time for different particle sizes**

This study repeated this calculation for a wide range of particle sizes at temperature 1500 K and 2000 K since the temperature significantly impact the surface reaction rate. Results are summarized in Figure 7. These calculation results match the experimental data of carbon particles smaller than 10 μm and larger than 100 μm [6] within a reasonable deviation, which supports the validity of this calculation. Both the experiments and calculation shows that for carbon particles smaller than 10 μm, the combustion process is controlled by the surface reaction. Also with the particle size shrinking, the difference between reaction rates and diffusion rates becomes larger. This is reasonable since the required diffusion path of oxygen is much shorter. The diffusion control becomes the limiting step for carbon particles larger than 100 μm.

For our interested particles – particle size between 10 μm and 100 μm, experimental data is currently not available while the calculation shows that it is difficult to state that there is a dominant step since the differences of combustion time is not obvious. Meanwhile, the exact surface temperature remains unknown, which adds great barrier to identify the key limiting step, the diffusion or the surface reaction.

However, no matter which step is limiting the combustion process, the particle size plays an important role in it. The combustion rates for a single carbon particle, $\dot{m}_c$ can be written as [3]:

$$\dot{m}_c \propto d_p^2 \qquad \text{[Eq.17]}$$

Then, the combustion rate for a certain dust cloud, $\dot{M}_c$, can be expressed as:

$$\dot{M}_c \propto \frac{1}{d_p} \qquad \text{[Eq.18]}$$

For larger particles, the combustion process are limited by diffusion. Then the relationship can be expressed as [3]:

$$\dot{m}_c \propto d_p \qquad \text{[Eq.19]}$$

And,

$$\dot{M}_c \propto \frac{1}{d_p^2} \qquad \text{[Eq.20]}$$

According to [eq. 18] and [eq. 20], further studies on combustion rate and particle size are needed to figure out the mechanism how reaction and diffusion interact in controlling the combustion process.

## 5. Conclusion

To study agglomeration's influence on particles' combustion and explosion behaviors, this study selected two groups of CNFs, including one milled sample and one unmilled sample in each group. MECs of these four samples were identified in a customized 36 L dust explosion vessel. It was found that MECs decreased after the milling process. This finding calls for attention to widely used processing method in nanomaterial industries. A comparison between the CNFs and micro-sized graphite particles is also discussed in this study, which could confirm the hypothesis that the size of agglomerates rather than that of single nanoparticles can be used as an index to estimate CNFs' MECs.

Theoretical analyses were performed for carbonaceous particles with different sizes at different surface temperature to explain how specific surface area affects particles' combustion and to find out the limiting step in combustion process. Combustion time calculation results using classic equations match well with available experimental data. For interested particles – particle size between 10 μm and 100 μm, experimental data is currently not available while the calculation shows that it is difficult to state that there is a dominant step since the differences of combustion time is not obvious. The unknown surface temperature makes it hard to identify the limiting step, the diffusion or the surface reaction

**Acknowledgements**

The authors gratefully acknowledge the financial support from the National Science Foundation Eager Program (No.1321581).

## 6. References

[1] Eckhoff, R.K., *Dust Explosions in the Process Industries: Identification, Assessment and Control of Dust Hazards*. Third ed. 2003: Gulf Professional Publishing.

[2] Wu, H.-C., R.-C. Chang, and H.-C. Hsiao, *Research of minimum ignition energy for nano Titanium powder and nano Iron powder*. Journal of Loss Prevention in the Process Industries, 2009. **22**(1): p. 21-24.

[3] Holbrow, P., et al., *Fire and explosion properties of nanopowders*. Health & Safety Executive: Buxton, UK, 2010.

[4] Eckhoff, R.K., *Influence of dispersibility and coagulation on the dust explosion risk presented by powders consisting of nm-particles*. Powder Technology, 2013.

[5] Wu, H.-C., et al., *Explosion characteristics of aluminum nanopowders*. Aerosol and Air Quality Research, 2010. **10**(1): p. 38-42.

[6] Bouillard, J., et al., *Ignition and explosion risks of nanopowders*. Journal of Hazardous Materials, 2010. **181**(1): p. 873-880.

[7] ASTM, *E1515: Standard Test Method for Minimum Explosible Concentration of Combustible Dusts*. 2007.

[8] Castellanos, D.Y., *The Effects of Particle Size and Crystallinity on the Combustion Behavior of Particulated Solids*, in *Department of Material Science*. 2013, Texas A&M University.

[9] Denkevits, A. and S. Dorofeev, Dust explosion hazard in ITER: Explosion indices of fine graphite and tungsten dusts and their mixtures. Fusion engineering and design, 2005. 75: p. 1135-1139.

[10] Turns, S.R., *An introduction to combustion*. Vol. 499. 1996: McGraw-Hill New York.

[11] Burke, S.P. and T.E.W. Schumann. *The mechanism of combustion of solid fuel*. in *Proc. 3rd Int. Conf. Bituminus Coal*. 1931.

[12] Van der Held, E.F.M., *13. The reaction between a surface of solid carbon and oxygen*. Chemical Engineering Science, 1961. **14**(1): p. 300-312.

[13] Graaf, D. and J.G.A., Brennst-Warme-Kraft, 1965. **17**: p. 227.

[14] Neufield, P.D., A.R.Jansen, and R.A.Aziz, The Journal of Chmeical Physics, 1972.

[15] Wilke, C.R. and C.Y.Lee, Industrial & Engineering Chemistry Fundamentals, 1955.

# 348963 Shock Interaction with Dust Layers for Different Mach Numbers and Dust Layer Depths

**Abstract:**

Secondary dust explosion is a serious industrial issue because it occurs under conditions corresponding to an increased quantity and concentration of dispersed, combustible dust when compared with the primary explosion. The problems of lifting and dispersing of a dust layer behind a propagating shock wave must therefore be understood to ensure safety regarding secondary dust explosion hazards. A new facility was developed recently by the authors for the study of shock waves over dust layers. A shock tube and high-speed (15,000 fps) shadowgraph system were used to film shock waves over limestone dust layers. New data were collected for image analyses over longer periods than found in prior works, where the longer observation time and higher frame rates led to the discovery of trends not previously observed by earlier studies. Further investigation is needed to aid the development of predictive models for shock wave/dust layer interaction.

Data were collected for various Mach numbers and different limestone dust layer depths. Within the current data, it was observed that the rate of dust rise increases linearly with increasing shock Mach number. This trend does not appear to hold true above Mach number 1.32 as the rise rate and overall rise height of Mach 1.32 and 1.40 are nearly identical. However, an increase in rise rate is observed as Mach number increases between 1.10 and 1.32. Further experiments at higher Mach numbers above 1.40 are required to fully observe the effects of shock Mach number on the dust layer rise rate and raise height. When comparing the effects of the dust layer depth, it was observed that increasing the dust layer depth resulted in an increased dust layer rise height but a slight decrease in the rate of the dust layer rise.

# 348967 Numerical Simulation of Cryogenic Boiling

**Abstract:**

The main source of hazard associated with cryogen is the loss of primary containment which can lead to vaporization, two-phase spreading and vapor dispersion. Numerous studies have been conducted in modeling vapor dispersion, but much less attention has been given to phenomena such as vaporization and two-phase spreading despite the fact that those are critical to the accurate prediction of the consequences of spill. Vaporization and two-phase spreading models are known as source term model since results of these models are used to perform the vapor dispersion calculations. As a result, LNG vapor cloud dispersion models are highly sensitive to the accuracy of the vapor generation rate estimated from the source term model. The vaporization rate modeling includes boiling and evaporation. During the initial spillage, the surface temperature of the substrate, such as concrete or insulated dyke material, is much higher than the boiling point of the cryogen thus boiling takes place mainly due to the heat conduction from substrate. When the temperature of the substrate falls close or equal to the boiling point, conduction heat transfer becomes insignificant in comparison to convective heat transfer from air to liquid thus the vapor generation is mostly due to the evaporation. In this study, our focus is to model the boiling heat transfer due to conduction for an accurate prediction of the vaporization rate that will subsequently be combined with evaporation model and two-phase spread model. A CFD based approach is considered rather than integral modeling approach for future unification of vaporization, two-phase spreading and vapor dispersion models.

The boiling of a fluid have 3 different regimes, i.e., film boiling, transition boiling and nucleate boiling. Each mode of boiling occurs over a range of excess temperatures (surface temperature exceeding the saturation temperature). Nucleate boiling is characterized by the formation of bubbles from randomly distributed nucleation sites over the surface whereas the film boiling is characterized by the formation of a continuous blanket of vapor film on the heating surface with bubbles leaving the surface at a regular intervals. Transition mode is much less understood but it connects nucleate and film regimes. Liu et al. proposed a volume-of-fluid (VOF) method with a sub-model of phase change to simulate cryogenic pool boiling using the commercial CFD software Fluent. A user defined interphase mass transfer model considering both homogenous and heterogeneous boiling was proposed. Three different boiling regimes were identified, and the boundaries of each regime were well predicted. It was also found that the simulation can reasonably predict the experiments in a film boiling regimes. However, the heat flux for nucleate boiling was under-predicted by Liu's model. In this work, the predictions in the regime of nucleate boiling will be improved by seeking a better sub-model to describe the homogenous model, which is supposed to be critical to predict nucleate boiling.

# 349002 Homogenous-Gaseous and Particle-Gas-Air Combustion in Turbulent Environment: Analytical Formulation and Experimental Validation

**Abstract:**

Analytical studies of premixed turbulent flame velocity, namely, the renormalization approach of Bychkov et al [Combust. Sci. Tech. 179 (2007) 137] and the spectral formulation of Chaudhuri et al [Phys. Rev. E 84 (2011) 026322] are reconciled, and validated by Hybrid-Flame Analyzer experiments on turbulent methane-air burning [Combust. Flame 160 (2013) 635]. Both the homogenous gaseous premixtures as well as the particle-gas-air systems containing coal-duct impurities are investigated. The turbulent flame velocity is tabulated as a function of various flame parameters (such as the laminar flame speed, thermal expansion coefficient, etc.) and flow parameters (such as the turbulent intensity, integral and Kolmogorov length scales, etc.) as well as that of the sizes and concentrations of the impurity particles. Specifically, it is shown that the particles typically intensify turbulent combustion. The formulation accounts for the flame-flow feedback due to thermal expansion. Besides, turbulence coupling to the intrinsic combustion instabilities is analyzed by means of the concept of the turbulence-induced Darrieus-Landau cutoff wavelength. It is eventually demonstrated how multi-phase impurities and turbulence-instability interaction modify the conventional turbulent combustion regimes diagram.

# 349400 Application of Leading and Lagging Indicators to Improve Laboratory Operation Safety

**Abstract:**

Unlike the chemical manufacturing process with fully automatic equipment, the laboratory operation in the chemical industry will still involve multiple types of manual handling for chemical analysis, reliability test, material characterization and failure analysis. The test operators or lab engineers have different levels of exposure to a lot of chemicals and physical hazards in multiple phases. In order to improve the operation safety and mitigate the potential risk, the study of situation awareness in the laboratory operation aims to investigate the perception of what is happening around the operation environment for both the normal situation and the abnormal situation, which has become an important part in the PSM(Process Safety Management). One of the effective methods to improve individual and team situation awareness and their individual responsibilities to accomplish the team's goals and objectives is the application of leading and lagging indicators, which have been recognized as the effective tools to drive process safety performance improvement. The processes and decisions to develop and implement the process safety leading and lagging indicators within the company involves the integration of the applied criteria, the differences between facility-level implementation and company-wide implementation, the potentially encountered barriers to success and the methodology to overcome, and the demonstrated benefits through the use of leading indicators. This paper will describe the straightforward application of leading and lagging indicators to identify the hazards and assess the risks of laboratory operations. The analysis will be deployed in the ways of manpower, machine, materials, method and environment. The analysis output can be regarded as the aid to safer laboratory design, operator training and real-time fault diagnosis. This will help to debottleneck and achieve operation safety and occupational health in the semiconductor industry.

# 349855 Graphic Visualization of IPL Status Enables Better Judgement of Inspection, Maintenance and Repair Priorities and Helps Day to Day Assessment of Operational Risk and Work Management Decisions

**Abstract:**

**With hundreds and even thousands of IPL components to be inspected, maintained or repaired at a facility, monitoring and managing their status is a challenge whose difficulty rises exponentially when dealing with ageing assets. It is not uncommon to have a situation where there is a degree of impairment on several prevention or protection systems but it is difficult to judge where the priority lies given constraints on maintenance and inspection resources. With a backlog of safety critical maintenance, dilemmas arise such as whether it is more important to action an overdue callibration on a relief valve or to inspect a set of gas detectors which should have been carried out last month. These practical day to day problems are very common in aging plants or production facilities in mature basin areas. Our maintenance management systems are excellent for setting up and managing an appropriate inspection / maintenance schedule for our critical elements but once a backlog has developed they are not so helpful in prioritising overdue work. Often the right decision is based on proximity of impairments so that you avoid a situation where multiple systems are impaired in the same location, however, the data is not configured to present in this manner.**

**This paper will describe these challenges and talk about new technology platforms which are now available to help in this area. They are expansions of existing work management technology solutions where work activities are shown graphically with icons on plant plot plans. These systems can now be used to track IPL impairments graphically showing the geographical location of the impairment allowing better decisions to be made on the cumulative local effect of impairment. They also enable better decisions to be made on day to day work whose activities may have an impact on overall Operational risk based on the status of IPL's in the vicinity of the work. For example if you were about to do maintenance work offshore on a main oil pump which required breaking containment and the tools you were going to use introduced a spark potential. Added to this the deluge system in the area is 3 months beyond its inspection due date. Showing this information graphically rather than in lists and schedules enables better assimilation of the facts and more informed decision making on whether to do the work or what additional precautions to take. The paper will look at how such information is gathered and presented and we will describe some realistic scenarios showing how these systems help with prioritising inspection and maintenance work and how they help with day to day work decisions. While this expanded technology platform is new, the original work management technologies have been around for over a decade and in common use today. We will include some comments on how these systems have evolved, who uses them and how successful they have been.**

**The growth in this area represents a broadening of the technology tool set which can be used for a risk based approach to inspection, maintenance and repair and helps to combine and share information and data from three disparate functional groups: Operations, Maintenance and Safety whose collaboration is essential for safe and efficient running of the plant.**

# 349881 2. Successes in Implementing PHA/HAZOP/LOPA in Major Capital Projects in a Steel Company

**Abstract:**

HADEED, the steel making affiliate of SABIC, began implementing a multiple phase approach for major capital projects 8 years ago. For complex processes, such as direct iron reduction, 4 sequential PHAs throughout the project have been required, including a PHA of the operating procedures before commissioning. For simpler processes, such as rolling mills for rod and sheet and plate, 2 to 3 PHAs have been required. This paper describes the approach and successes to date achieved by HADEED using this approach, including discussing the gains in startup efficiency and operational cost savings that more than pay for the PHA efforts during the project. The paper also illustrates the versatility of the phased approach to performing PHAs in non-chemical processes, such as steel making.

<<NOTE: If HADEED cannot get approval from their management to take the lead on this paper, then we will withdraw this submittal >>

# 349909 Liquid Fuels Release Rate Calculation in Transport Pipelines with Complex Topographical Conditions


**Abstract:**

Pipelines are the most efficient and economical mean of hazardous materials (HazMat) transportation through long distances. In the particular case of hydrocarbons, production sites are often far away from processing facilities and/or final users, therefore, this industry uses the greatest amount of transmission pipelines. The main concern related to the operation of these systems is the loss of containment events (LoC), due to the fact that the fluid itself constitutes a source of hazard. Such events, despite their low frequency, might have severe consequences on the surrounding communities, the environment and the assets of the operating company or any other company in the vicinity of the pipeline.

In order to ensure a safe and reliable operation of these systems, and to support decision-making processes in a risk management framework, it is required to perform risk analysis. This analysis will allow establishing possible consequences of the accidental events and the required measures to reduce the frequency of events and/or the effects magnitude. One of the main inputs to risk analysis is the release rate, i.e., the rate at which the material is released to the surroundings once the loss containment takes place. Modeling the release rate is particularly important in countries with complex topographic conditions, in which a pipeline undergoes several geographical changes along its route. The amount and manner in which a HazMat is released to the surroundings will define the possible accidental events and therefore the consequences on the elements of interest, such as population, environment or infrastructure. Therefore, it is necessary to properly quantify the available volume for release at each segment of the pipeline and estimate the related release rate profile. As a result, several models have been developed to model hydrocarbon release scenarios, including phenomenological approaches that consider one (axial) and two-dimensional (axial and radial) flow patterns, with or without phase change, transient effects, and more. Although the phenomenological models offer the best representation of reality, their computational cost is extremely high, with computing times on the order of days.

This work presents a tool that supports decision making processes during the design, operation and/or revamping of hydrocarbons transmission pipelines. The tool allows a simple, non-phenomenological calculation for a quick estimation of available volume for release, and the related release profile. The tool considers the presence of active-automatic safety barriers (such as emergency shutdown valves and check valves) in pipelines with complex topographical conditions, given a failure in any point of the pipeline. These calculations are performed at a very low computational cost, with computing times on the order of minutes. The results provided by this tool become a key input to perform *a priori* risk analyses of pipeline systems to identify its critical segments (those with the highest release rates and times). Such information is crucial to define the layers of protection to be implemented (e.g., determine segmentation requirements, application of other safety barriers) and for the strategic design of emergency response plans in case a failure occurs (e.g., establish the critical areas of the pipeline, location of sensing devices, emergency response brigade).

# 349929 Decision Tree to Optimize NFPA 30 Criteria in Fire Protection Systems Applied in Oil and Gas Industry

**Abstract:**

In accordance with risks and hazards associated to oil and gas industry operations, criteria must definitely be developed in order to use NFPA 30 code (Flammable and Combustible Liquids Code) in an adequate way. To meet this purpose, ECOPETROL has developed a decision tree so that both engineers and technicians can use NFPA 30 criteria more easily and more efficiently tailoring the decision making process to the most outstanding and recurrent hazards in west.

# 349931 Sil Determination of High Integrity Pressure Protection System (HIPPS)

**Abstract:**

High Integrity Pressure Protection Systems (HIPPS) are Safety Instrumented Systems (SIS) implemented to address overpressure scenarios in lieu of a pressure relief valve (PRV). HIPPS essentially replaces the PRV for those scenarios that the SIS is designed to prevent. HIPPS applications are generally pipeline and pressure vessel overpressure protection.

The fourth edition of API 521 allows credit for a favorable response of the instrumented systems that prevent over-pressure and/or over-temperature. The recommended practice refers to these systems as high integrity protection systems and states that they should be at least as reliable as a pressure relief device.

The Pertamina Hulu Energy Offshore North West Java (PHE ONWJ) intents to install new YYA NUI (Normally Unmanned Installation) platform will be sent to the existing unmanned KLB platform via 12 inch three phase pipeline. The KLB platform has fix flare system and has limited space to increase its capacity due to the additional production from the YYA.

The overpressure scenario has been identified in the HAZOP study and need to be mitigated by either release the pressure through PSV to flare or installing another layer of protection to prevent overpressure scenario. In this case of the releasing pressure to existing KLB flare is not sufficient to handle the gas relief from YYA wells. Therefore HIPPS was selected to replace the PSV in preventing the potential overpressure at the KLB.

The LOPA workshop was conducted for SIL determination of the HIPPS system and attended by discipline engineer and operator to identify initiating cause, the protection layer and severity level. It was identified 5 initiating events related to the overpressure scenario above. The independent protection layer also identified the existing pressure switch high which can be assumed to have a credit for SIL 0. The consequences or severity level for safety, environment, cost and business image was determined as impact of the scenario happened. The modifier factor was also included for occupancy (NUI) while the ignition probability was not credited as the ultimate consequence ultimate scenario is rupture.

The assessment of SIL determination resulted IL target IL1 for safety, IL2 for environment and IL0 for commercial. From this LOPA method, maximum IL2 with PFD 6.1E-03 was selected for IL target. The next step is to design HIPPS and verify the SIL level after gathering data of PFD from the vendor.

# 350028 Comparison of Different Methods to Determine the Activation Energy of Flammable Dusts Mixed with Inert and Inhibitory Materials


**Abstract:**

**Comparison of different methods to determine of the activation energy of flammable dusts mixed with inert and inhibitory materials**



**C. Wanke, B.Binkau, P.Rettig, U.Krause**


Porous flammable media often undergo spontaneous combustion and the spread of smoldering flame fronts. Adding inert or inhibitory substances, the self-ignition and combustion behavior can be influenced. In a series of experiments the activation energy of flammable dusts (lignite, carbon) mixed with inhibitors (ammonium sulfate, ammonium phosphate) was determined and compared to the pure flammable dusts.

To determine the activation energy self-heating methods (isoperibolic and adiabatic hot storage tests) and temperature programmed methods (thervogravimetric (TG) and differential scanning calorimeter (DSC) tests) can be used. The common method in Europe is the isoperibolic hot storage test according to DIN EN 15881. This method is very time and material consuming, due to the fact, that the self-ignition temperature of three to four different sample sizes must be analyzed. An alternative is the adiabatic hot storage test where the activation energy can be determined from just one single experiment. The disadvantage is the difficult analysis of the resulting Arrhenius plot, where the activation energy is the steepest slope of the logarithmic temperature rise. To find this value a Matlab tool was developed.

The advantage, using TG and DSC is the small amount of sample material. The analysis of the mass loss rate from the TG experiment results in an Arrhenius plot as well, which can be evaluated by the mentioned Matlab tool. Using the DSC, three experiments with different heating rates have to be executed. In spite of the slow heating rates, this method is still faster than the isoperibolic test.

Summing up, the activation energy of pure substances, as well as mixtures, can be calculated by the methods above. The isoperibolic test lead to the highest activation energies, the DSC and TG experiments lead to the lowest values. This can be traced back to the different experimental procedures. Isoperibolic tests start at the self-ignition temperature (SIT), adiabatic tests at the onset temperature, which is below the SIT, and DSC and TG start at ambient temperature. The reactions below the starting temperature can't be taken into account for the determination of the activation energies, therefore the results of DSC and TG are lower.

# 350059 Economical Approach Quantification of Impacts in Major Accidents

**Abstract:**

In the industry field, the nature of certain processes has always represented an inherent threat not only to humans but also to the environment in which they are executed. Although the frequency of scenarios that lead to the occurrence of major accidents in facilities is considerably low, their consequences are usually catastrophic. Experience has demonstrated that high level impacts occurred during the development of industrial activities imply significant financial losses. Therefore, one of the main interests for risk assessment results in the economical estimation of these consequences.

Usually, damage costs are estimated once the potential consequences have been already materialized. Nonetheless, in order to generate mechanisms for prevention, mitigation and control over these negative impacts, the implementation of a pre-emptive quantification of potential leakages allows the estimation of useful indicators as incomes for decision making analysis. For this purpose, qualitative, semi quantitative and quantitative methodologies have been developed to support risk analysis.

Within this context the present work pretends to propose an *a priori* quantitative methodology that allows carrying out the economical conversion of impacts originated by major accidents through the integration of two methods. The first method is defined to estimate the impact level over a specific system while the second establishes the transformation of impacts into monetary terms. The suggested methodology is applied later to a particular case of study.

A quantitative vulnerability analysis has been defined for the estimation of impact level. The model used is a classical approach that implements *probit* analysis which gives the level of impact. The fact that the result corresponds to a quantitative estimation, makes easier its conversion into monetary terms. Even so, as a limitation of the method, it must be considered that each type of major accident, and subsequently their specific cases, has a unique equation that represents the system of study. Accordingly, the application of this method for a particular situation requires the definition of the appropriate expression that relates the *probit* variable with the dose associated to the specific event.

Once the level of impact is estimated for a system of interest, the second method is applied in order to express these impacts in monetary terms. In this case, the methodology of CEPAL is used. It defines the types of effects produced by an accident: Direct, indirect and macroeconomic. Since the main interest of this work is not the estimation of impacts effects of an accident at socioeconomic level, only the direct and indirect effects over the system are considered. For direct costs, this method defines a set of parameters for each type of system affected (infrastructure, humans, services). These must be considered for the breakdown of costs including sometimes the appraisal over a period according to the item evaluated. In the case of indirect costs, the execution of two analyses is necessary. As a first step, a projection of each item is done to estimate certain economic parameters dependant of the system type. Then a comparison of the values that will be reached in the supposed accidental scenario, with respect to the ones that would be reached if the accident did not happen is developed. With these values, finally the net incomes for each case are estimated.

An explosion in an energy supply facility is considered as a case study. The correspondent vulnerability analysis with probit method is done in order to estimate the damages caused over the structure. As a result, costs of reparation / reconstruction for damaged infrastructure taking into account the appraisal over the period of time that reparation / reconstruction of the facility takes, costs of equipments and costs of raw materials are determined. Also, operation and capital costs are estimated taking into account the time required for reparation / reconstruction of facility. Capital costs include equipment acquisition, operation, labor force and materials. Finally the net incomes for the accidental and no accidental scenarios are estimated in order to obtain the approximate net indirect costs.

The appropriate use of methodologies, as the one proposed in the present work, defines non conventional paths for generation of indicators that support decision making analysis. Due to the different considerations and approaches used within the methodology, the indicators that could be generated contain valuable information different from the one given by indicators that do not consider an economic perspective of effects. The information provided by these indicators offers new perspectives and more accurate points of view for project evaluation.

# 350080 Application of Consequence Analysis in the Development of Emergency Response Plans for Accidental Events in Liquid Fuels Transport Pipelines

**Abstract:**

Pipelines are the world's most common operation for hazardous materials transportation due to its low cost. Oil and gas industry (upstream, mid-stream and downstream) concentrates the greatest amount of pipelines of the world, due to the fact that production fields are normally too far away from final users, export facilities or processing plants. The main concern regarding the operation of these systems is the events of loss of containment (LoC), which might take place for several reasons such as corrosion, external impacts, etc. Such events, despite their low occurrence frequency, could have severe consequences to surrounding populations, environment and assets of the operating company or any other company. The consequences of accidental events might be reduced by application of different protection layers (e.g., construction materials, alarms, shutdown valves). Emergency response corresponds to the last layer of protection, and its objective is to control, if possible, an accidental event, or to reduce its consequences in the case the event develops completely.

In a risk management framework, it is imperative to perform risk analysis to determine possible consequences of accidental events and required actions to reduce the frequency of events and/or the magnitude of effects, and thus, support the decision-making process. As a result, many risk analysis techniques have been developed, which can be qualitative, semi-quantitative or quantitative. The selection of a specific technique depends on the particular characteristics of the system under study (e.g., complexity of the operation, neighboring population, sensitive environment). Typically, emergency response plans have been developed from the application of qualitative risk analysis techniques such as Hazid and Hazop. This approach produces "generic" emergency response plans, which do not differentiate most likely scenarios or possible consequences of an accidental event.

This work presents the application of a quantitative approach of risk analysis, through consequence analysis , to determine major factors in attention of emergencies such as preparedness activities (e.g., drills), critical attention points, potential safety barriers required, shelters, evacuation routes and times, etc., in loss of containment events during pipeline transmission of liquid fuels in complex topographical conditions. These factors will allow the development of emergency response plans that guarantee effective attention of accidental events in pipelines. The methodology developed in this work, given the characteristics of the material being transported and the pipeline, guides the user in the construction of analysis scenarios, the selection of source, effects and consequence models. To illustrate this, a hypothetical pipeline is studied and the most important results are presented.

# Shale Oil and Gas – Wellhead to Fuel

Dr. Lawrence Kremer - Baker Hughes
AIChE - Hilton New Orleans Riverside
April 1, 2014

14AIChE
Spring Meeting, New Orleans, LA

BAKER HUGHES

## Outline Unconventional Shale Oil

- Properties of shale oils
- Problems
  - Transportation
  - Refining

14AIChE
Spring Meeting, New Orleans, LA

BAKER HUGHES

## General Characteristics of Shale Oils

- API gravity typically varies 30-55+
- Very paraffinic
- Low sulfur
- Can have hydrogen sulfide and mercaptans
- Low TAN
- Very low asphaltenes
- Light crude with very little residual material
- Many samples show great quality variation

| Factor | Bakken | WCS |
|---|---|---|
| API | 39.7 | 21.6 |
| S% | 0.25 | 3.44 |
| Light ends | 2.6% | |
| LSR | 3.5% | 3% |
| Light Nap | 12.4 | 3.0 |
| Heavy Nap | 9.7 | 5.0 |
| Jet | 10.5 | 10.0 |
| LFO | 27.6 | 12.0 |
| Gasoil | 27.2 | 37.0 |
| VacTar | 6.7 | 29.8 |



## Shale Oil Property Comparison

| Parameter | Eagle Ford | Bakken |
|---|---|---|
| API | 52 | 40.8 |
| TAN (g KOH/g) | < 0.05 | 0.09 |
| Sulfur (wt %) | <0.2 | 0.304 |
| Asphaltene (wt %) | 0.1 | 0.41 |
| Resin (wt %) | 1.6 | 4.95 |
| Filterable Solids (PTB) | 225 | 76 |

## Utica Shale Crude - Three Phase

14 AIChE
Spring Meeting, New Orleans, LA

BAKER HUGHES

## Eagle Ford Crude Samples

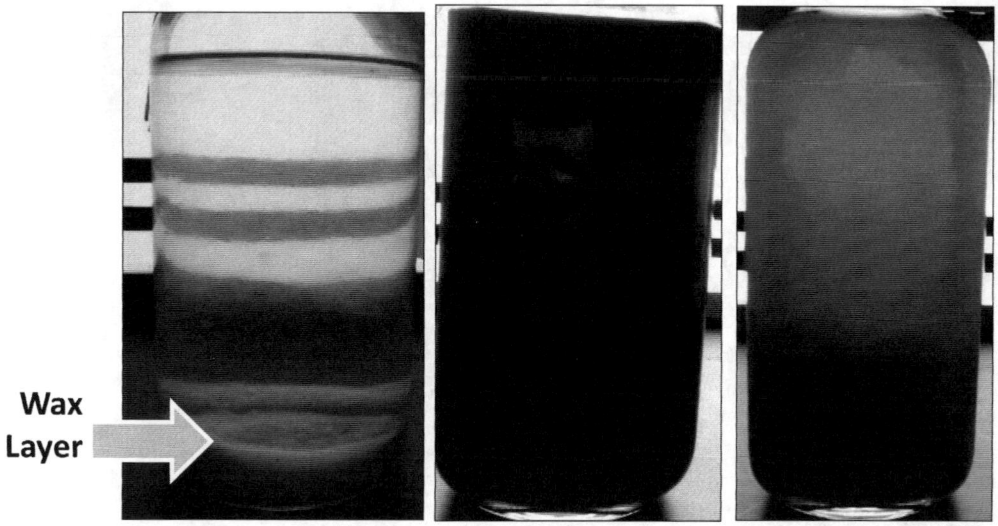


14 AIChE
Spring Meeting, New Orleans, LA

BAKER HUGHES

# Analytical Breakdown

| Value | Yellow | Red | Black |
|---|---|---|---|
| API | 55.0 | 44.6 | 52.3 |
| TAN | <0.05 | 0.07 | <0.05 |
| Sulfur | <0.20% | <0.2% | <0.2% |
| Na | 1.0 ppm | 1.6 ppm | 1.6 ppm |
| K | 0.3 ppm | 0.4 ppm | 0.5 ppm |
| Mg | 3.4 ppm | 2.9 ppm | 3.0 ppm |
| Ca | 2.6 ppm | 2.8 ppm | 3.8 ppm |
| Asphaltene | 0% | 0% | 0.1% |
| Resin | 0.5% | 3.2% | 1.6% |
| Filterable solids | 176 PTB | 295 PTB | 225 PTB |
| P | 5.3 ppm | 4.6 ppm | 5.0 ppm |

14 AIChE Spring Meeting, New Orleans, LA

BAKER HUGHES

# Micrograph of Sludge

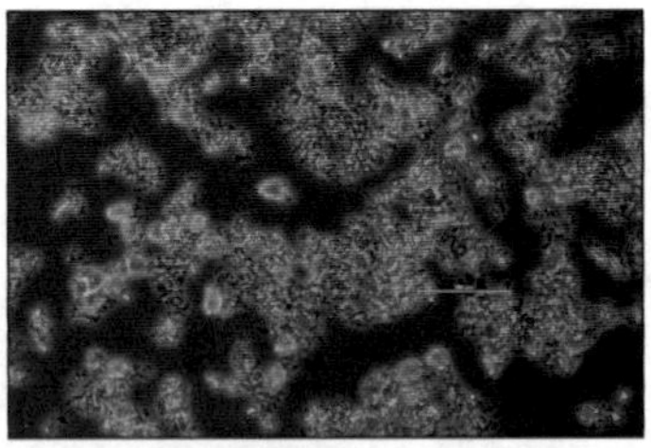

100 X

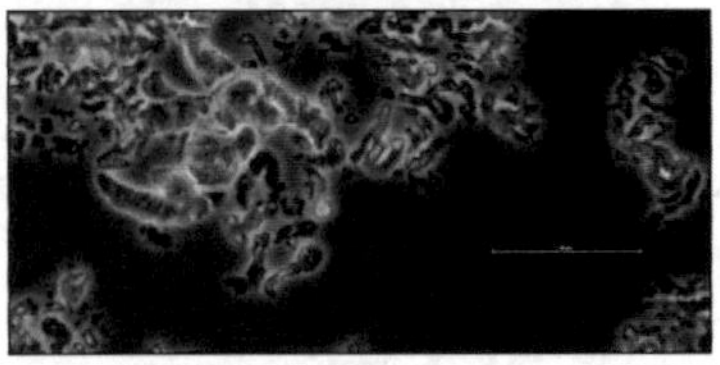

400 X

14 AIChE Spring Meeting, New Orleans, LA

BAKER HUGHES

# Paraffin Analysis of Sample

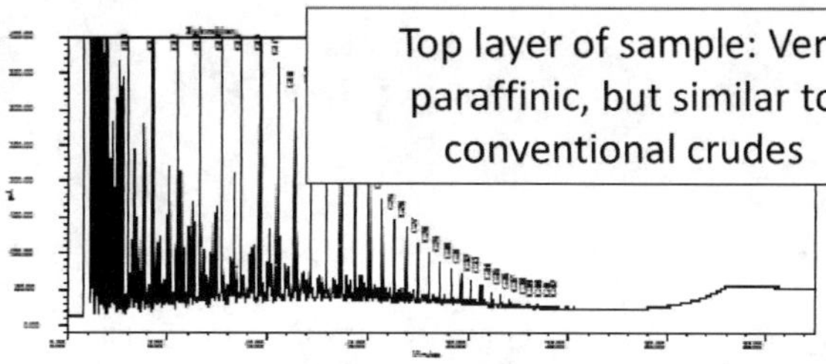

Top layer of sample: Very paraffinic, but similar to conventional crudes

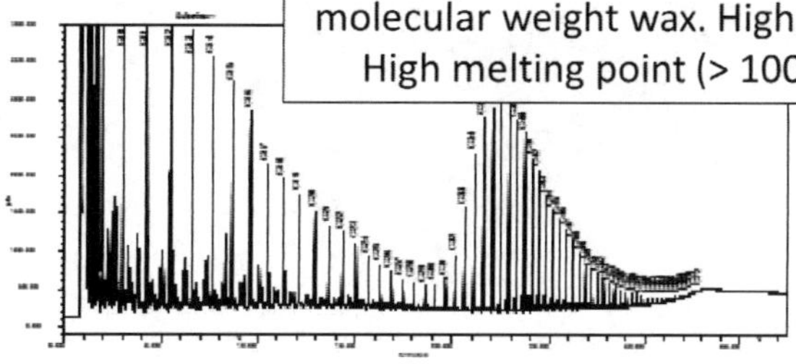

Bottom layer: Considerable high molecular weight wax. High MW = High melting point (> 100 °C)

14 AIChE
Spring Meeting, New Orleans, LA

BAKER HUGHES

# Transport of Shale Production

- Infrastructure does not exist to meet needs
  - Most shale oil moved by rail or trucks
  - Limited pipelines expanding to meet transportation needs
- Unit trains to East Coast, West Coast, and Gulf Coast
- Problems:
  - Tankers and barges with heavy wax deposits ROB
  - Tight emulsions that look like oil or white condensate but high BS&W
  - Cars not uniformly treated for $H_2S$
  - Corrosion in rail cars

14 AIChE
Spring Meeting, New Orleans, LA

BAKER HUGHES

## Viscosity of ROB Before Treatment

ROB from untreated compartment

14 AIChE Spring Meeting, New Orleans, LA

BAKER HUGHES

## Eagle Ford Shale Oil Wax Dispersion Testing

Several products were effective in dispersing the paraffin wax in an Eagle Ford shale oil sample at ambient temperature for over a month.

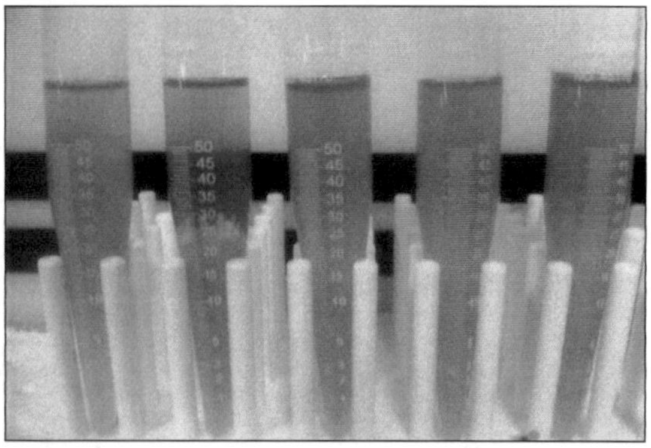

14 AIChE Spring Meeting, New Orleans, LA

BAKER HUGHES

## Product 'A' Effect on Wax Deposition (Utica)

**Blank** **200 ppm** **400 ppm**

14 AIChE Spring Meeting, New Orleans, LA

BAKER HUGHES

## Refinery Issues

- Crude Blending
- Fouling
- Corrosion
- Product quality

14 AIChE Spring Meeting, New Orleans, LA

BAKER HUGHES

# Crude Unit Fouling

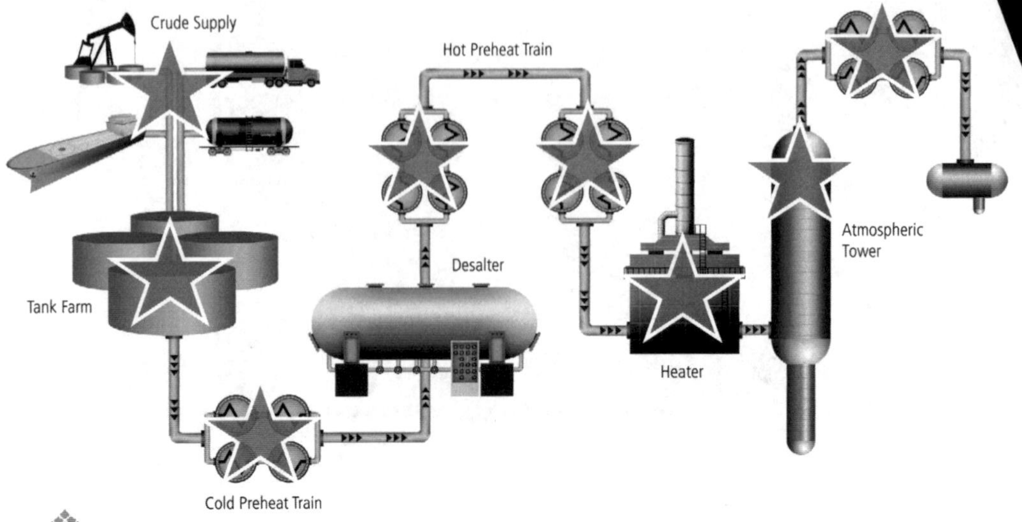


★ = Susceptible to Fouling



# Asphaltene Precipitation

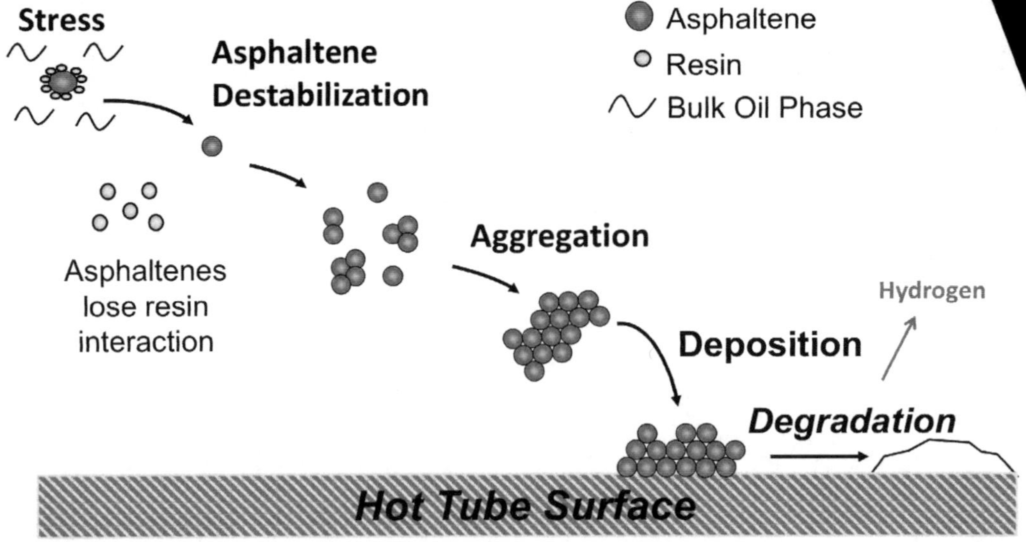

# Asphaltene Stability Index Test

Field ASIT services™ technology measures the onset of the flocculation of the asphaltenes with high accuracy by inducing the asphaltene precipitation via titration with a paraffinic solvent.

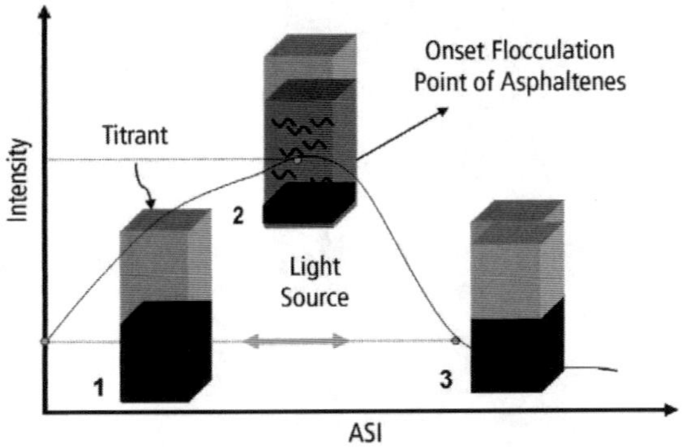


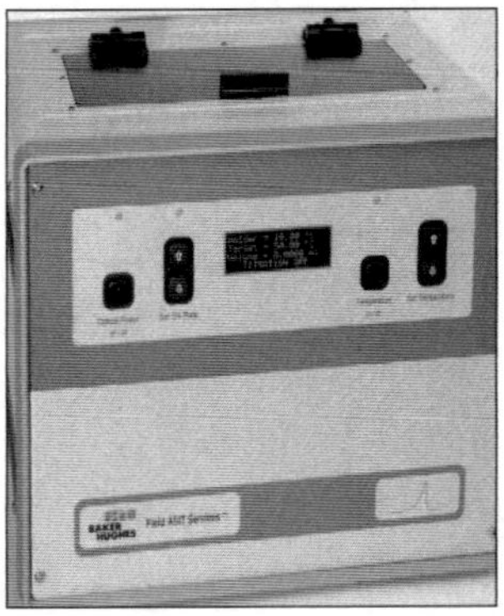

14 AIChE® Spring Meeting, New Orleans, LA

Field ASIT services is a trademark of Baker Hughes Incorporated

BAKER HUGHES

# ASI Shows Eagle Ford Can Destabilize Crude

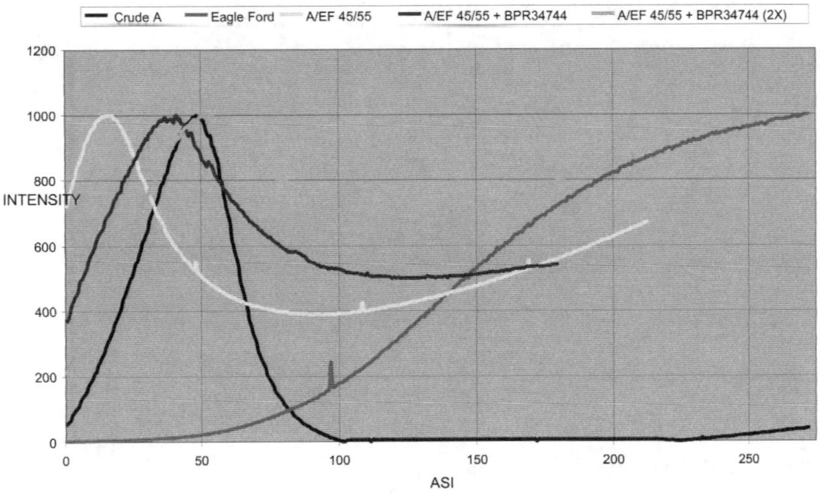


14 AIChE® Spring Meeting, New Orleans, LA

BAKER HUGHES

## Tower Tray Fouling - Phosphorus

- Seen by refineries running Bakken, Eagle Ford, and WTI
- Operationally manifested as increasing tower dP
  - Typically in the kero/jet zone
- Tower inspection shows deposits on trays in kero/jet zone
  - Some deposition may also show a few tray levels above and/or below kero/jet zone
  - Deposit scale typically tightly bonded to tray surface restricting flow and valve function

Above kero/jet zone

In kero/jet zone

Below kero/jet zone

14 AIChE® Spring Meeting, New Orleans, LA

BAKER HUGHES

## Shale Oil Impact – MEA Contamination

**Corrosion and Fouling**

- Shale oils transported by truck, rail, barge.
- Many shipments are treated with $H_2S$ scavengers:

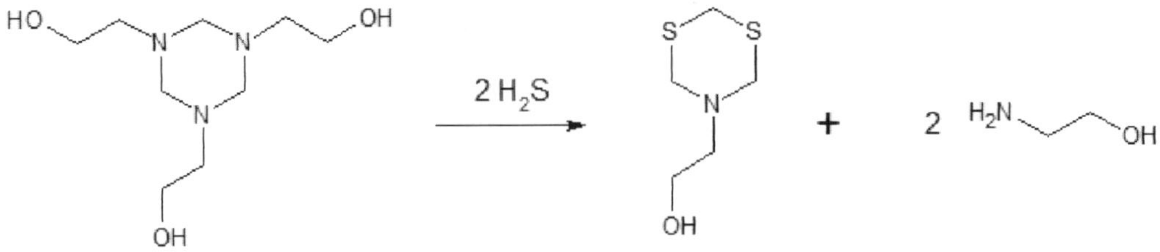


- Potential Impacts:
  - $H_2S$ scavengers leave byproduct amine in crude oil (MEA, MA).
  - Tramp amines distill to OH system, create amine-HCl salts:
  - Corrosive to system metallurgy.

14 AIChE® Spring Meeting, New Orleans, LA

BAKER HUGHES

# Impact of Tramp MEA on OH Water

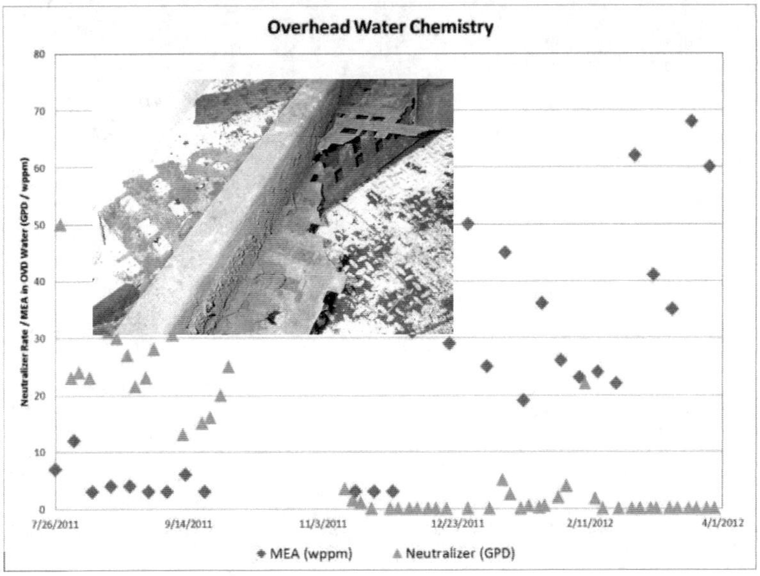


14AIChE® Spring Meeting, New Orleans, LA

BAKER HUGHES

# Salt Risk w/MEA Concentration

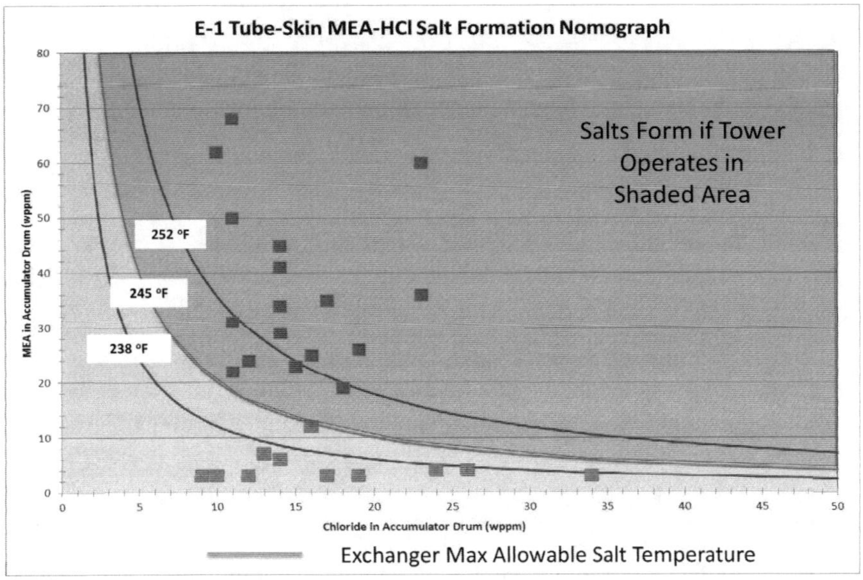


14AIChE® Spring Meeting, New Orleans, LA

BAKER HUGHES

# Shale Oil – Refinery Process Concerns and Monitoring

BAKER HUGHES

| CHARACTERISTIC OF SHALE OIL | CONCERN IN REFINERY | PROCESS DATA | SAMPLES AND ANALYSES Samples | Analysis | OUTPUT DATA | FREQUENCY | BHI PROGRAMS |
|---|---|---|---|---|---|---|---|
| Crystalline Wax | Tank bottom sludge | • Tank Sludge Depth | • Tank bottom | • Visual sludge<br>• GC analysis | • Tank Sludge Depth<br>• Crystal Wax Presence | • Monthly | • XERIC™ Pretreatment |
| | Cold Preheat Fouling | • Desalter Temperature | | | • Temp vs. Time | • Update 2x/mo. | • LIFESPAN™ Wax Dispersants |
| | Desalter emulsions | • Emulsion layer growth | • Triline emulsions | • Visual wax<br>• GC analysis | • Crystal Wax Presence | • After Shale Oil begins<br>• When emulsion grows | • XERIC Demulsifiers |
| Asphaltene Stability in Blends | Desalter emulsions | • Emulsion layer growth | • Triline emulsions | • Precipitated Asphaltenes | • Precipitated Asphaltenes | • After Shale Oil beings<br>• When emulsion grows | • XERIC Asphaltene Stabilizers<br>• Field ASIT services™ |
| | ColdPreheat Fouling | • Desalter temperature | • Crude blend | • ASIT | • Temp vs. Time<br>• ASIT # | • Pre-Shale Oil<br>• Update 2x/mo. | • LIFESPAN Antifoulants<br>• Field ASIT services |
| | Hot Preheat Fouling | • Furnace inlet temp<br>• Furnace tube temps<br>• Furnace fuel usage<br>• Preheat DP | • Crude blend | • ASIT | • Temp vs. Time<br>• Fuel use vs. Time<br>• ASIT # | • Pre-Shale Oil<br>• Update 2x/mo. | • LIFESPAN Antifoulants<br>• MILESTONE™ Furnace Fouling Control<br>• Field ASIT services |
| Paraffinic | Furnace Coking | • Furnace tube temps<br>• Furnace fuel usage | • Crude blend | • HLPS | • Coking tendency | • Pre-Shale Oil<br>• W/Shale Oil | • MILESTONE Furnace Fouling Control |
| Tramp MEA | Tower/OH corrosion (Early Warning) | • OH Neutralizer rate<br>• EXCAL usage rate | • OH water<br>• Desalter brine | • OH pH<br>• Brine pH | • Early Sign of Amine<br>• Early sign of Amine | • Pre-Shale Oil<br>• W/Shale Oil | • TOPGUARD™ Overhead Corrosion Control<br>• EXCALIBUR™ Contaminant Removal Technology |
| | Tower/OH corrosion (Quantitative) | • Inputs for CRM | • OH water<br>• Desalter ww<br>• Desalter brine | • IC Amines | • Confirm MEA in Crude<br>• OH MEA ppm<br>• Salt Risk Tower<br>• Salt Risk OH System | • Pre-Shale Oil<br>• W/Shale Oil<br>• Update 2x/mo. | • TOPGUARD Corrosion Control<br>• EXCALIBUR Technology<br>• TOPGUARD Corrosion Risk Monitor (CRM) |
| Phosphorus | Tower P Fouling | • Tower DP | | | • Tower DP Increase | • 2x/mo. | • LIFESPAN Antifoulants, Dispersants |
| | | | • Crude | • ICP for P | • P in Crude | • 2x/mo. | |
| | | | • Kerosene | • Filt Solids<br>• XRF Filter P<br>• ICP for P | • Incr. Kero Solids<br>• P in Solids<br>• P in Kerosene | • 2x/mo. | |
| | | | • Kerosene<br>• Diesel | • ASTM D86 | • Incr. in Diesel IPB – Kero EP overlap | • 2x/mo. | |

EXCALIBUR, FIELD ASIT SERVICES, LIFESPAN, MILESTONE, TOPGUARD and XERIC are trademarks of Baker Hughes Incorporated.

# AUTHOR INDEX

# AUTHOR INDEX